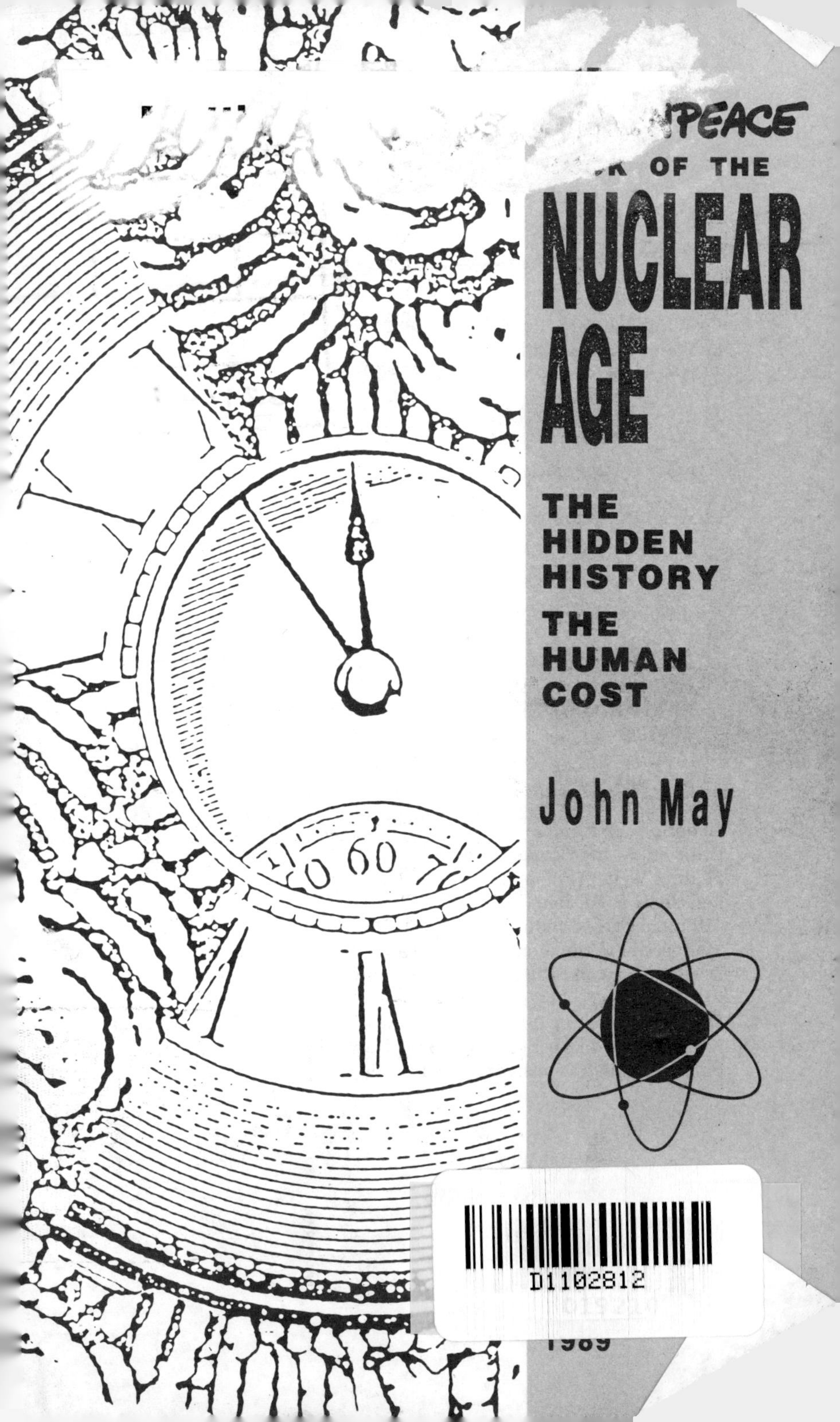
PEACE
OF THE
NUCLEAR
AGE
THE
HIDDEN
HISTORY
THE
HUMAN
COST
John May
D1102812

First Published in Great Britain 1989
by Victor Gollancz Ltd.
14 Henrietta Street, London WC2E 8QJ

A Gollancz Paperback Original

British Library Cataloguing in Publication Data
May, John
The Greenpeace book of the nuclear age.
1. Nuclear power. Environmental aspects
I. Title II. Greenpeace
333.79'24

ISBN 0-575-04567-1

Greenpeace Books
Principal Author/Project Director: John May
Feature Writer (Palomares, Thule, Three Mile Island and Chernobyl): John Trux
Additional Text/Production Manager: Ian Whitelaw
Manuscript Production: Tanya Seton
Design, Maps and Diagrams: Andy Gammon

Desktop publishing: Stella Cardus and Judy Bentley/Desktop Design
Index: Richard Raper et al./Indexing Specialists, Hove
Part-title Illustrations: Matthew Cooper

Printed and bound by
The Guernsey Press Co Ltd, Guernsey

Contents

Contributors and Consultants

William M. Arkin is author of numerous books on military and nuclear matters, co-editor of the *Nuclear Weapons Data Books*, a member of the editorial board of the *Bulletin of the Atomic Scientists* and disarmament adviser to Greenpeace.

Simon Carroll is a Greenpeace nuclear campaigner, based in Brussels.

Damian Durrant is a Greenpeace campaigner on the Nuclear Free Seas issue, based in London.

Geoff Endacott is a freelance science writer and broadcaster.

Shaun Gregory is a Research Fellow at the Department of Peace Studies at Bradford University. His new book, *The Hidden Cost of Deterrence: Nuclear Weapons Accidents*, is published by Brassey's Defense Publishers Ltd.

Helmut Hirsch is a physicist who works in Hanover with the Gruppe Ökologie as an independent consultant dealing with matters of nuclear safety. He has also advised the Austrian Federal government and the State government of Lower Saxony.

Dr David Lowry is a Research Fellow at the Energy and Environment Research Unit at the Open University, Milton Keynes, and director of the European Proliferation Information Centre in London. Dr Lowry has written and broadcast widely on nuclear issues.

Barbara Moon is Senior Editor of *Saturday Night* magazine in Toronto.

Robert S. Norris is a Senior Research Associate at the Natural Resources Defense Council in Washington DC.

Paul Rogers is a Senior Lecturer at the Department of Peace Studies at Bradford University.

Mycle Schneider is an independent consultant, researcher and author on nuclear issues working from the Paris office of the World Information Service on Energy.

Steve Sholly is an independent analyst on nuclear issues working with MHB Technical Consultants in California.

Andy Stirling is the international coordinator of Greenpeace's nuclear campaigns, based in Amsterdam.

Dr Jinzaburo Takagi is an independent nuclear scientist and a director of the Citizens Nuclear Information Centre in Tokyo.

Ralph Torrie is an independent consultant on nuclear and energy issues working from Ottowa in Canada.

John Trux is a freelance science writer and Features Consultant to the Science Photo Library in London.

First Words

This is a book about accidents and risk, about the nature of chance and the oppressive weight of secrecy, about invisibility and intrigue, about tragic events, about causes and effects, about official lies and the true human cost of atomic energy.

The world described in this book exists all around us, like an alternative reality. It is a world where human error meets sophisticated technology, where people make a sequence of logical decisions for the right reasons only to find they have created chaos. Where there are many versions of the truth.

This book will introduce you to the black briefcase in the red shack, take you on a helicopter ride over a burning nuclear core, into a damaged reactor in which life-forms are breeding, and through a realm of black humour and mystical coincidence.

Because the effects of radiation take decades to reveal themselves, almost all the stories in this book are current, even though some of them begin in the mid-1940s. In this context, information develops its own half-life — the amount of time it takes for official truth to leak out of the canisters in which it is contained.

There is a new geography here, with a new set of significant points and a complex network of interconnections. Nuclear material is constantly on the move around the globe, it is stacked in caverns deep in the earth, and dumped in deep-sea trenches. Above our heads nuclear-powered satellites produce high-resolution images of nuclear navies. While we sleep, radars scan the horizon for the heat signature of incoming missiles.

Souvenirs of our journey round this planet might include sand alchemically transformed by the heat of a nuclear blast, a belt buckle from a drowned Soviet submariner, an irradiated restaurant table leg, a nuclear warhead dropped by accident in someone's back garden.

A $25-million three-stage Trident 2 missile, fitted with dummy warheads, cartwheels out of control only four seconds after being launched from the nuclear submarine USS *Tennessee*, submerged about 50 miles off Cape Canaveral in Florida, on 21 March 1989. This was the fifth Trident 2 rocket test to end in failure.
(Credit: Associated Press)

The route runs from Alamogordo to the Z-9 buildings at the Hanford Reservation, via desert test sites on the edges of Mongolia, Pacific coral reefs, the Antarctic, a Swiss cave, the Himalayas and the inner reaches of outer space.

The stories in this book bear witness to the daily risks we are running and the night-time dangers inherent in the nuclear dream. They lift the shroud that cloaks an enormous industrial enterprise designed to defend rival ideologies: an atomic papacy of Byzantine splendour and colossal scale with deep and historic institutional roots and a secretive bureaucracy.

Against this are ranged the unborn, the innocent and the unsuspecting. The widows of the ex-servicemen who happily posed for pictures in the Pacific sunshine on board irradiated battleships in 1947, and collected a cancer along with their discharge medal. The children of Kiev, who were not evacuated until *after* their traditional May Day parade had been dusted with radiation. The Navaho shepherds, the Mexican steel workers, the Marshallese islanders, the Australian Aborigines — all have been on the receiving end of this invisible threat.

In these stories, valves never work, computer programmes glitch, barrels leak, weather conditions are never right: the unexpected, the bizarre, the last-thing-anyone-would-have-expected, has the annoying habit of happening. Murphy would understand.

'There is a latent fear, almost an intuitive folk wisdom belief...that something just has to go wrong in anything that complicated...people believe in Murphy's Law.' (P. Bracken, *The Command and Control of Nuclear Forces*, Yale University Press, 1983.)

Murphy's Law was coined in California's Mojave Desert in 1949. Major John Paul Stapp had just risked his life on an experimental rocket sled that pushed him beyond 31 Gs — 31 times the force of gravity. He survived only to discover that none of the G-measuring devices had worked.

Captain Edward Aloysius Murphy Jr was called in to find out what had gone wrong and found that somebody had installed each of the six G-measuring devices backwards. 'If there's more than one way to do a job and one of those ways will end in disaster, then somebody will do it that way,' Murphy remarked.

Thus Murphy's Law was born.

Murphy was later to comment: 'My original statement was to warn people to be sure that they cover all the bases because if you haven't, you're in trouble. It was never meant to be fatalistic.'

Murphy's Law has since achieved lasting popularity, and has made several people, but not Murphy, quite rich.

Homo sapiens retains a natural scepticism towards nuclear matters, and it is encouraging to note that radiophobia is one of the fastest growing human attitudes in the world, in *all* societies. This global phenomenon is a sign that ordinary people instinctively understand that radiation damages living cells, and an indication that they are not prepared to suffer its consequences any longer. This book is written in the belief that the information it contains should be public knowledge; in the hope that sense will prevail. Our common human future is at stake: such odds make this the most important issue of our times.

John May
Lewes, July 1989

Introduction

The Greenpeace Book of the Nuclear Age is the most comprehensive popular account of both civil and military nuclear accidents produced to date. It is not a work of academia, a scientific treatise, or a book of pure propaganda. It *is* a hidden history written for the best journalistic reasons — strong stories and an even stronger reason to tell them.

This book does not document all nuclear accidents. That would be both pointless and encyclopaedic. Our selection procedure for accidents involving nuclear reactors was based on the **Bertini** report, which used the following categories: 'The accidents selected for inclusion fulfil at least one of the following conditions: 1. caused death or significant injury; 2. released a significant amount of radioactivity offsite (e.g., many times the maximum permissible concentration for extended periods of time); 3. resulted in core damage (melting and/or disruption), or core damage was suspected although it did not actually occur; 4. resulted in severe damage to major equipment; 5. caused inadvertent criticality; 6. was a precursor to a potentially serious accident; 7. resulted in significant recovery cost (e.g., greater than half a million dollars).'

We added a few more: 8. strange but true; 9. significant for the survival of life on this planet; 10. unbelievable.

The military have their own euphemisms to describe significant incidents:

> **Broken Arrow:** An unexpected event involving nuclear weapons or radiological nuclear components that results in any of the following situations where a risk of outbreak of nuclear war does *not* exist: 1. nuclear detonation; 2. non-nuclear detonation or burning of a nuclear weapon or radiological nuclear weapon component; 3. radioactive contamination; 4. seizure, theft, or loss of a nuclear weapon or radiological nuclear weapon component, including jettisoning; 5. public hazard, actual or implied.
>
> **Bent Spear:** An unexpected event involving nuclear weapons or radiological nuclear weapon components which does not fall into the nuclear weapon accident category but: 1. results in evident damage to a nuclear weapon or radiological nuclear weapon component to the extent that major rework, complete replacement or examination or recertification by the US Department of Energy (DOE) is required; 2. requires immediate action in the interest of safety or nuclear weapon security; 3. may result in adverse public reaction (national or international) or premature release of classified information; 4. raises suspicion that a nuclear weapon has been partially or fully armed; 5. could lead to a nuclear weapon accident and warrants that high officials of the signatory agencies be informed or take action.
>
> **Faded Giant:** An uncontrolled reactor criticality resulting in damage to the reactor core or release of fission products from the reactor core to the atmosphere or surrounding environment.
>
> **Dull Sword:** An occurrence that results in the loss of control of radioactive material presenting a hazard or potential hazard to life, health or property.

We've included all those known accidents that fall into these categories and many others that remain officially unannounced or conveniently forgotten.

In the case of many of the military nuclear accidents it is not clear whether nuclear weapons were on board or not. Our reasons for including these accidents have been best expressed by Lloyd J. Dumas in 'National insecurity in the nuclear age', *Bulletin of the Atomic the Scientists* (May 1976). He writes:

'The justification for including accidents in which the presence of nuclear weapons was denied or unspecified is threefold. First, wherever it is possible to deny or gloss over the involvement of weapons of mass destruction in an accident, it is a virtual certainty that this will be done regardless of the actual facts, because of the desire to avert fear and adverse public reaction.

'Second, there is some specific evidence that the distinction between "nuclear capable" delivery systems and systems which actually carry nuclear weapons is more apparent than real. As retired Admiral Gene La Rocque has testified relative to the US Navy: "My experience...has been that any ship that is capable of carrying nuclear weapons, carries nuclear weapons. They do not off-load them when they go into foreign ports such as Japan or other countries. If they are capable of carrying them they normally keep them aboard ship at all times except when the ship is in overhaul or in for major repairs."

'Finally, even if we unquestioningly accept that nuclear weapons were not present except where their presence was specifically confirmed by authorities, we must still consider accidents of this type to be serious and relevant to the nuclear accidents problem. The absence of these weapons at the time of accidents can be considered good fortune, but we must realise that the presence of nuclear weapons would not have prevented the accidents.'

There are many disciplines, technologies, nations, and substances bound up in this book. We have tried to pick our way through this thicket of obscurity and to offer a cogent, informative account of the issues which will be of interest to the general reader.

The book covers incidents in the USA more thoroughly than those in other countries for one very good reason: the Freedom of Information Act. We have been mindful of this unavoidable bias and have endeavoured to paint a global picture of this most global of topics.

It should be pointed out that the management of nuclear power in those countries which do not have a Freedom of Information Act is unlikely to be better — indeed it is probably worse; is merely harder to monitor than in the US. Throughout the brief history of the nuclear enterprise, 'no news' has never been good news.

The stories in this book are constantly developing and changing their shape as fresh data adds to our knowledge or alters our perspective. This is unlikely to be the final word on *any* of these subjects, and it may well be that some of the facts in this book will later be proved incorrect for just this reason. This is an area dominated by secrecy and falsehood. We have been consistently deceived and constantly misinformed by nuclear authorities in the past. Under such circumstances the 'truth' is always questionable.

This book begins with a series of opening essays by specialists in their individual fields. They deal with the 'security' supposedly provided by nuclear weapons; the political history of civil nuclear power; the dangers of nuclear accidents and the possibility of accidental nuclear war. The next section, *Fundamentals*, presents the basic science and technical details of nuclear power and nuclear weaponry, as well as explaining the effects of radiation on living organisms. For easy reference, we give an index of key terms at the end of the chapter (p. 50). This is followed by explanations of abbreviations and units of measurement.

The main body of the book is divided into the five decades since man-made nuclear energy first made its presence felt. Each section contains detailed accounts of the principal nuclear incidents of that decade (the geographical locations of which are indicated on maps on pages 52 to 59), as well as a compendium of shorter additional stories. A marginal symbol denotes the type of accident and a key to these is given on p. 50.

A concluding chapter gives Greenpeace's own perspective on the current state of the Nuclear Age and the dawning possibility of a planet no longer menaced by the threat of nuclear disaster — civil or military.

In compiling the material, a number of sources have been referred to constantly. These appear in bold type throughout the text and full bibliographical details are given on p. 352. Each main story is individually sourced at the end of the book, while sources for the shorter additional stories appear at the end of each entry.

We have passed the information in this book through a matrix of experts to whom we give due acknowledgement. It remains to be said that any mistakes that remain are the responsibility of the author.

The Illusion of Security

One cannot open a newspaper, tune in to the radio or turn on the TV without being reminded of the atom bomb. Its presence is all-pervasive; the dangers inherent in its existence are consistently alarming. This book bears witness to the fact that nuclear accidents occur with frightening regularity. Missiles explode, tests backfire, bombers crash, ships and submarines collide and bomb factories leak. The many accidents become that much more important when they are juxtaposed with a nuclear weapons system that is always wired and primed, awaiting the provocation that will activate it and destroy everything.

The existence of the bomb not only threatens *our* existence directly. It also threatens our economies with bankruptcy. New generations of ever more deadly nuclear weapons — Midgetman missiles, Stealth bombers, Trident submarines — come with obscene price tags. Star Wars solutions to the nuclear peril promise only further financial ruin. We now learn that the shopping spree of the Reagan era has been accompanied by widespread waste and corruption; far from being satisfied with their increased budgets and forces, the generals and admirals continue to press for more funding.

As tensions between the North Atlantic Treaty Organisation (NATO) and the Warsaw Pact decline, there is little effect on the arms spending race. There seems to be no end point in sight when the military will accept that their nuclear arsenals are large enough and that the technology they possess needs no further improvement. They will never volunteer to redirect their military budgets towards more humane and productive ends. Their weaponry will never make the world a safe enough place for them. They cannot promise us real security or assure us meaningful defense.

Not with the bomb. The very nature of the bomb demands secrecy and strengthens autocratic power, thus creating inevitable conflicts with the public interest. When Cruise and Pershing missiles were successfully installed in Europe, NATO viewed this as a victory over the mass peace movements. The signing of the Intermediate-range Nuclear Forces (INF) Treaty and the completion of the Strategic Arms Reduction Treaty (START) were seen by the nuclear advocates as moves that would nullify public opposition to nuclear weapons once and for all. The attempts to force New Zealand to modify its anti-nuclear stance — by threatening trade sanctions and by throwing it out of an intimate and long-standing defence alliance — were designed to serve as an example to other governments that they had better not question their adherence to nuclear policies.

Time and again, governments and the military-industrial complexes are found to have lied to the public about the real dangers of the nuclear enterprise. These dangers are constantly on the increase as nuclear proliferation spreads the technology and know-how to a wider group of nations, most of whom are situated in the most unstable political regions of the world.

Even in a new era of *détente* and *rapprochement*, the bomb continues to confound the optimists. An arms control agreement is signed to eliminate a class of weapons and others are redesigned to take their place. A new lock is invented to make weapons harder to use without authorisation, a new procedure is developed to repair some flaw in the war plans, yet new problems continue to be uncovered. No matter how many failsafe gadgets and gizmos are installed, no matter how many assurances are given that the bomb is safe, there is always another defect, another deception.

Despite all the attempts of the nuclear advocates to preserve the bomb's anonymity, it constantly seeks and demands our attention. Despite their best efforts to establish the bomb as an acceptable and respectable fact of modern life, public opposition hasn't died. The public remains as worried, as uncomfortable and as contrary in its views on the subject as ever. Public opinion may be temporarily subdued and inert when times are good but when crises erupt, or superpower jousting becomes too cavalier, public nervousness is quickly reawakened. Being against nuclear weapons is the seventh sense.

It is not so much our fear of imminent world destruction that creates an atmosphere of permanent unease but rather the day-to-day revelations of technological failure, mismanagement and close calls that have brought home the true meaning of what living with the bomb is all about.

For four decades, hundreds of thousands of engineers and scientists have been employed to increase the bomb's reliability and performance, and aid in its anonymity. Strategists and theorists have exhaustively examined scenarios of destruction and simulated crises in a vain attempt to control the unpredictable. Governments have cynically manipulated information to try and sanitise the bomb and neutralise opposition to it — all without success.

In the 1980s the public became even more profoundly distrustful of nuclear weapons in an era when the Reagan Administration's hawkish policies and the breakdown of US-Soviet relations threatened to lead to global conflict.

With his Star Wars dream, Reagan further undermined faith in the nuclear system. He told the public that the promise of the bomb — that it would prevent war — was no longer true and that we now required a defensive shield if we were to be safe.

The rhetoric may have been intended to mobilise the public in support of another new and expensive mega-programme but the message was read in a different way: it seemed that technology had rendered the deterrence argument obsolete; nuclear weapons were no longer perceived as a means of guaranteeing peace.

With the arrival of Gorbachev, a complete reversal of the global political picture has taken place and now the nuclear priesthood and the Cold War warriors are busily promoting every counter-argument to disarmament. They say we should be cautious about developments in the Soviet Union and they predict — almost wish — that Gorbachev will fail in his bold attempt to reduce the military burden in his country and eliminate the nuclear threat. As far as they are concerned, the nuclear bomb has provided more than 40 years of 'peace'. They believe that changing existing nuclear policies will upset the international order; that to tamper with the bomb is not only risking some amorphous catastrophe, but also risks handing control of the entire nuclear enterprise over to the public.

But, for ordinary people, disenchantment with the concept of nuclear defence has set in, and those that govern us will not find it easy to reverse this process. We will not return to our earlier position of unknowing trust and naïve faith in nuclear weapons.

Moreover, the issue is not just whether or not we retain the bomb: arguments over how public funds should be allocated in our society; what priority we are going to give to protecting the environment; and what kind of government we need all arise out of the nuclear debate.

This is no longer an argument about levels of relative risk. We can no longer accept the presence of nuclear weapons as a necessary evil if we want to live in a free and civilised world. The risks that we share and the sacrifices we are being asked to make are justified by the contention that there is no alternative.

An open record of nuclear accidents, incidents and abuses represents the greatest danger to those who support the nuclear system and want to maintain the status quo. What we need is a fuller airing of the true atomic history: of the true cost of nuclear testing and nuclear research; of the real nuclear strategies and plans; an accounting of the full extent of the pollution and the problems caused by nuclear waste. The compilation of this book is just one stage in that process.

Nuclear weapons have been the subject of the most intensive campaign of secrecy in human history. But when all the obscurities and obfuscations, all the bogus facts and rhetoric are stripped away, we must conclude that nuclear weapons are wrong. This is why they must be eliminated. Let us hope that the final chapter of the history of the nuclear age will describe how we nearly went wrong, rather than how we did.

William M. Arkin
Washington DC, July 1989

Reactor Risk

The achievements of atomic science and engineering, if looked at purely in terms of the speed with which massive military production programmes were designed and developed, are quite extraordinary. Sadly, the enormous human genius gathered to build the 'bomb' and its support infrastructure did not attend to comprehensive safety measures, the need for which would surely have been recognised had not the pressure of wartime, mixed with the exhilaration of discovery, led to corners being cut. A clear legacy of the short-term thinking of the 1940s is the massive clean-up programme, costing well in excess of $110 billion, currently being undertaken at the US military nuclear materials production plants (see p. 301).

By the early 1950s, post-war euphoria had led to unbridled atomic optimism, whereby the bounties tied up in the atom would be liberated, making life a heaven on earth. The prevailing mood was amply demonstrated by the 'Atoms for Peace' programme, launched by Eisenhower at the United Nations (UN) in 1953. Designed to bring to world attention the positive uses of atomic energy, 'Atoms for Peace' was not an entirely philanthropic venture. The US government had two ulterior motives. Firstly to ensure that American industry captured a major advantage in global sales of nuclear power plants (though none were yet built in the US). Secondly, to control the global supply of nuclear fuels, for the US was worried that other countries might wish to join the exclusive weapons 'club', of which only themselves, the Soviet Union and Britain were at the time members.

In the drive to commercialise nuclear energy, the US Congress amended its extremely draconian Atomic Energy Act, originally drafted in 1946 to protect the secret of the atom for America. In 1954, Congress attempted to entice private industry into collaboration with the federal atomic energy establishment by offering partnership with the US Atomic Energy Commission (AEC). The drawbacks of the technology were barely mentioned in public; but this did not mean they were not recognised in private.

For instance, in March 1955 Professor George Weil wrote in *Science*, the prestigious weekly journal of the American Academy for the Advancement of Science: 'The beneficial prospects associated with the development of nuclear energy have been widely publicised. On the other hand, discussions of the unpleasant aspects have been limited almost exclusively to technical meetings and publications.'

Weil went on to explain the almost uniquely dangerous characteristics of nuclear reactors with their potential for major accidents. He was not saying this from the point of view of an anti-nuclear antagonist; he had in fact been the scientist who extracted the first fuel rod from the Chicago 'pile' reactor in 1942.

Another warning was issued by Dr James Conant, one of Weil's colleagues in the atomic bomb effort of the 1940s. He became President of both the American Chemical Society and Harvard University. In 1952 Conant predicted that atomic power would flounder because of the virtually insoluble problem of radioactive waste disposal. Such doubts were rare; and their message fell on deaf ears. As Carroll L. Wilson, who was the General Manager of the US AEC from 1947-51, recalled in an article in the *Bulletin of the Atomic Scientists*, June 1979: 'Chemists and engineers were not interested in dealing with waste. It was not glamorous, there were no careers, it was messy and nobody got "brownie points" for caring about nuclear waste.'

The risk of atomic energy was recognised by the atomic insiders, but doubts were kept private except on rare occasions. One example is the testimony given by the nuclear 'guru', Dr Edward Teller — the so-called father of the H-bomb — who appeared before the US Congressional Joint Committee on Atomic Energy (JCAE) in 1953. Teller headed the US AEC's Reactor Safeguards Committee, established in 1947, and admitted candidly that 'up to the present time we have been extremely fortunate in that accidents in nuclear reactors have not caused any fatalities. With expanding application of nuclear reaction, and nuclear power, it cannot be expected that this unbroken record will be maintained.'

Despite this admission, Teller did not urge caution over the development of atomic power. Rather, he promoted it. 'No legislation will be able to stop future accidents and avoid completely occasional loss of life. It is my opinion,' he said, 'that the unavoidable danger which must remain after all reasonable controls have been employed must not stand in the way of rapid development of nuclear power.'

It was deemed unacceptable that private doubts about nuclear safety should jeopardise the whole atomic energy programme. The secrecy that had surrounded the development of the atom bomb was transferred, shroud-like, to the atomic power projects spawned in its post-war wake. It was understood by the small coterie of insider scientists that nuclear technology posed two fundamental risks:

- It produced low level radioactive effluents that could be discharged into the environment in gaseous and liquid form; high level nuclear liquid wastes that must be stored in isolation for very long periods of time; and low and intermediate level solid wastes which required a disposal route.
- A reactor accident might cause the release of large quantities of deadly fission products into the environment.

However, owing to the policy of secrecy, in no country which embarked on nuclear power programmes for electricity production did an environmental risk-benefit debate take place, not even in the US where the political culture should have permitted it. Instead, the US adminstration's nuclear establishment determined to forge ahead with its plans, and promoted the benefits of atomic energy by practically force-feeding its Power Reactor Demonstration Program, launched in 1955, with tax-dollars, as part of a ploy to seduce private industry into partnership.

The US AEC also took every opportunity to sell the 'nuclear dream' abroad. A few months after the launch of the demonstration programme, US atomic technology was on sale at the first global fair of the commercial atomic age, at the UN-sponsored Peaceful Uses of Atomic Energy Conference in Geneva. Reactor risk was relegated to impolite asides, as salesmanship was the main business.

The development of the nuclear safety regulatory regime in the United States is central to an understanding of the global agreement developed on reactor risk because, in commercial terms, it was US-designed and US-licensed reactors that were to dominate the world market. The story of how science was politicised and perverted for the sake of expediency is a salutary one. Almost anything was acceptable if it did not interrupt the expansion of the atomic empire worldwide.

In 1956 the JCAE, recognising that atomic power would not be attractive to private industry unless financial liability was covered by government, instructed the US AEC to prepare a study of the full implications of a reactor accident. The final report, prepared by Brookhaven National Laboratory on Long Island, New York, was published in March 1957 in a sanitised form. Codenamed WASH-740, the Brookhaven report on 'the theoretical possibilities and consequences of major accidents in large nuclear power plants,' came up with some startling figures.

It established the chance of a major accident as being between in 100,000 and one in 1,000 million per year per reactor. Were such an accident to occur, the report estimated that it would result in 3,400 fatalities, 43,000 injuries and around $7,000 million-worth of property damage. As there was literally no operational experience of commercial-scale nuclear reactors, the entire report was based not on reality but on a theoretical model. Nonetheless, with such a prognosis, Congress was forced to legislate for a limited liability for the prospective nuclear operators. The Price-Anderson Act provided for a limit of $560 million in a federal fund, to be paid out in compensation following any atomic accident. (In 1987 Congress increased the limit to $7 billion.) The nuclear industry was to have no residual liability; the public would have no common-law right to bring any claim against the builders or operators of any accident-damaged plant. The Act was passed after perfunctory debate in the House of Representatives and none in the US Senate. The very existence of the Price-Anderson Act indicated that the nuclear industry was calling into question the reliability of its own reactor-safety propaganda.

The Price-Anderson Act had an initial 10-year lifespan. In 1964 the US AEC began to re-evaluate WASH-740 to prepare for the Act's forthcoming renewal in 1967. This review proved even more controversial than the original Brookhaven Report. It was being prepared by the AEC at a time when its own chairman, Glenn Seaborg, was telling conferences in Europe that 'great strides' had been taken with nuclear technology and that the prevailing optimism was 'rooted in experience'.

The real experience was somewhat different. The WASH-740 update had analysed a larger reactor type, which was being planned for commercial use. In a letter to the AEC's *ad hoc* reactor-safety steering committee (which was preparing the update) Stanley Szawlewicz of the AEC's Division of Reactor Development warned of the dangers of publishing a revised WASH-740. He pointed out that the public would be worried by the conclusion that the reactor accident hypothesised could have consequences that would be more serious than an atomic explosion. Extrapolations to this effect could be made from the technical data in the analysis and they might have very serious implications when trying to obtain site approvals for future reactors.

The new Brookhaven study had warned that the larger nuclear plants would hold as much long-lived radioactive material as would be released by about 1,000 Hiroshima-sized atomic bombs. It added that should even a small fraction of the radioactive inventory be released, 'the possible size of the area of such a disaster might be equal to that of the State of Pennsylvania.' Fifteen years later, this prophesy was nearly fulfilled with terrible irony during the Three Mile Island accident at Middletown, Pennsylvania.

The Brookhaven scientists told the AEC that the revised study had shown there was nothing inherent in the design of current reactors or safeguards systems which guaranteed 'either that reactor major accidents will not occur or that protective safeguard systems will not fail. Should such accidents occur and the protective systems fail, very large damages could result.' The AEC was appalled, not so much by the potential for an accident, but by the prospect of losing public and industry support for nuclear power, so carefully cultivated over a decade.

Despite the fact that when curious Congress people enquired about the new study they were informed by the AEC that it was not completed, Seaborg sent a letter to the JCAE in 1965 recommending that the Price-Anderson Act be extended. He stated that the likelihood of major nuclear accidents was 'still more remote', although the consequences could be greater. The Act was extended.

The nuclear industry's own lobby organisation, the Atomic Industrial Forum, advised the AEC in 1965 that the revised version of WASH-740 'should not be published in any form at the present time'. Taking the advice, Seaborg thereafter insisted that the study was never completed. Yet in an internal AEC summary of the project written in 1969, one of the members of the committee that prepared the review, Forrest Western, stated that: 'an important factor in the decision not to produce a complete revision of WASH-740 along the lines proposed by the Brookhaven staff was the public relations considerations.' Once again public relations were given priority over public safety.

From the mid-1960s, when the AEC was covering up the real risk of reactor accidents, the global sale of nuclear plants rose rapidly. Had the truth been widely publicised it seems very doubtful that this important expansion would have been possible.

By the early 1970s another nuclear risk issue had begun to break in the United States, this time over the effectiveness of the Emergency Core-cooling System (ECCS), essential for light-water reactors, which have high power densities. The big difference in this dispute was that, for the first time, outside critics of nuclear energy led the charge.

The political pressures generated were such that Congress agreed to special hearings on the ECCS, which began in January 1972. The quality assurance, promoted as being of the highest standard by the nuclear proponents, came under heavy fire. So, with serious internal expert criticism also being voiced, the AEC decided, in March 1972, to initiate a further major study on accident probabilities. The new AEC chairman, James Schlesinger, decided that what was needed was an in-house study; but to give it an objective and independent appearance, the AEC sought an outsider as a titular director.

After some false starts, the AEC got their man. His name: Norman Rasmussen, a physics professor at the prestigious Massachusetts Institute of Technology (MIT). Rasmussen was not a nuclear reactor specialist, and had never published any papers on reactor safety. However, he had published a short opinion article in January 1972, in the industry magazine *Nuclear News*, where he had described the risk of a major accident at a nuclear plant as 'very small' and 'insignificant'.

Rasmussen accepted the appointment immediately, and on the very same day enthusiastically wrote back to the AEC with his outline on how to conduct the study. He said, in a joint memorandum with his MIT colleague Dr Benedict, that the AEC would have control over the final report so that the conclusions would be put 'into a form which the AEC were willing to issue'. They added: 'The sensitive nature of these studies will require careful control of all information releases.' Once more the consequences of the radiation releases were considered secondary to the consequences of damaging information releases.

The Rasmussen Reactor Safety Study (RSS) developed a so-called fault-free analysis, setting out the accident options and consequences on diagrams that resembled family trees. Fault-free analysis is a way of presenting complex events in a streamlined fashion: it is well suited to incorporating hardware failures, but not design deficiencies. The RSS did not review the validity of the basic computer codes used by the nuclear industry, having neither the time nor the experts at hand to carry this out. Instead, the RSS relied on data from the plant operators.

After a series of carefully planned leaks to the press in the first six months of 1974, giving reassuring hints about the findings of the RSS, a draft version of the Rasmussen Report was released in August. The AEC had organised its own internal review group to analyse the report. Their comments ran to over 200 pages, but were not made public. These comments were critical of various flaws in the methodology adopted by Rasmussen. Despite knowledge of the criticisms from their own experts, the AEC promoted Rasmussen's study with unabashed enthusiasm. A massive publicity campaign, involving pre-recorded television messages, was launched across the US. The message was that nuclear plants were proved to be safer than just about anything else in modern life.

Scientists were not so sanguine. The October 1974 issue of the *Bulletin of the Atomic Scientists* warned that the RSS was 'essentially an in-house study by an agency under heavy pressure to get critics off its back. In no sense is it an independent evaluation...instead it is a defense of the US AEC's policies.' The most damning criticism came in the summer of 1975, when the American Physical Society (APS) published an evaluation from its study group on light-water-reactor safety. WASH-1400, as the RSS was coded, began to be seen as a whitewash, lacking scientific integrity and credibility. A revised version, NUREG 75/014, was released in October 1975.

In June 1976, the US Environmental Protection Agency (EPA) published a trenchant critique of the revised WASH-1400. Amongst other concerns, the EPA asserted that Rasmussen had underestimated the latent cancer risks from a reactor accident by a factor of between two and 10.

Far from convincing the US public, the prospective investors in nuclear energy, or the US scientific community, that reactors were relatively risk free, Rasmussen's nine-volume RSS, running to 2,300 pages, had further undermined confidence in an industry that by the mid-1970s was waning due to economic uncertainties. Further reviews of the RSS were conducted by the Nuclear Regulatory Commission (NRC), which replaced the old AEC in 1975. The NRC's review group concluded that Rasmussen's report had probability calculations containing so many uncertainties as to make it virtually meaningless. Many of the calculations were 'deficient when subject to careful and probing analysis.' Cogent comments from critics had gone unacknowledged or were evaded.

In January 1979, after four months of evaluating the review group's criticisms, the NRC accepted them and publicly repudiated the Rasmussen Report. The NRC stated tersely that it was withdrawing any explicit or implicit past endorsement of the Executive Summary of WASH-1400. Two months later, barely two miles away from the family farm where Norman Rasmussen had been brought up alongside the Susquahenna River, the Three Mile Island plant suffered its serious accident.

Such are the ironies of nuclear risk. The NRC got to the truth just in time to save face, but not to save Three Mile Island. Nuclear risk remains ever present.

David Lowry
Milton Keynes, July 1989

Sources

Bulletin of the Atomic Scientists, 1947 and 1951
Science (4.3.55)
R. Nader & J. Abbotts, *The Menace of Atomic Energy* (W. W. Norton, 1977)
N. Moss, *The Politics of Uranium* (Andre Deutsch, 1981).
D. Ford, *The Cult of the Atom: the Secret Papers of the Atomic Energy Commission* (Simon & Schuster, 1982)

Command and Control

Nuclear weapons, like all complex high technologies, are subject to failure. It is part of common expectation that, however competent safety measures may be, aircraft will crash, power grids overload, nuclear power plants leak, and satellites stray out of orbit. While the consequences of such events may disrupt normal behaviour, or claim lives, the prospect of a nuclear weapons accident is perceived to be of a different order of seriousness. Images of the devastation at Hiroshima and Nagasaki provide a mental standard against which the potential destruction of a nuclear weapons accident can readily be measured.

It is no surprise therefore that, almost since the dawn of the nuclear era, measures have been taken by nuclear governments and military authorities to reduce the chances of nuclear weapons accidents and the risks of unauthorised or unintentional use of nuclear weapons which might precipitate nuclear war. The measures taken to control nuclear weapons operations, and ensure only the authorised and intentional use of nuclear weapons, can loosely be described by the blanket term 'command and control'.

For safety purposes the nuclear powers have adopted a variety of procedural, physical and electronic measures, firstly to reduce the chances of an accident having serious repercussions — such as a nuclear detonation or the widespread dispersal of radioactive contamination — and secondly to reduce the chances of an accident happening. The extent to which these measures have guaranteed, and will continue to guarantee, the public safety remains the subject of debate.

The overriding principle of nuclear weapons safety has been to construct nuclear weapons in such a way as to reduce almost to zero the chance that an accident might detonate a nuclear weapon, no matter how severe the circumstances of the accident. Although there has never been an accidental nuclear detonation — a record with which the nuclear powers can justifiably be pleased — this provides no firm guarantee for the future. Moreover, for many, the potential effects of such an accident are so grave that the only acceptable long-term risk is zero. No one in either the government or the armed forces has been prepared to state that there is no risk at all.

Procedural safety measures are designed to ensure that no unauthorised or irrational person will be able to cause the use of a nuclear weapon. Measures such as the two-person rule (which means that all nuclear weapons operations — from the highest level of command, to the soldier in the field — cannot be effected without involving at least two individuals) are designed to prevent incidents such as that which took place at RAF Sculthorpe, a USAF base in the UK, in October 1958. On that day Master Sergeant Leander Cunningham went berserk following a nervous breakdown, locked himself in a nuclear weapons store and threatened to commit suicide by detonating one of the weapons (Campbell, 1984). Similar aberrant behaviour can result from drug and alcohol abuse which, as research by Herbert Abrams (1988) has shown, is presently running at high levels in the nuclear armed services.

Physical safety measures include a variety of devices such as environmental sensing switches: these render the nuclear weapon inoperable until they sense some specified environmental change, e.g. the zero gravity of free-fall. If a nuclear weapon is fitted with such a device the switch will remain open, and the weapon 'safe', while it is on the ground, but if it accidentally falls from an aircraft it will 'think', having detected zero gravity, that it is being used against a target and the switch will close, thereby threatening to detonate the weapon. Similar switches which detect altitude or barometric pressure, or use radars, are often incorporated in weapons in series. The crash of a B-52 at Goldsboro on 29 January 1961 (see p. 140), caused a nuclear weapon to fall from the aircraft and five of its six switches closed before it hit the ground, providing clear evidence that even complex safety systems can be circumvented by accident.

The threat of irrational or unauthorised (such as terrorist) use of nuclear weapons has also been countered by electronic measures. The most important of these is probably the Permissive Action Link (PAL). PALs can be understood as electronic 'locks' which render nuclear weapons inoperable unless the correct electronic 'key' is inserted to unlock the weapon (Caldwell, 1987.) PALs were first fitted to American nuclear weapons in the early 1960s following a number of scares in Europe where nuclear weapons almost fell into unauthorised Allied hands. (Stein and Feaver, 1987.) By keeping PAL 'keys' in relatively few senior military hands it has been possible to exercise a high degree of control over nuclear weapons deployed at any number of stage facilities. PALs have proven so successful that the US has taken the unprecedented step of offering information about their design and use to the UK, France, and the Soviet Union (Kleine and Littel, 1969), in an effort to encourage their wider adoption and thereby improve global safety and security.

It needs to be added that PALs are expensive and early designs were crude and relatively easy to circumvent. The fact that improved designs are installed mostly on new weapons and relatively few new PALs are retrofitted on older weapons means that many nuclear weapons are presently not as well protected as they could be. Moreover, the navies of the US and UK have eschewed the use of PALs altogether on their submarine-launched ballistic missiles, and also on short-range naval nuclear weapons. This is probably also true for the other nuclear powers.

As well as the systems designed to try and prevent accidents from occurring and to reduce their potential consequences, one ought to mention the organisations which exist to respond in the event of a nuclear weapons accident. Both the US and UK have standing units deployed ready to respond to accidents around the world. The US has Accident Response Teams (ARTs) deployed at over 450 sites around the world (NARCL, 1987) and Britain is thought to have some units both in the UK and in countries where British nuclear weapons are routinely stored (at present West Germany). ARTs usually have personnel specially trained in security, radiological monitoring, medicine, fire control, weaponry, construction, and other activities necessary to control the accident site and eventually restore it to normal usage. Not all ARTs have all these facilities and hence it is usual for several ARTs to respond to an accident, bringing together specialists in a range of related fields. As many as 500 personnel may eventually assemble at an accident site for periods as long as several months (NARP, 1984). The Soviets, French and Chinese probably have similar arrangements, although access to the evidence is less clear.

History does, however, show that, despite considerable efforts, nuclear weapons accidents persist as a feature of the nuclear arms race. Given that this is the case, every effort must be made to reduce further the incidence of accidents and mitigate their potential effects.

Aside from the problems of retaining command and control over individual nuclear weapons, perhaps even more serious difficulties become apparent when one considers the early-warning, communications, and computer systems which enable the weapons to be used. The worst military nightmare of the past 20 years has been the prospect of an adversary, using accurate nuclear weapons with short flight-times, attacking vital government and military headquarters, communications systems and nuclear weapons bases in an unexpected 'bolt from the blue' attack, and thereby 'winning' the nuclear war (Steinbruner, 1981/2). However implausible one considers such a scenario, it has been taken sufficiently seriously by nuclear governments and military planners to alter profoundly the nature of nuclear weapons command and control, and has greatly increased the likelihood of the accidental or unintentional use of nuclear weapons in the process.

The possibility of rapid destruction — these days, missiles from either superpower may reach their adversary's territory in under 15 minutes — has meant that many nuclear weapons on both sides have been placed on ever higher levels of alert, and computers have increasingly been used to interpret sensor data and provide attack warning, to facilitate prompt retaliation. At the extreme, a 'launch on warning' (LOW) policy has been advocated, and may have been adopted as an option by the nuclear powers. This proposes that forces be readied for use at all times so that they can be launched *on warning of attack* merely. Obviously, this depends on extremely accurate and reliable early-warning technology since any error might cause a massive launch of nuclear missiles (Aldridge, 1983).

The close relationship between nuclear weapons and computers is part of the increasing trend towards automating nuclear-weapons systems (Borning, 1987), a trend which finds its most startling expression in the Strategic Defense Initiative (SDI) programme (Roberts and Berlin, 1987). The SDI project — now a far more modest programme than that envisaged by President Reagan in 1983 — depends on response times to attack warning which are so short as to preclude meaningful human control. It is too early to say that computers control nuclear weapons, but the increasing passivity of human operators working from data provided and presented by computers is a cause for real concern.

Computers are not infallible. Since their initial use for early warning in the US SAGE air defence system in the 1950s, and in the BMEWS radar early-warning system in the early 1960s, many serious errors have occurred. The generation of false alerts of nuclear attack is undoubtedly the most serious manifestation of these errors. In 1960 a BMEWS computer undergoing trials erroneously indicated that a large-scale nuclear attack had been launched against the USA. The computer had accidentally lopped two zeros off the distance indicator of the radars and as a result showed a large target at 2,500 miles (which was taken to be a massed Soviet attack) instead of at 250,000 miles (which was in fact the moon). (Morrison, 1984.) Twice in June 1980 a malfunctioning computer chip, valued at just 46 cents, caused an attack-warning computer to indicate that a Soviet missile attack against the USA had begun (see p. 255). The fault was soon identified, but not until elements of the US nuclear retaliatory force had been readied to strike back (Hart and Goldwater, 1980). Following the 1980 incidents, a number of congressional investigations took place which concluded that false alerts were likely to occur again. The situation may be even worse on the Soviet side because, in general, they have far less sophisticated and reliable computers than the West.

In spite of this, many eminent scholars have argued that the risk of a nuclear weapons launch as the result of a false alert is, in fact, very remote in peacetime. This view trusts to the checks which presently have to be made by human operators after any indication of attack warning — including checking with more than one sensor system — and to the general perception that, in peacetime, a nuclear attack is highly improbable. It may indeed be true that warnings of nuclear attack are assumed to be false in peacetime and though some steps are taken to prepare nuclear forces to retaliate, these do not proceed too far before being held or stood down. But in periods of international tension or conventional war, dependence on such a benign interpretation of false alerts may no longer be possible. Perceptions may be revised — may be led to *expect* a nuclear attack — and forces will undoubtedly be placed on even higher alert-levels for immediate retaliatory launch. As the likelihood of nuclear attack grows, so too will the risks of a false alert.

Moreover, in periods of international tension the plans for nuclear war might be implemented in their early stages and so the risk of nuclear weapons being used unintentionally also increases. This could happen either as a result of the blurred distinction between nuclear and conventional systems or through the loss of centralised control of nuclear weapons — which might prove inevitable in wartime — and the resulting escalation of the conflict.

As a crisis situation escalates, adversaries will watch each other's military activities ever more closely, using a variety of intelligence-gathering apparatus. Steps taken to mobilise forces and prepare for defence might be interpreted as provocative by the adversary; a situation exacerbated by the difficulty of distinguishing defensive from offensive preparations. Detecting adversarial moves to prepare for war might encourage counter actions and stimulate a spiral of alert and counter-alert which might become mutually reinforcing and lead to the breakout of war. Parallels with World War I, in which the mobilisation of forces ground inexorably and uncontrollably towards war, have been made with the present situation in Europe (Van Evera, 1984).

In preparing for war the nuclear powers will have to take a number of important decisions about their nuclear forces. If nuclear weapons are left in storage (particularly where storage depots are close to the battlefield, as is the case in Europe) they run the risk of being overrun by advancing enemy forces or becoming the prime targets of those forces. To avoid this, nuclear weapons may be taken from storage very early in a conflict, and dispersed in the battle area. This poses a further threat If weapons are deployed without being readied for use (e.g. without having PALs unlocked) it may later prove very difficult to reach those weapons with the correct PAL 'keys', or to communicate with them.

The situation in a battle area will undoubtedly be complicated by the chaos of battle; the jamming of communications systems, and so forth. Militarily, this scenario argues for the unlocking and readying of weapons *before* they are dispersed. This would, in effect, devolve control of nuclear weapons on to many, perhaps hundreds, of officers in the battle area who might themselves come under direct conventional (or nuclear) fire. The number of fingers on nuclear buttons which could be pressed, either by accident or in response to attack, would thereby be greatly increased. Moreover, once some weapons have been used, the disruption caused — and the huge psychological impact of crossing the nuclear threshold — may be sufficient to precipitate the widespread use of nuclear weapons, a process which, once started, would prove extremely difficult to stop (see Bracken, 1983).

To a greater or lesser extent the nuclear powers are aware of these difficulties. The steps they have taken to deal with them have often only gone as far as responding to dangers as they occur. The need for direct communications between the superpowers in order to clarify political and military actions, and counter misunderstandings and misapprehensions, might have been apparent to anyone who gave thought to the problems of conducting superpower relations in a competitive and volatile world. Not until the Cuban Missile Crisis of October 1962, however, did the need for a 'Hot-Line' between the Soviet and American leaders become obvious to those leaders. Even now the 'Hot-Line' links military rather than political centres in the US and USSR and remains an inadequate tool for sensitive and critical discussions between political leaders. Moreover, it has on many occasions been subject to corruption and breakdown.

Following a spate of nuclear weapons accidents in the 1960s, particularly those involving US B-52 aircraft at Goldsboro (see p. 140), Palomares (see p. 148), and Thule (see p. 162), the superpowers agreed the Nuclear Accidents Treaty in September 1971 which committed them to immediate notification of missile warning; notification of any accidents involving a nuclear detonation or launch; advanced notification of any missile launches that might be perceived as threatening; and the use of the 'Hot-Line' to exchange information. (Goldblat, 1982.) These measures were intended to ensure that even a serious nuclear weapons accident would not be construed as the first blow of a nuclear attack.

Many areas of risk, however, are entrenched in the nature of relations between NATO and the Warsaw Pact and may not lend themselves as readily to mutual solution. Of particular concern are the present counterforce capability of nuclear forces, the high-alert postures of weapons, and the abundance of nuclear munitions in most regions of the world. The present climate of glasnost might, however, provide a way out of the impasse. A substantial reduction of the strategic nuclear arsenals will probably reduce the incidence of accidents, improve superpower relations and add to strategic stability. It is this improvement in relations between East and West and the increasing trust between all parties which should provide the climate for seeking mutual solutions to common dangers.

Nuclear safety and security will remain areas of common interest irrespective of the climate between East and West. Just as the sharing of PAL technology proved a universally beneficial measure in the 1960s so might the sharing of other safety and security technologies benefit all parties in the future. Optimistically, it should be possible to move from the present situation to one where there are fewer, less provocatively deployed, and better protected nuclear weapons as an important half-way house to further disarmament.

Shaun Gregory, Bradford, July 1989

Sources

H. Abrams, 'Inescapable risk: human disability and "accidental" nuclear war', *Current Research on Peace and Violence*, vol 1/2, 1988

R. Aldridge, 'The counterforce syndrome', Institute for Policy Studies, 1983

L. R. Beres, *Apocalypse: Nuclear Catastrophe in World Politics* (The University of Chicago Press, 1980)

A. Borning, 'Computer system reliability and nuclear war', In: D. Bellin and G. Chapman (Eds), *Computers in Battle: Will They Work?* (HBJ Inc, 1987)

P. Bracken, *The Command and Control of Nuclear Forces* (Yale University, 1983)

D. Caldwell, 'Permissive action links', *Survival* (May/June, 1987)

D. Campbell, *The Unsinkable Aircraft Carrier* (Michael Joseph Ltd, 1984)

Center for Defense Information, 'Nuclear weapons accidents: danger in our midst', *The Defense Monitor*, vol 10, No. 5, 1981

Defense Nuclear Agency, Joint Nuclear Accident Co-ordination Center (JNACC), *Nuclear Accident Response Capability Listing (NARCL) Manual*, DNA 5100.52.1L, (22.9.87)

Defense Nuclear Agency (USA), *Nuclear Accident Response Procedures (NARP) Manual*, DNA 5100.1 (January 1984)

J. Goldbalt, *Agreements for Arms Control: A Critical Survey*, SIPRI, (Taylor and Francis, 1982)

G. Hart and B. Goldwater, 'Recent false alerts and the nation's missile attack warning system', Report for the Senate Armed Services Committee, US Senate (9.10.80)

E. Kleine and R. Littell, 'Shh, let's tell the Russians', *Newsweek* (5.5.69)

P. Morrison, 'An absence of malice: computers and armageddon', *Prometheus*, vol 2, No. 2, December 1984

E. Roberts and S. Berlin, 'Computers and the strategic defense initiative', In: Bellin and Chapman (Eds), op cit.

P. Stein and P. Feaver, 'Assuring the control of nuclear weapons', CSIA Occasional Paper Series No. 2, Harvard University, 1987

J. Steinbruner, 'Nuclear decapitation', *Foreign Policy* (Winter 1981/82)

S. Van Evera, 'The cult of the offensive and the origins of the First World War', *International Security*, vol 9, No. 1, 1984

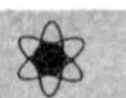

Fundamentals

What is Radioactivity?

Radioactivity is a naturally occurring phenomenon in which certain atoms change their structure. Our relatively recent understanding of the process has enabled us to make use of it in many ways.

The greatest amount of radiation to which we are exposed comes from natural sources — from space, rocks, soil, water and even from the human body itself. This is called background radiation and levels vary considerably from place to place, though the average annual dose is fairly constant. The largest single contributor to background radiation is radon gas, formed mainly by the decay of radioactive materials in the soil or in certain building materials.

The radiation which concerns us most here is that which results from human activities. Sources include medical applications of radioactive substances, fallout from atmospheric nuclear weapons tests conducted largely before the Partial Test Ban Treaty in 1963, discharges from the nuclear industry, and radioactive waste. While artificial radiation accounts for a small proportion of the total, its effects can be disproportionate. Some of the radioactive materials discharged as a result of human activities are not found in nature — e.g. plutonium — while others which *are* found naturally may be discharged in different physical and chemical forms, allowing them to be dispersed more readily into the environment, or perhaps to accumulate in the food-chain.

Doses vary considerably. Areas near a discharge, for instance, may receive much higher levels of radiation than the national average.

For all these reasons, simple comparisons of background and artificial radioactivity may not reflect the relative hazards. Equally important, it has never been shown that there is such a thing as a safe dose of radiation and so the fact that we are progressively raising global levels should be of as much concern to us as the possibility of another major nuclear disaster. We are taking an additional and unnecessary health risk.

A SCIENTIFIC EXPLANATION

The nucleus of every atom contains protons (positively charged) and neutrons (with no electrical charge) surrounded by a cloud of negatively charged electrons. Normally atoms

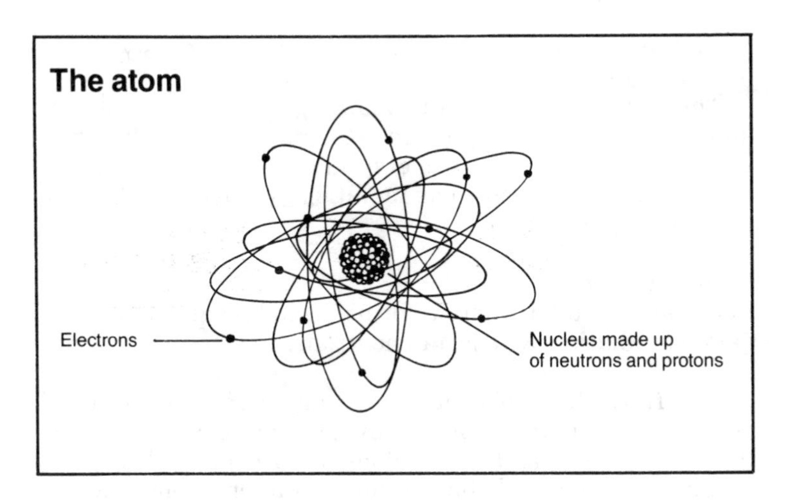

have the same number of protons and electrons balancing each other out so that the atom is electrically neutral. Adding or removing electrons leaves the atom with a net electric charge and the resulting particle is known as an ion.

All atoms of the same chemical element have the same number of protons. This is known as the atomic number of the element. But atoms of the same element can have different numbers of neutrons and are called isotopes of the element. They are identified by a number (also called the mass number) which is the total number of protons and neutrons in the nucleus. Uranium, for example, has two common isotopes: uranium-235 (92 protons and 143 neutrons) and uranium-238 (92 protons and 146 neutrons).

There are 92 naturally occurring elements of which hydrogen (one proton) is the lightest and uranium (92 protons) is the heaviest. Each element can have several natural isotopes, so there are several hundred different atoms. Elements which have more than 92 protons can be produced artificially in particle accelerators and nuclear reactors. And these techniques can also generate new isotopes of lighter atoms.

Although the positive charge of the protons means they are constantly pushing away from each other, the nucleus is held together by a counter force generated between the protons and the neutrons. Most nuclei are therefore held in a stable state, but some combinations of protons and neutrons cannot maintain the balance and will spontaneously change (decay) to reach a more stable form.

These changes are what is known as radioactivity, and the atoms are radioactive isotopes. In reaching a more stable state, some of the atom's energy must be released and it is this energy that is known as radiation, which can be harmful to living organisms.

RADIATION TYPES

Radiation can take different forms, including alpha, beta and gamma radiation, X-rays and neutrons.

Alpha Radiation: if a nucleus is alpha radioactive it will decay by expelling a rapidly moving alpha particle comprising two neutrons and two protons. As a result of losing two protons the atom will change into a different element which has an atomic number lower by two. Alpha radiation normally occurs in heavy elements.

Beta radiation: in beta decay, a neutron is converted to a proton (or vice versa) and a beta particle is expelled in order to balance the electrical charges and release excess energy. The atom becomes an element one higher or lower in the progressive series. For example, uranium-239 (92 protons and 147 neutrons) decays by emitting a beta particle to become neptunium-239 (93 protons and 146 neutrons).

Gamma rays: the emitting of particles in alpha and beta decay does not *always* leave the nucleus in its most stable state and the remaining excess energy can be released as gamma rays. (These are a form of electromagnetic radiation, as are X-rays, light, radiowaves and microwaves.)

Neutron emission: this occurs when the nuclei of very heavy, unstable atoms undergo spontaneous nuclear fission, breaking up into two large fragments and releasing several free neutrons.

All these forms of radiation may interact with surrounding material to produce electrically charged atoms known as ions. These can either break chemical bonds or create new ones and when this occurs in the cells of living organisms it may cause damage. Different types of radiation react with the matter through which they pass in different ways and some types are more penetrating than others.

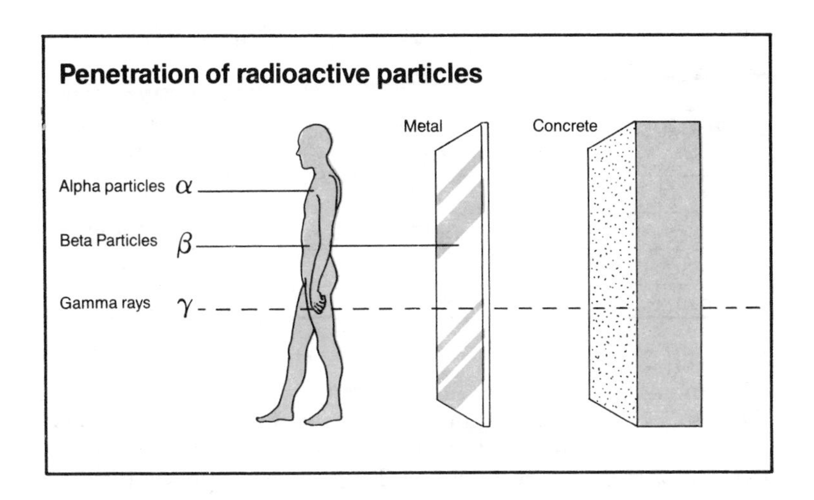

THE PROPERTIES OF RADIATION TYPES

Alpha particles: these particles are relatively large and heavy. They lose their energy as they move — and therefore slow down — in two main ways. Being electrically charged, they lose energy as they pass other charged particles. Secondly, they collide with other particles, thereby losing energy. They can therefore only travel short distances through air and cannot penetrate the human skin or a piece of paper. The concern for humans is that should an alpha-emitting substance be ingested or inhaled, the particles emitted can do great damage in a concentrated region of living tissue.

Beta particles: these have less mass and are faster moving. They are therefore more penetrating than alpha particles but they can still be stopped by thin sheets of aluminium or perspex. The more energy the particle carries, the greater its depth of penetration. Beta particles can penetrate the skin, but they deposit their energy in a larger volume of tissue and therefore cause less concentrated damage than alpha particles.

Gamma rays and neutrons: gamma rays, X-rays and neutrons have no electrical charge and therefore lose their energy more slowly. They can therefore travel long distances in air and are very penetrating.

The relative hazards of different types of ionising radiation can be summarised as follows:

Radiation	Internal Hazard	External Hazard
alpha	high	low
beta	moderate	moderate
gamma	low	high
neutron	low	high

HALF-LIFE

The moment at which an individual radioactive nucleus decays is completely random. However, for a given radioactive isotope, there is a characteristic average rate at which a sample of the material decays. This can be expressed as the half-life - the time taken for half of the total number of atoms in a sample to decay. It then takes the same amount of time for half the remainder to decay, and so on. Half-lives can vary from a fraction of a second to millions of years. For example, the half-life of uranium-238 is more than four billion years, whereas that of bismuth-214 is less than 20 minutes.

MEASURING RADIATION

Radioactivity, radiation and its effects require different types of measurement.

Activity is the number of nuclear disintegrations per unit of time in a quantity of radioactive substance. The modern unit of activity is the becquerel. One becquerel represents one decay per second. For example, a sample of plutonium-239 with an activity of 2,000 megabecquerels will emit an average of 2,000 million alpha particles every second. Activity was formerly measured in curies. One curie = 37,000 million becquerels.

Absorbed dose indicates the amount of radiation deposited in the material through which it is passing. The unit of absorbed dose is the gray, which is the dose of radiation that will cause one kilogram of material to absorb one joule of energy. Absorbed dose was formerly measured in rad. 100 rad = one gray.

Dose equivalent allows for the fact that some forms of ionising radiation are more harmful than others. The dose equivalent is the absorbed dose in grays, multiplied by an appropriate 'quality factor'. For example, high-energy neutron are given a quality factor of 10. Alpha radiation is ascribed a quality factor of 20. (This quality factor is only an approximation, because our knowledge of radiation effects is far from complete.) The unit of dose equivalent is the sievert. It was formerly measured in rem. 100 rem = one sievert.

Effective dose equivalent Because different parts of the body are affected differently by the same amount of radiation, a weighting factor is needed to calculate overall risk to the body from exposure of a single part. This weighting factor multiplied by the organ dose expresses the equivalent whole-body exposure. It is also measured in sieverts.

The Effect of Radiation on Health

WHAT DOES RADIATION DO TO LIVING THINGS?

The functions of a living organism are performed by its cells. Ionising radiation can break the links between atoms inside the molecules of living cells. It can also change the nature of the atoms themselves. Such damage to a cell may change the way it functions, or even kill it.

Cells have the ability to repair some levels of damage but if it is too extensive, or in a particularly vulnerable part of the structure, then it may be irreparable.

In general, the effects of radiation can be divided into those which affect the individuals who are exposed and those which affect their descendants. Somatic effects are those which appear in the irradiated individual. These include leukaemia and cancer. Hereditary or genetic effects are those which arise in subsequent generations.

There are three principal effects which radiation can have on cells: 1. The cell may be killed; 2. The cell multiplication may be affected, resulting in cancer; 3. Damage may occur to cells in the ovaries or testes, leading to the development of a child with an inherited abnormality.

In most cases, cell death only becomes significant when large numbers of cells are killed. Most organs contain more cells than are necessary to maintain normal function. The effects of cell death therefore only become apparent at comparatively high dose levels.

If a damaged cell is able to survive radiation dose, the situation is rather different. In most cases the effect of the

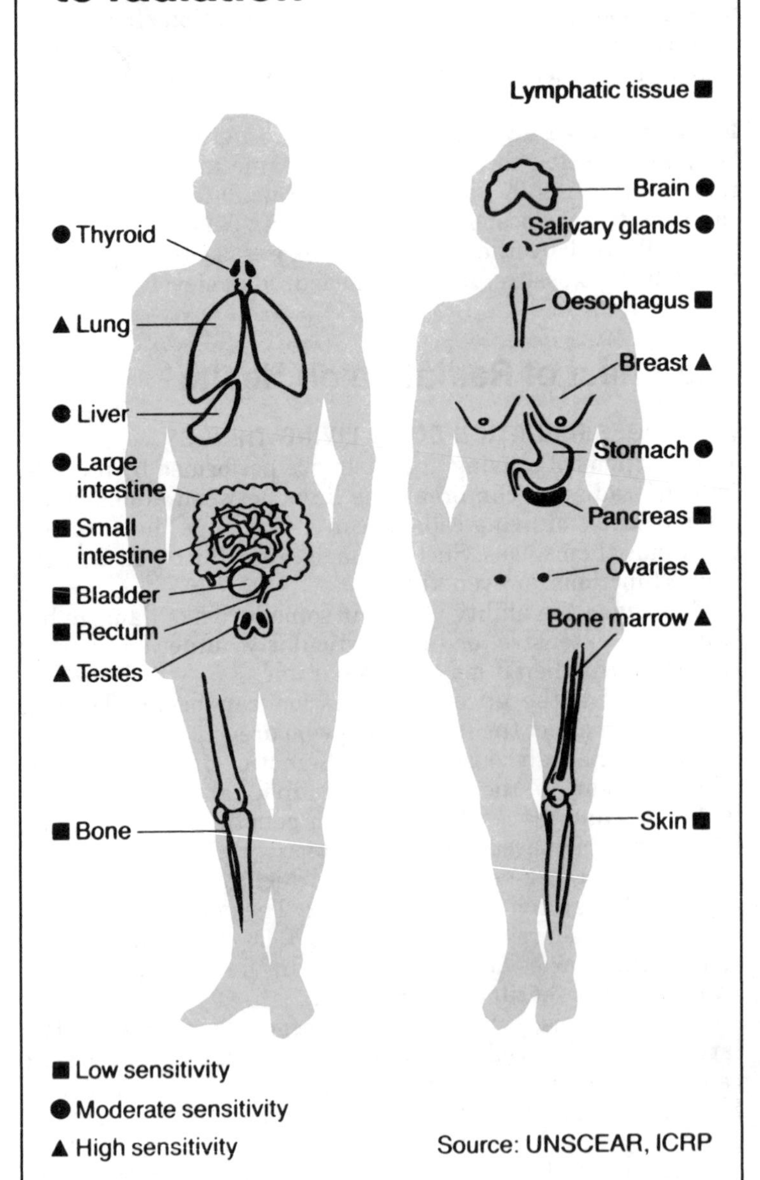
Sensitivity of human organs to radiation
Lymphatic tissue ■
● Thyroid
▲ Lung
● Liver
● Large intestine
■ Small intestine
■ Bladder
■ Rectum
▲ Testes
■ Bone
Brain ●
Salivary glands ●
Oesophagus ■
Breast ▲
Stomach ●
Pancreas ■
Ovaries ▲
Bone marrow ▲
Skin ■
■ Low sensitivity
● Moderate sensitivity
▲ High sensitivity
Source: UNSCEAR, ICRP

cell damage may never become apparent. A few malfunctioning cells will not significantly affect an organ where the large majority of cells are still behaving normally.

However, if the affected cell is a germ cell within the ovaries or testes, the situation is somewhat different. Ionising radiation can damage DNA, the molecule which acts as the cell's 'instruction book'. If that germ cell later forms a child, all of the child's cells will carry the same defect. The localised chemical alteration of DNA in a single cell may be expressed as an inherited abnormality in one or many subsequent generations.

In the same way, when a somatic cell in body tissue is changed in such a way that it or its descendants escape the control of processes which normally control cell replication and proliferation, the group of cells so formed may continue to have a selective advantage in growth over surrounding tissues. It may ultimately increase sufficiently in size to form a detectable cancer and in some cases cause death by spreading locally or to other parts of the body.

ACUTE RADIATION SYNDROME

Following a high radiation dose, many cells will be killed directly, or damaged in such a way that they will not be able to replicate or reproduce themselves successfully. These mechanisms lead to forms of radiation sickness. Cells which are in the process of growing and dividing rapidly are particularly vulnerable, e.g., the cells of an embryo or a young child, the lining of the intestinal tract, red bone marrow and reproductive cells.

High doses of up to 30 sieverts damage the central nervous system, causing nausea, severe vomiting, disorientation, coma and possibly death within hours. Lower doses, of between 10 and 30 sieverts, damage the gastro-intestinal tract, leading to nausea and vomiting in the first few hours. Internal haemorrhages, diarrhoea and septicaemia follow. Fluids are lost and infection may enter from the bowel. Death may occur within a few weeks, due to a breakdown of cell replacement within the lining of the gut.

Doses of between one and 10 sieverts induce nausea and vomiting, but this is short-term and is followed by a period of general well-being. However, the radiation will have affected the blood-forming cells in the red bone marrow.

The effects of high doses of radiation can therefore be broadly summarised as follows:

100 Sv: Death within days due to central nervous system damage.
10-50 Sv: Death after one or two weeks from gastro-intestinal damage.
3-5 Sv: Half of the exposed group will die within one or two months, from damage to the red bone marrow.

THE EFFECTS OF LOW-DOSE RADIATION

While there is now broad agreement about the effects of high-level radiation, there is vehement controversy about the long term effects of low-level doses. Data interpretation is complicated by: the length of time it can take for effects to become apparent; the fact that the populations under study (bomb survivors, people exposed to fallout from weapons testing, workers in the nuclear industry) are small and exact dosages are hard to calculate; and the fact that, for a number of reasons, it is difficult to compare one study with another.

One result of this low-dose controversy is to open to criticism predictions about the effects of any given dose. A growing number of scientists point to evidence that there is a disproportionately high risk from low doses of radiation. Others assume a directly proportioned relationship between the received dose and the risk of cancer for all levels of dose, while there are some who claim that at low doses there is a disproportionately low level of risk.

RADIATION PROTECTION STANDARDS

The harmful effects of radiation have been recognised since the first report on X-ray damage was published in 1896, and in 1928 an International X-Ray Radium Protection Committee (IXRPC) was established to protect those people regularly exposed in the course of their work. This Committee was re-formed in 1950 into the International Commission for Radiological Protection (ICRP).

Accurate information about the effects of different radiation doses is, of course, essential to the assessment of risk but there is great debate about data interpretation, particularly about low-level radiation exposure.

The United Nations Scientific Committee on the Effects of Atomic Radiation (UNSCEAR) produces its own assessments, as does the US Academy of Science's Committee on the Biological Effects of Ionising Radiation (BEIR). The latest study (1988) indicates that the internationally accepted figure for the risk of cancer induction may have been underestimated by a factor of between five and seven.

The single most important body of information on the effects of radiation comes from the study of Hiroshima and Nagasaki survivors. The interpretation of this data has provided the basis for international radiation protection standards, but, as the level of our scientific knowledge increases, it has become apparent that many of the assessments have been inaccurate. It appears, for instance, that the amount of radiation released had been calculated wrongly and, as a result, the harmful effects per unit dose may need to be doubled.

New studies have calculated general risk estimates using this and other recent data and all agree that the risks from ionising radiation are anything from three to 15 times higher than is recognised in safety regulations. The safety standards of most countries follow the recommendations of the ICRP. It has been argued that the membership of the committee is too narrow; it is drawn largely from the nuclear industry, its regulatory bodies and medical radiologists. Historically, the largest proportion of its membership has been physicists, while geneticists, pathologists, epidemiologists and, strangely enough, biophysicists have rarely appeared. Their recommendations have been increasingly incorporated into national radiation protection legislation and regulations. As more has been learned about the nature and scale of the risks, the committee has been forced to react by lowering its maximum permissible levels of whole-body radiation exposure. It has not, however, revised its estimates since the 1970s, claiming that the evidence that risks have been consistently underestimated remains inconclusive. In 1900, the recommended maximum dose was the equivalent of 100 mSv per *day*. In 1925, the level was 10 mSv per *week*. Since 1977, the ICRP's occupational dose limit for workers regularly exposed to radiation has remained effectively at 50 mSv per *year* (less than one mSv per week).

The first recommended limit for the general public was introduced in 1952 at 15 mSv per year. In 1959, this was lowered to five mSv per year, where it has remained ever since. A number of national authorities have adopted stricter levels than those suggested by the ICRP.

Fission, Fusion & the Nuclear Fuel Cycle

THE NUCLEAR REACTION

Energy can be liberated through nuclear processes by either nuclear fission or fusion. This can be an uncontrolled release of energy such as that in nuclear weapons, or a controlled release of heat energy in nuclear reactors to create electricity.

In splitting the nucleus of an atom into two or more parts (nuclear fission) an enormous amount of energy is released. Only a few isotopes of heavy elements are capable of being fissioned easily — e.g., uranium-235 and plutonium-239. To initiate fission the nuclei are bombarded by neutrons and, as they fly apart, they emit two or three neutrons of their own. If there are other fissionable atoms nearby then these neutrons may cause further atoms to fission, resulting in more neutrons and more fission products. This continuing process is called a chain reaction. The two major fragments of the split nucleus are now different chemical elements, mostly highly radioactive and including such isotopes as iodine-131, caesium-137 and strontium-90.

Nuclear fusion is in some respects the opposite of nuclear fission. It is the combining or fusing together of nuclei. This process offers the potential for releasing an enormous amount of energy. Getting nuclei close enough together to make them fuse is, however, extremely difficult because all nuclei are positively charged and like charges repel each other. The closer the nuclei approach, the greater the repulsive force.

Only the nuclei of 'light' atoms (below iron — atomic number 26 — in the periodic table) produce a net amount of energy when they fuse. (In practice, the only isotopes used in fusion are the hydrogen isotopes deuterium and tritium.) In addition, the smaller the nuclei involved, the lower the amount of energy required to overcome the repulsive forces between two nuclei. Sheer force can be used to make nuclei fuse — for example, the extremely high temperatures and pressures created in a nuclear (fission) explosion.

THE NUCLEAR FUEL 'CYCLE'

The complete set of activities, starting with the mining of uranium (or thorium) ores, through their use in reactors, to the subsequent handling of radioactive wastes, has become known as the nuclear fuel cycle. It may be broken down into several main stages, although with different fuel types, or

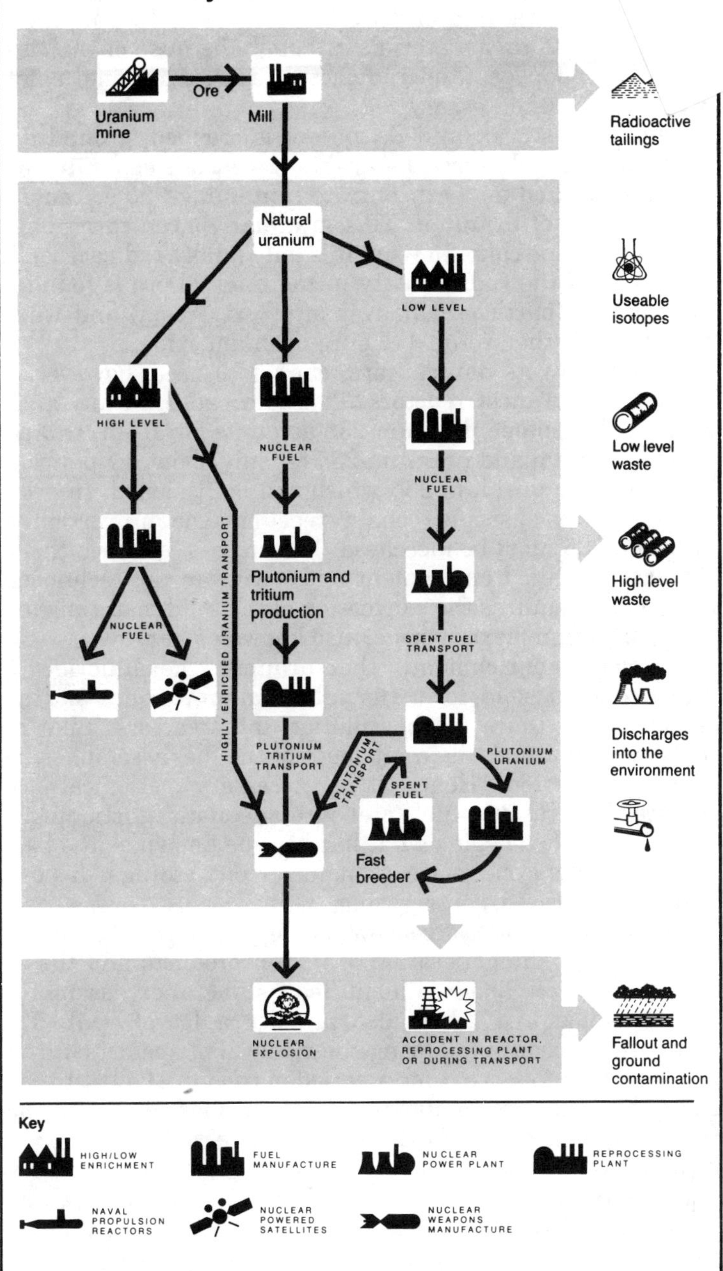
The nuclear cycle
Ore
Uranium mine
Mill
Radioactive tailings
Natural uranium
LOW LEVEL
Useable isotopes
HIGH LEVEL
NUCLEAR FUEL
NUCLEAR FUEL
Low level waste
HIGHLY ENRICHED URANIUM TRANSPORT
Plutonium and tritium production
High level waste
NUCLEAR FUEL
SPENT FUEL TRANSPORT
Discharges into the environment
PLUTONIUM TRITIUM TRANSPORT
PLUTONIUM TRANSPORT
PLUTONIUM URANIUM
SPENT FUEL
Fast breeder
NUCLEAR EXPLOSION
ACCIDENT IN REACTOR, REPROCESSING PLANT OR DURING TRANSPORT
Fallout and ground contamination
Key
HIGH/LOW ENRICHMENT
FUEL MANUFACTURE
NUCLEAR POWER PLANT
REPROCESSING PLANT
NAVAL PROPULSION REACTORS
NUCLEAR POWERED SATELLITES
NUCLEAR WEAPONS MANUFACTURE

different government policies, not all fuel passes through every stage of the cycle.

Mining of uranium ore involves the same techniques as the mining of other materials although, because of the risks posed by radon gas and radioactive dust, additional safety precautions are necessary.

Once it has been mined, the ore is crushed, ground and then leached to dissolve the uranium, which is separated out and precipitated as a concentrate containing 90 per cent, or more, oxides of uranium. This granular concentrate is also known as yellowcake. In itself it is only mildly radioactive. In fact, most of the radioactivity in the original ore is found in the tailings (particles left over after extraction) and waste streams. It is then refined to pure uranium oxide.

Uranium in its natural form cannot be used, however, in weapons or in most reactors. This is because, in any given sample, the isotope uranium-238 accounts for about 99.3 per cent of the total and uranium-235 for only about 0.7 per cent. But it is only uranium-235 which can be fissioned. In order to produce a sustained chain reaction, the proportion of uranium-235 must be increased. The process by which this is achieved is called enrichment and, owing to the technology involved, it requires the conversion of the solid materials to a gas — uranium hexafluoride, also known as hex.

Following enrichment, the uranium hexafluoride is returned to a solid form (usually uranium oxide) and assembled into units for installation into reactors. Like all fuels, in time the fissile component of the fuel assemblies becomes exhausted. Hence it is necessary periodically to replace the uranium fuel to ensure adequate functioning of the reactor. Now begins what has become known as the back end of the fuel cycle — waste management, storage, disposal and, in some cases, reprocessing.

Spent fuel from a reactor is very different from the original fuel. The generation of fission products and the effect of neutron bombardment leaves the fuel assemblies highly radioactive. They also produce a lot of heat. The standard procedure following the removal of spent fuel from the reactor is to store it for a period of time to allow it to cool somewhat and also to allow some of the short-lived radioactive isotopes to decay. At the end of this period there are two basic options — some form of long-term storage, or reprocessing. Reprocessing involves the break-up of the fuel assemblies and then the chemical separation of the different components in the spent fuel.

Arguments for reprocessing include a technical need (due to the long-term instability of spent-fuel stores under water), waste management (it is claimed that reprocessing allows for better management of the wastes produced) and better utilisation of raw materials (by recycling fissile materials for further use in reactors). Several countries have fuel cycles that do not involve reprocessing, preferring instead to manage the fuel in the form in which it leaves the reactor. Reprocessing generates large volumes of radioactive liquids and gases and leaves the problem of management of the wastes, in particular of the long-lived radionuclides, unsolved. Historically, reprocessing has been responsible for some of the world's worst radioactive pollution problems. Furthermore, it requires the movement of large amounts of intensely radioactive spent fuel and the consequent risk of serious accidents during transport between reactor sites and the plant where the reprocessing is to take place. Were it not for reprocessing, there would be virtually no civil transport of plutonium. The recycling of the fissile materials from reactor fuel has its origins in the military reactor cycle and the search for plutonium for weapons. Reprocessing increases the volume of radioactive waste that will need long-term management. From the environmental and occupational health aspects then, reprocessing creates a large number of additional problems. The Organisation for Economic Co-operation and Development (OECD) has pointed out that the reprocessing of spent fuel is less economic than long-term storage.

Radioactive wastes are divided into three categories, depending upon their level of activity. High-level Waste (HLW) comprises most of the heavy elements and fission products from nuclear reactors. These will remain radioactive for many thousands of years and must therefore be consigned to some form of long-term storage. At present, much of this waste is stored in liquid form in tanks. Long-term plans include vitrification — incorporation of the waste into glass blocks which would then be buried deep underground.

Intermediate-level Waste (ILW) comprises such materials as the cladding from spent reactor fuel cans. It is less radioactive than HLW, but is produced in larger volumes. ILW is at present stored in silos. Long-term plans include incorporating it in concrete for eventual burial.

Low-level Waste (LLW) is produced by medical and industrial users of radiation, as well as the nuclear power industry. Anything which has come into contact with radioactive material is regarded as potentially suspect and classified as LLW. This category includes such items as used protective clothing and discarded paper towels, and also residues from the treatment of radioactive waste water from nuclear power plants.

Nuclear Reactors

ENGINEERING PRINCIPLES

A nuclear reactor is designed to allow the fission chain reaction to proceed at a controlled rate, so that the heat released can be channelled into electricity generation. This means controlling some characteristics of the fission reactions inside the core of the reactor, so that each fission event leads to *exactly* one more. At the point at which this occurs the reactor is said to be critical.

Inside the core, neutrons are both released during fission and 'lost'. Loss occurs either through leaking from the reactor into the surroundings or through being 'captured' by various atoms with which they come into contact. To maintain the correct balance, control rods are used, composed from materials such as cadmium and boron, which have a very high neutron-absorbing capability. Should an excess of neutrons be allowed to build up, then the chain reaction could escalate out of control. Taking the control rods out of the core slightly will allow the neutron population, and thus the power, to rise. Inserting them further into the core causes a fall in power.

It is not just the numbers of neutrons that need to be controlled but also their speed, because in order to fission uranium-235 most efficiently the neutrons must be slowed down. This process is called moderation and the substance used is known as a moderator. The moderator is incorporated into the core of the reactor and as neutrons pass through it, they undergo many collisions, thereby losing speed.

The main materials used as moderators are light (ordinary) water, heavy water, graphite (carbon) and beryllium. Heavy water is water in which the hydrogen atoms have both a neutron and a proton in the nucleus. This hydrogen isotope is known as deuterium.

To start up a reactor the first step is to provide a small source of neutrons, which releases a very small population of neutrons in the core of the reactor. The fission process is then used to multiply neutrons through many generations from this initial population, so bringing the reactor up to the required full power. This is a gradual process.

All these processes are carried out in the core of the reactor — the fuel is surrounded by the moderator, and the control rods are moved into and out of the core as required. When the reactor is operating, fission of the uranium produces fission products which deliver up their energy as heat when they are stopped in the surrounding material. A coolant, which may also be used as the moderator, conveys heat away from the reactor core. (Coolants include water, heavy water, air, carbon dioxide, helium and sodium.) The heat energy is then converted into electrical energy via a system of heat exchangers and turbines. This is exactly the same method employed by coal-fired power stations. In a nuclear power station, only the heat source is different.

DIFFERENT REACTOR DESIGNS

As with technologies employed in some other stages of the nuclear cycle (enrichment, reprocessing), commercial reactors of the present generation have been developed from design-types originally formulated for military purposes. (Some still serve a dual civilian-military role. Pressurised-water reactors, for example, were developed as a compact propulsion reactor for naval vessels. Gas-graphite reactors, such as Magnox, were developed from the atomic piles designed to produce plutonium for weapons.) Commercial reactors did not, therefore, have as their primary consideration the safest and most economic production of energy for civilian purposes. As a result, all those now in commercial operation are bound by the basic constraints of decades-old design principles. Within these constraints, the designs have developed to a point where they have now reached the limits of their potential safety or their economic performance.

There are a number of different ways of arranging the fuel, coolant, moderator etc., in a reactor which have led to a number of fundamentally different designs.

LIGHT-WATER REACTORS

These are the most common power reactors in operation worldwide. They were developed in the United States and use lightly enriched uranium. There are two major types — the boiling-water reactor and the pressurised-water reactor. In a boiling-water reactor (BWR) the coolant water circulates through the core, and boils. The steam turns a turbine and produces electricity. In a pressurised-water reactor (PWR)

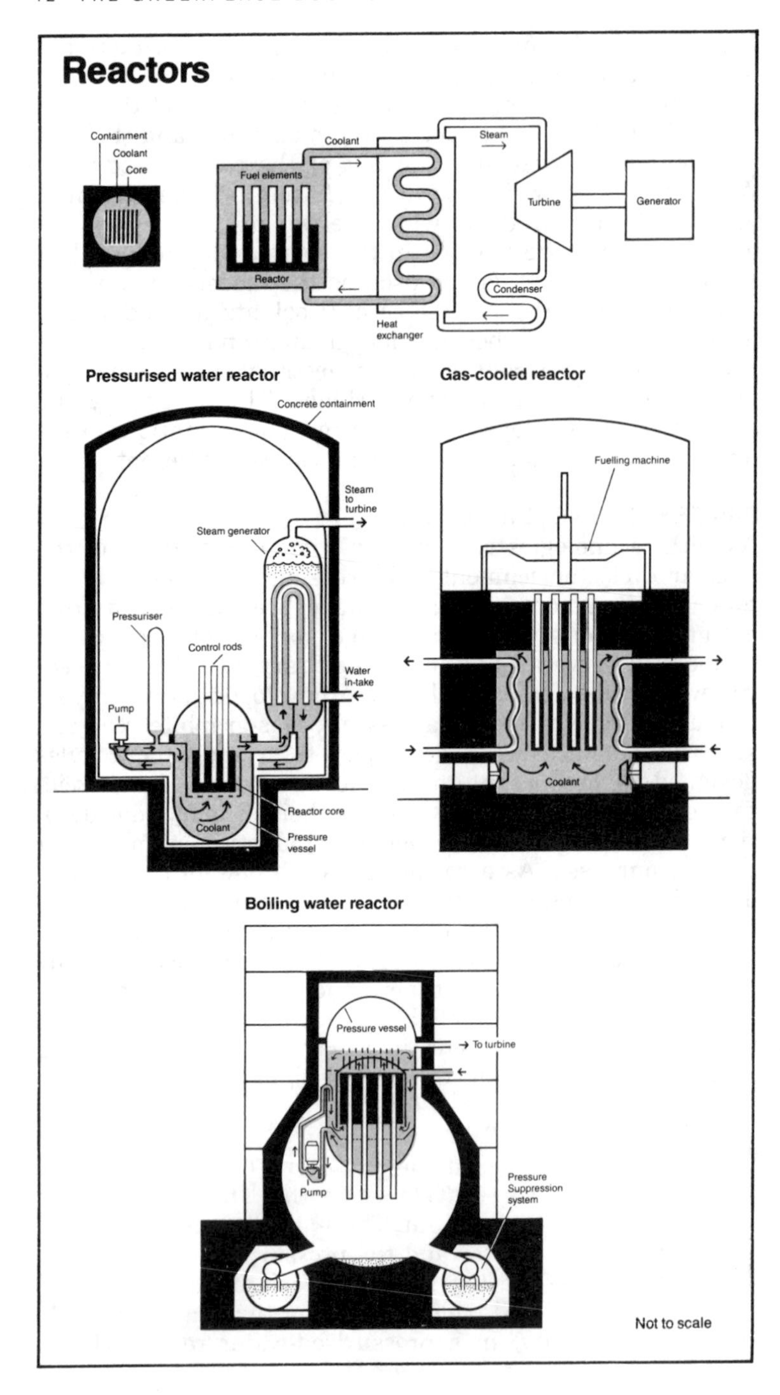
Reactors
Containment
Coolant
Core
Coolant
Fuel elements
Reactor
Steam
Turbine
Generator
Condenser
Heat exchanger
Pressurised water reactor
Concrete containment
Steam to turbine
Steam generator
Pressuriser
Control rods
Water in-take
Pump
Reactor core
Coolant
Pressure vessel
Gas-cooled reactor
Fuelling machine
Coolant
Boiling water reactor
Pressure vessel
To turbine
Pump
Pressure Suppression system
Not to scale

the coolant water is under sufficiently high pressure not to boil and an additional (secondary) water circuit extracts the heat from the coolant and is used to generate electricity.

In the BWR, the reactor cooling circuit is less complex than that of the PWR. There are no steam generators, no pressurising mechanisms for the coolant and the pumping capacities are lower. Operating pressures are lower, while the temperature is almost equally high. The whole circuit, including the turbines, is radioactively contaminated to a certain degree. The pressure vessel for a BWR must also house the steam separators and driers for the turbines and is therefore larger than that of a PWR. In both types of reactor, the coolant water is also used as a moderator.

The safety problems of PWRs are particularly related to the dependence on the integrity of the reactor pressure vessel and the complex active safety systems which are an inherent part of the design. The BWR, while sharing the same basic problem as the PWR of reliance on the pressure vessel, complicates matters by having a much larger pressure vessel volume and more complex internal structure. The safety systems are more complex than those of the PWR and, as the control rods are introduced from below, these cannot rely on gravity to bring them into the reactor core.

The graphite-moderated pressure tube boiling-water reactor (RBMK) is a Soviet design. It is light-water cooled and uses low-enriched uranium fuel. As in the BWR, the water boils in the reactor core. Heated graphite reacts with oxygen, so that the reactor space is filled with helium and nitrogen. The reactor cooling circuit consists of two parallel loops. The complexity of the pressure tube design makes maintenance and inspection difficult.

The capacity of the reactor to be refuelled while it is running, originally designed to facilitate weapons-grade plutonium production, presents its own hazards. Under normal operating conditions, there is what is known as a 'positive void coefficient', the implications being that a loss of coolant to the reactor core results in a rapid escalation of power levels inside the core. This can lead to melting of the fuel and the possibility of loss of integrity of the reactor containment. This effect, coupled with an explosion of hydrogen inside the core, resulted in the accident at Chernobyl.

HEAVY-WATER REACTORS

These use heavy water as both moderator and coolant. Heavy water is a much better moderator than ordinary water (it slows neutrons very effectively and with less tendency to absorb them) and such reactors do not need to use enriched uranium as fuel.

The main commercially available heavy water reactor is the Canadian-designed CANDU reactor. (CANDU = CANadian Deuterium Uranium reactor.) Unlike the light-water reactors, in which the water circulation is through one large pressure vessel containing all the fuel rods, coolant in the CANDU flows through hundreds of separate pressure tubes containing the fuel. These pressure tubes are surrounded by unpressurised heavy water acting as moderator. Due to this pressure tube design, it is possible to remove and insert fuel while the reactor is operating (on-load refuelling).

CANDUs have a secondary circuit which uses light water. The CANDU suffers from some of the design faults of the Chernobyl-style RBMK reactors, having a positive void coefficient under some operating conditions, an on-load refuelling capacity and the possibility of hydrogen-evolving reactions in the core. The complexity of the pressure tube design creates its own problems for effective containment.

GAS-COOLED REACTORS

Gas-cooled reactors use gases, such as carbon dioxide or helium, in place of water as coolant. These reactors were developed in France and the United Kingdom, the earliest type being the Magnox reactors. In these the fuel is made from natural uranium metal sheathed in an alloy of magnesium, from which the name of the reactor is derived. Carbon dioxide coolant circulates under pressure through the reactor core. The core itself consists of the fuel rods and blocks of graphite moderator, with coolant channels running through the graphite. The gas passes through heat exchangers to produce steam for the generators.

The Advanced Gas-cooled Reactor (AGR) was also developed in the UK. It uses enriched uranium and operates at higher temperatures. In both the AGR and the Magnox reactor types, on-load refuelling, originally incorporated into the design type to facilitate plutonium production, adds to the risks of reactor operation. The earlier Magnox reactors are particularly vulnerable to loss of coolant accidents, whereas the AGRs have a major design flaw which could lead to water intrusion into the core and to a violent steam-graphite reaction, with the possibility of containment breach.

A third, distinctive type of gas-cooled reactor is the high-temperature gas reactor. This uses highly enriched uranium oxide or uranium carbide in sintered pellet form. The pellets themselves are coated in graphite moderator. (Other fuels like thorium-233 have also been used.) The core can withstand very high temperatures and the helium coolant is heated to about 800 °C.

Fast-breeder reactors have a core consisting of both fissile uranium-235, plutonium-239 and non-fissile uranium-238, surrounded by a blanket of uranium-238. The fast neutrons which escape from the core can be captured by the uranium-238 nuclei and form more fissile material in the form of plutonium-239. The fast-breeder reactor uses enriched uranium. The high power density in the core means that cooling requirements are greater, ruling out the use of water. Helium has been considered, but all fast-breeder reactors use liquid sodium in the primary circuit. The problem with liquid sodium in the primary circuit is that it is highly inflammable with air, and reacts explosively with water.

Nuclear Weapons

NUCLEAR DEVICES, WARHEADS & WEAPONS

The terms nuclear device, nuclear warhead and nuclear weapon are often confused. A nuclear explosive device is an assembly of nuclear and other materials and fuses which could be used in a test, but generally cannot be reliably delivered as part of a weapon. A nuclear warhead implies further refinement in design, resulting in a mass-produced, reliable and predictable nuclear device which may be carried by missiles, aircraft or other means. A nuclear weapon is a nuclear warhead fully integrated with its delivery system. Delivery systems carry the nuclear warhead to its target. These systems include missiles, artillery shells, atomic demolition mines, aircraft and nuclear bombs. Nuclear weapons are further classified as strategic, theatre, or tactical, depending upon their intended military function.

FISSION WEAPONS

An explosion is the release of a large quantity of energy in a small volume over a short period of time. A nuclear weapon is a device in which most or all the explosive energy is derived from either uncontrolled fission, fusion or a combination of the two processes.

Nuclear weapons therefore contain, if we consider both fission and fusion, the fissile isotopes uranium-235, plutonium-

239, uranium-233, or a combination of these. Plutonium is more expensive to produce and must be made artificially, but gives a higher yield-to-weight ratio than uranium. There is a minimum amount of material (critical mass) needed to sustain a chain reaction. Any less than this and too many neutrons leak out. A mass that is less than critical is called subcritical, whereas a greater mass is called supercritical.

It is advantageous to achieve critical mass in as small a quantity of fissile material as possible and critical mass can be lowered in several ways. It can be surrounded by a shell of other material which reflects neutrons that would otherwise escape. Secondly, the density of the material can be increased (see 'implosion technique' below). Once fission begins, the mass can become subcritical before an uncontrollable chain reaction is produced. To avoid this, it is held together by a surrounding tamper of heavy material.

FISSION EXPLOSION DESIGN

There are two basic designs for fission weapons. These are the implosion technique and the gun assembly technique.

The implosion technqiue is the more sophisticated, complex and efficient. Here a peripheral charge of chemical high explosive is uniformly detonated in a manner designed to compress or squeeze (implode) a subcritical mass into a supercritical one by increasing its density. This technique is used in weapons where the fissionable material is plutonium-239, uranium-235, or a combination of the two. It was used in the first nuclear test (Trinity, 16 July 1945) and also for Fat Man, the nuclear bomb dropped on Nagasaki.

The gun device involves the assembly of two (or more) masses of fissionable material, each less than critical mass and hence incapable of starting uncontrolled fission. They are then driven together by conventional explosives. This technique was used in Little Boy, the nuclear weapon dropped on Hiroshima.

Implosion devices can generally be made with higher efficiency, so most fission weapons in the US nuclear stockpile employ the implosion technique. Both methods of starting fission require an initial burst of neutrons at the right moment — a neutron initiator — such as a polonium-beryllium sphere, or an external high-voltage neutron initiator.

FUSION WEAPONS

Fusion or hydrogen weapons are usually defined as atomic weapons in which at least a portion of the release of energy

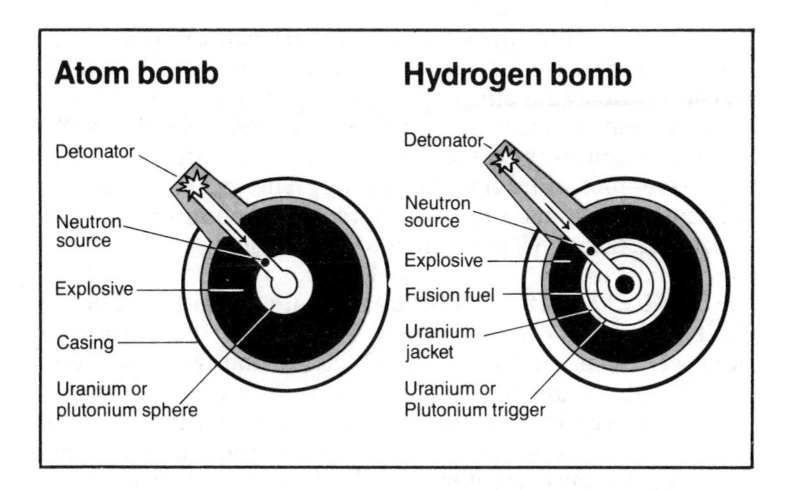

occurs through the fusion of light atomic nuclei such as the hydrogen isotopes deuterium and tritium (which is produced from lithium).

To overcome the repulsive forces between the nuclei, a great deal of energy is needed. The high temperatures and pressures necessary in a thermonuclear weapon are achieved with a large, uncontrolled fission explosion. In fact, US thermonuclear weapons depend upon two explosions, one to compress the fusion fuel and one to ignite it. As fusion releases neutrons it can create fission. Including fissile uranium is a cheap way to multiply the yield of the weapon. In general, the energy released in the explosion of a large themonuclear weapon stems from three sources — a fission chain reaction (first stage), burning of thermonuclear fuel (second stage) and fission of uranium-238, where it is included in the device (third stage).

There are also thermonuclear-boosted fission weapons. These use a thermonuclear fuel, typically deuterium and tritium gas, either in, or directly next to, the core of fissile material. This increases or 'boosts' the explosive power. It works by adding an extra quantity of free neutrons which increase the number of fissions.

Tritium decays quite rapidly with time, so the effectiveness of the boosting process can degrade with the age of a weapon. This requires the tritium to be replaced periodically in stockpiled nuclear weapons.

ENHANCED RADIATION WEAPONS (NEUTRON BOMBS)

The neutron bomb is a thermonuclear device designed to maximise the lethal effects of high-energy neutrons produced by the fusion of deuterium and tritium, and to minimise the blast. The neutrons released would destroy living organisms but leave most buildings standing. The burst of nuclear radiation (neutrons and gamma rays) is enhanced by minimising the fission yield relative to fusion yield.

PHYSICAL EFFECTS OF NUCLEAR WEAPON EXPLOSIONS

What form does all the energy released by the rearrangement of these nuclei in the fission-fusion-fission process take? What are the physical effects on the environment?

After a very intense burst of neutrons and gamma rays there is a silent wave of intense heat and flash of light, hundreds of times brighter than the Sun. A shock wave of extreme pressure follows, travelling like an expanding ring, followed by intense winds of thousands of km per hour. As the fireball lifts slowly from the ground explosion, it can lift millions of tons of vapourised earth which cools, condenses and falls towards the ground. Winds of the upper level of the atmosphere sweep the huge cloud downwind from the point of detonation. Some of this material will later return to earth as radioactive fallout. Billions of oxygen and nitrogen molecules in the air are combined in heat to form nitrogen oxides, which rise to the upper levels of the atmosphere.

One of the more alarming effects of a global nuclear war is the so-called nuclear winter. Massive amounts of dust, smoke, soot and ash would contaminate the atmosphere and the Sun could be obscured, resulting in dramatic climate changes, falling temperatures and consequent damage to remaining plant, animal and human life.

NUCLEAR WEAPONS ACCIDENTS

The hazards posed by accidents involving nuclear weapons are unlikely to be those of a nuclear explosion. Instead, the most likely weapons accident involves a nuclear warhead being consumed by fire. In the words of the US Defense Nuclear Agency (DNA), '[Nuclear weapons] accidents have occurred...which released radioactive contamination because of fire or high explosive detonations.' Such an accident could result from an external fire on board a ship or crashed aircraft setting fire to the fissile material in the warhead. The danger can be exacerbated by the independent detonation of the conventional high explosive, which can also be detonated by shock in the event of an accidental bomb-drop. In the case of a missile, the accident can be further worsened by the burning of solid or liquid propellant. Plutonium will burn readily under these conditions, and will create a toxic, radioactive 'plume' of plutonium particles that can contaminate a wide area and be inhaled by humans. The danger of fire to nuclear weapons is widely documented by the US Department of Defense (DOD), but no government studies of potential consequences and casualties have been carried out.

Greenpeace has, however, commissioned several studies which examine the human consequences of a nuclear weapon accident involving fire. For example, a study carried out by Dr Jackson Davies (of the Environmental Studies Institute, University of California at Santa Cruz) looked at the possible consequences for New York of a three-hour shipboard fire involving the incineration of a single nuclear warhead containing five kg of plutonium. According to the US Department of the Navy, 'The possibility of a fire at a nuclear weapons accident is very prominent. This will present an extremely hazardous environment because of radioactive, toxic, caustic and explosive materials contained in nuclear warheads and components.' (OPNAVIST 3440.15, Washington DC).

Dr Jackson Davies estimated the radioactive release, traced the pathways of dispersion of a radioactive cloud under probable weather conditions, and estimated the level of absorption by humans and the number of resultant casualties. He predicted that at distances of up to 200 km from the site of the accident, contamination levels would exceed the limits set by the US NRC. He also concluded that up to 30,442 long-term cancer fatalities would occur in the city.

Nuclear weapons and reactors that are lost at sea also present a long-term environmental risk. Greenpeace and The Institute for Policy Studies (IPS) have shown that there are 60 nuclear weapons and six nuclear submarines, with 10 nuclear reactors, on the ocean bed as the result of naval accidents, plane crashes and deliberate dumping (**Neptune III**). Nuclear weapons and reactors, breached by deep ocean pressures, can rapidly release their radioactive contents. At best, long-term corrosion will cause a gradual release. These materials, the majority of which will emit radioactivity for thousands of years, will eventually enter the marine food chains and will have a measurable effect on human populations unless the weapons and reactors are recovered.

Geoff Endacott, Simon Carroll, Damian Durrant. July 1989.

Index of Key Terms

Symbols

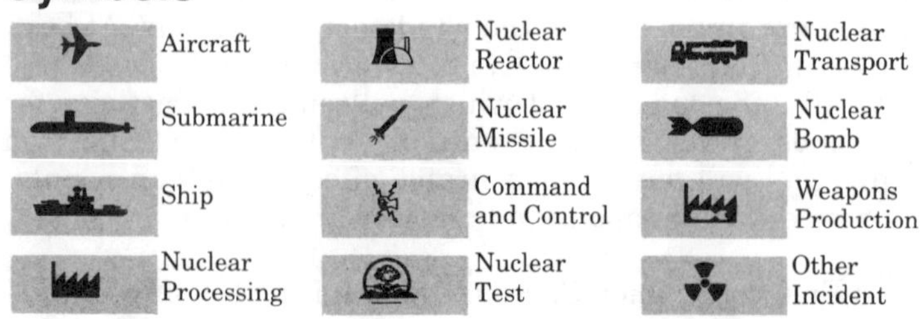

Abbreviations and Acronyms

AEA Atomic Energy Authority
AEC Atomic Energy Commission
AFB Air Force Base
AGR Advanced Gas-cooled Reactor
BMEWS Ballistic Missile Early Warning System
BNFL British Nuclear Fuels Ltd
BWR Boiling-Water Reactor
CANDU Canadian Deuterium Uranium reactor
CEA Commissariat à l'Energie Atomique
CEGB Central Electricity Generating Board
CEP Centre d'Expérimentation du Pacifique
CIA Central Intelligence Agency
DNA Defense Nuclear Agency
DOD Department of Defense
DOE Department of Energy (US)
DoE Department of the Environment (UK)
EDF Electricité de France
EEC European Economic Community
EPA Environmental Protection Agency
FOIA Freedom Of Information Act
GAO General Accounting Office
IAEA International Atomic Energy Agency
ICBM Inter Continental Ballistic Missile
ICRP International Commission on Radiological Protection
INF Intermediate-range Nuclear Forces
kt kiloton
kW kilowatt
Mt megaton
MOD Ministry of Defence
MW megawatt
NATO North Atlantic Treaty Organisation
NAAV National Association of Atomic Veterans
NASA North American Space Agency
NII Nuclear Installations Inspectorate
NORAD North American Aerospace Defense Command Center
NRC Nuclear Regulatory Commission
NRPB National Radiological Protection Board
NSA National Security Agency
NTS Nevada Test Site
OECD Organization for Economic Co-operation & Development
PTBT Partial Test Ban Treaty
PWR Pressurized-Water Reactor
RBMK graphite moderated boiling water reactor
RORSAT Radar Ocean Reconnaissance Satellite
SAC Strategic Air Command
SALT Strategic Arms Limitation Treaty
SDI Strategic Defense Initiative
SLBM Submarine Launched Ballistic Missile
SNAP Space Nuclear Auxiliary Power
THORP Thermal Oxide Reprocessing Plant
UKAEA UK Atomic Energy Authority
USAF US Air Force
WHO World Health Organization

Metric Prefixes

tera	T	1,000,000,000,000
giga	G	1,000,000,000
mega	M	1,000,000
kilo	k	1,000
milli	m	1/1,000
micro	μ	1/1,000,000
nano	n	1/1,000,000,000
pico	p	1/1,000,000,000,000

Units of Measurement

distance
1in = 2.54cm
1ft = 0.3m
1 mile = 1.6km
1cm = 0.39in
1m = 3.28ft
1km = 0.62 miles

area
1 sq in = 6.45 sq cm
1 sq ft = 0.09 sq m
1 sq mile = 2.59 sq km
1 sq cm = 0.16 sq in
1 sq m = 10.76 sq ft
1 sq km = 0.39 sq miles

weight
1oz = 28.35g
1lb = 0.45kg
1 ton = 1.02 tonne
1g = 0.03oz
1kg = 2.2lb
1 tonne = 0.98 ton

volume
1 cu in = 16.39 cu cm
1 gal = 4.54 litres
1 cu cm = 0.06 cu in
1 litre = 0.22 gal

MAP 1

Asia and the Pacific

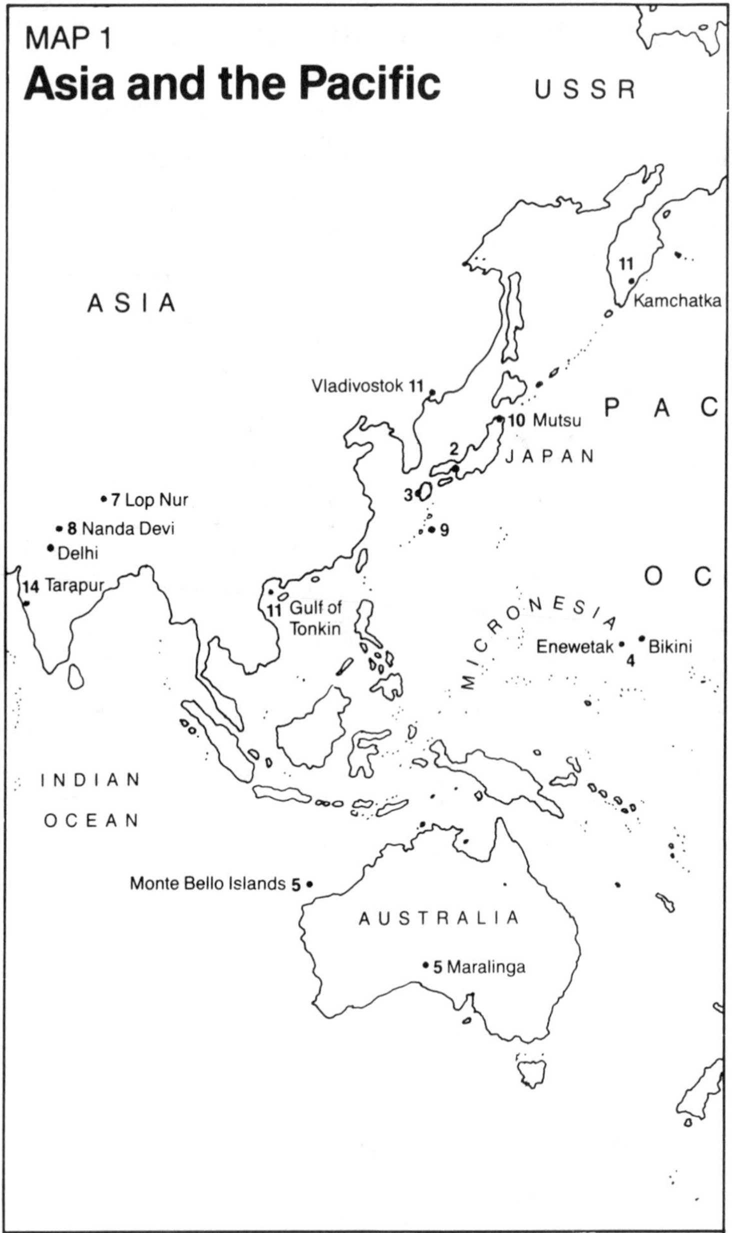

USSR
ASIA
11 Kamchatka
Vladivostok 11
10 Mutsu
PACIFIC
2
JAPAN
3
7 Lop Nur
8 Nanda Devi
Delhi
9
14 Tarapur
OCEAN
11 Gulf of Tonkin
MICRONESIA
Enewetak 4 Bikini
INDIAN OCEAN
Monte Bello Islands 5
AUSTRALIA
5 Maralinga

1 Los Alamos	**4** US Nuclear Tests	**7** Chinese Nuclear Tests
2 Hiroshima	**5** UK Nuclear Tests	**8** Operation Hat
3 Nagasaki	**6** Nukey Poo	**9** USS Ticonderoga

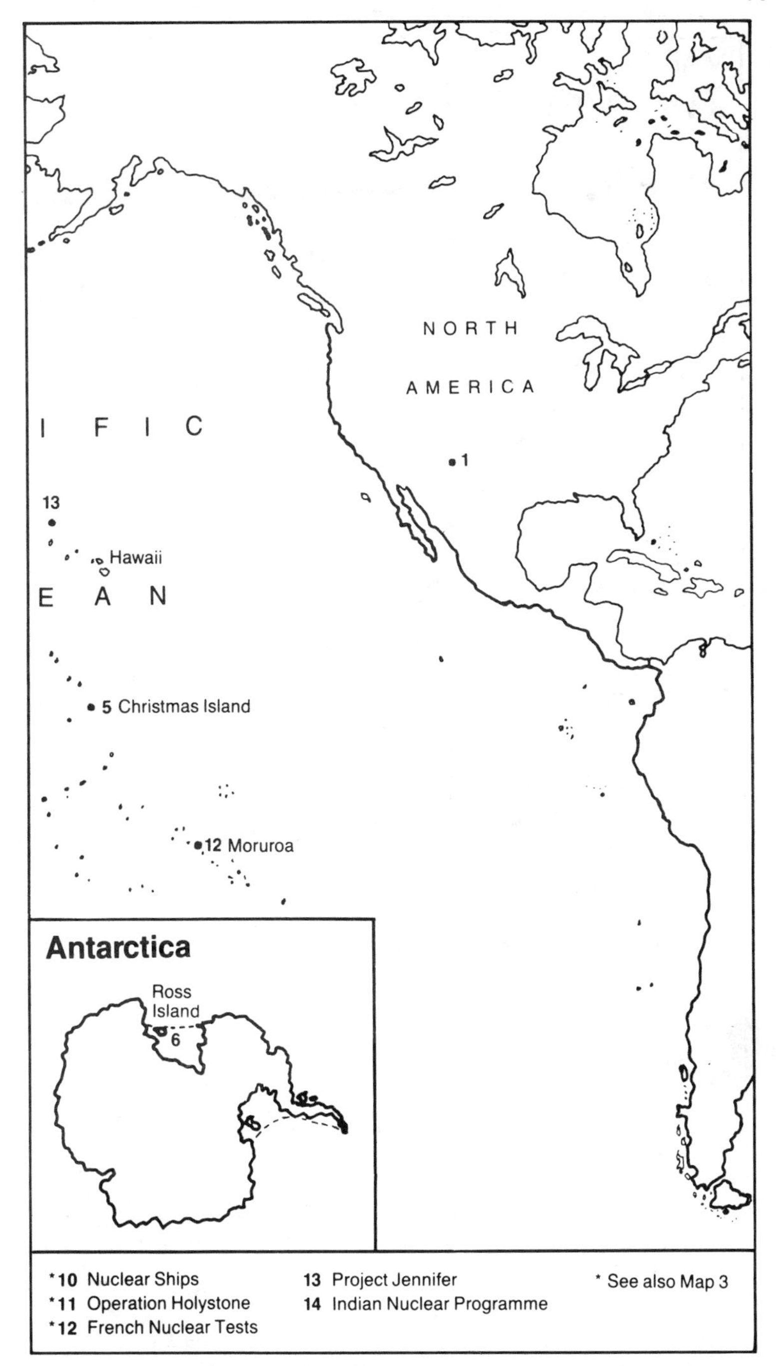

*10 Nuclear Ships
*11 Operation Holystone
*12 French Nuclear Tests
13 Project Jennifer
14 Indian Nuclear Programme
* See also Map 3

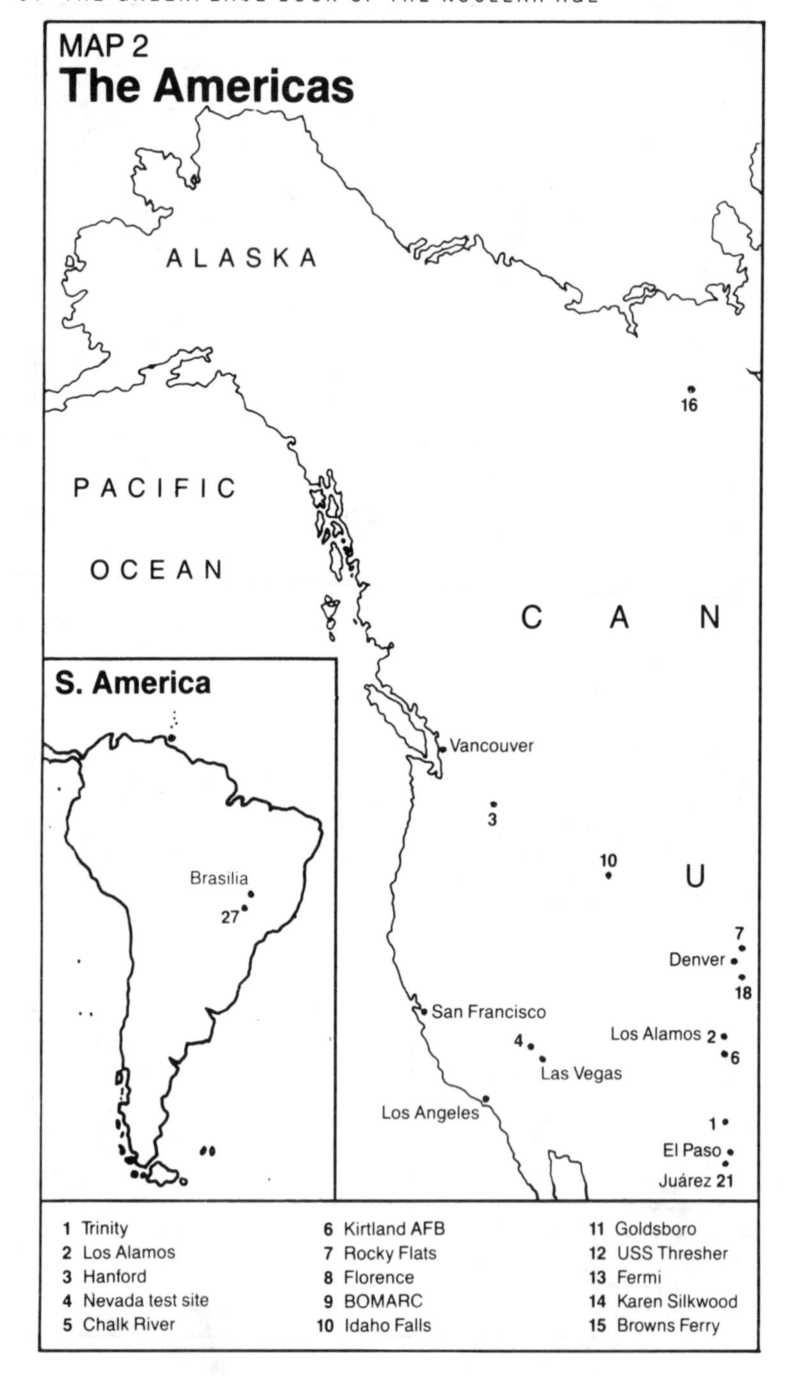
MAP 2
The Americas
ALASKA
PACIFIC
OCEAN
C A N
U
16
Vancouver
3
10
7
Denver
18
San Francisco
4
Las Vegas
Los Alamos 2
6
Los Angeles
1
El Paso
Juárez 21
S. America
Brasilia
27
1 Trinity
2 Los Alamos
3 Hanford
4 Nevada test site
5 Chalk River
6 Kirtland AFB
7 Rocky Flats
8 Florence
9 BOMARC
10 Idaho Falls
11 Goldsboro
12 USS Thresher
13 Fermi
14 Karen Silkwood
15 Browns Ferry

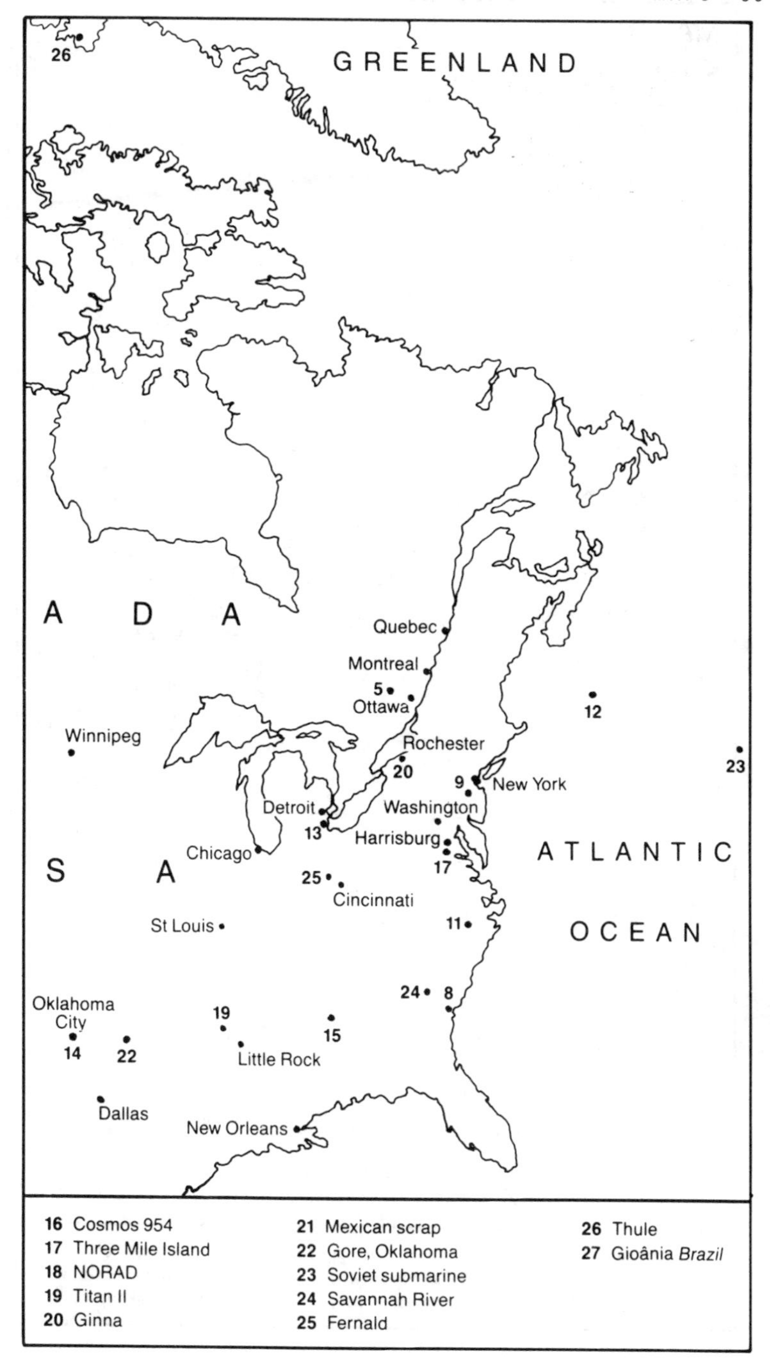
GREENLAND
26
A D A
S A
Quebec
Montreal
5
Ottawa
12
Winnipeg
Rochester
20
23
9
New York
Detroit
Washington
13
Harrisburg
Chicago
17
ATLANTIC
25
Cincinnati
St Louis
11
OCEAN
24
8
Oklahoma City
19
15
14
22
Little Rock
Dallas
New Orleans
16 Cosmos 954
17 Three Mile Island
18 NORAD
19 Titan II
20 Ginna
21 Mexican scrap
22 Gore, Oklahoma
23 Soviet submarine
24 Savannah River
25 Fernald
26 Thule
27 Gioânia *Brazil*

MAP 3

Europe

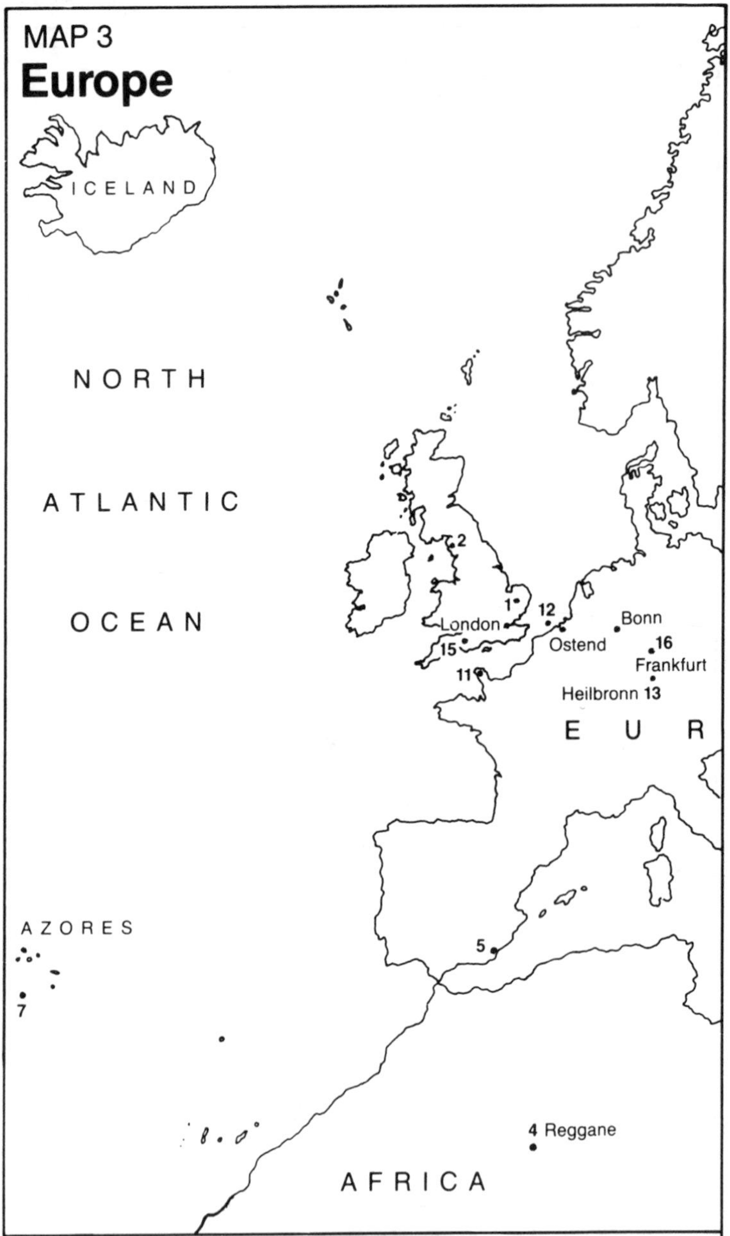

ICELAND

NORTH

ATLANTIC

OCEAN

2

1

12

London

Bonn

Ostend

15

16

Frankfurt

11

Heilbronn 13

EUR

AZORES

7

5

4 Reggane

AFRICA

1 Lakenheath AFB	*4 French Nuclear Tests	7 USS Scorpion
2 Windscale, Sellafield	5 Palomares	*8 Operation Holystone
3 Chelyabinsk-40	*6 Nuclear ships	9 USS Belknap

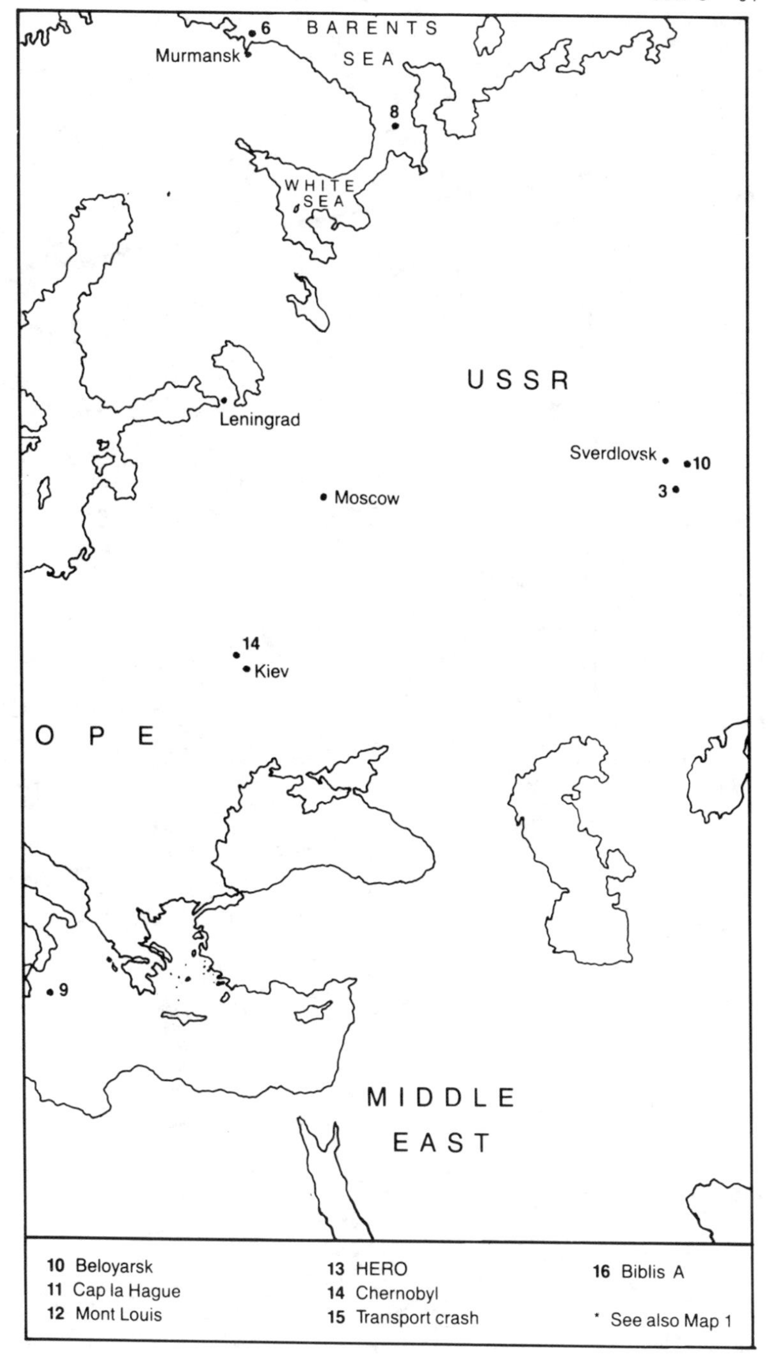

10 Beloyarsk
11 Cap la Hague
12 Mont Louis
13 HERO
14 Chernobyl
15 Transport crash
16 Biblis A

* See also Map 1

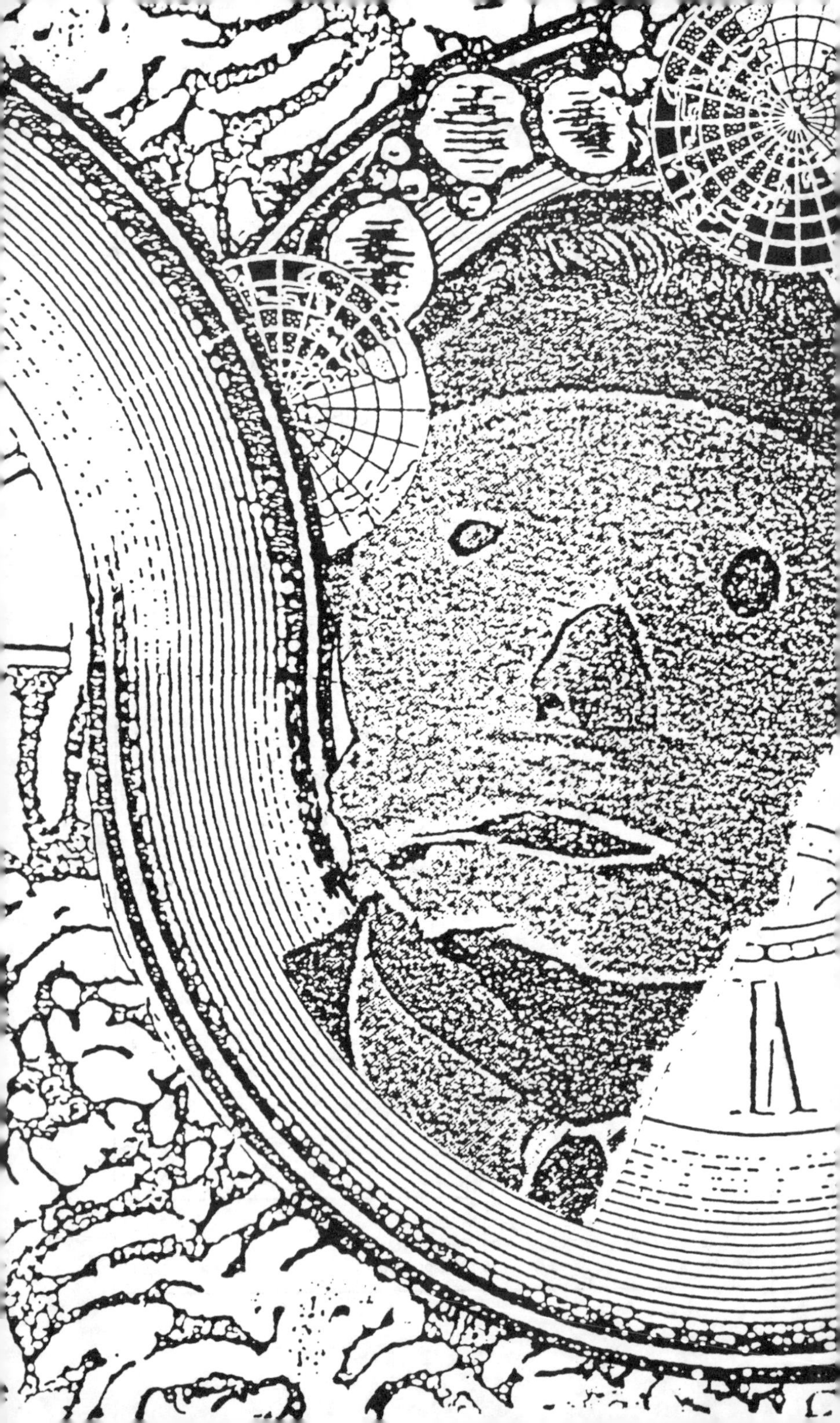

40S

16 July 1945

Trinity Test, New Mexico

The first test of the first atomic bomb — codenamed Trinity — was set to take place in July 1945 at a remote 90-mile stretch of high desert in New Mexico known as the Jornada del Muerto — The Journey of Death — a name it had been given centuries before by the conquistadores.

This area of land was already owned by the federal government, who had been using it for test bombing since shortly after Pearl Harbour. The combination of remoteness and federal ownership fulfilled all the requirements for siting such a secret project.

So it was that, in an 18-mile by 24-mile section of the Alamogordo Bombing Range, preparations for the detonation of this new weapon began to take shape. The Manhattan Project (the US and Allied effort to develop the first atomic bomb) was nearing its goal, under the direction of General Leslie Groves.

Then two months before the test, new calculations on the effects of the blast suggested that fallout from the explosion might force the evacuation of hundreds of civilians in the area around the test site. This was of particular concern to Colonel Stafford Warren, the Manhattan Project's medical director, who was also General Groves's personal physician. On his advice, the army began making serious plans for a potential evacuation of the surrounding area. Bets were laid as to both the yield of the bomb and the amount of fallout it would produce. To add to these unknowns there was the question of the weather.

Ignoring the advice of meteorologists, the test was set for 16 July, right in the middle of a predicted period of thunderstorms. The timing was important for purely political reasons. Truman was attending a crucial conference of war-time leaders at Potsdam (near Berlin) from 17 July to 2 August. Groves later wrote (in his autobiography, *Now It Can Be Told*, 1962): 'I was extremely anxious to have the test carried off on schedule. I knew the effect [it] would have on the issuance and wording of the Potsdam Ultimatum.'

Given the likelihood of changeable weather, Groves called New Mexico's governor to inform him that he might have to declare martial law throughout the central region of the state. He also had a cover story prepared for the press which claimed an ammunition dump had exploded. It was in this atmosphere of tension and uncertainty that Trinity exploded at exactly 05:29:45 local time on 16 July 1945.

I. I. Rabi was one of the eyewitnesses. He later recalled (*New Yorker,* 21.10.75): 'At first I was thrilled. It was a vision. Then a few minutes afterward, I had goose flesh all over me when I realised what this meant for the future of humanity. Up until then, humanity was, after all, a limited factor in the evolution and process of nature. The vast oceans, lakes and rivers, the atmosphere, were not very much affected by the existence of mankind. The new powers represented a threat not only to mankind but to all forms of life: the seas and the air. One could foresee that nothing was immune from the tremendous powers of these new forces.'

Within 15 minutes, the fallout split into three parts, with the main cloud drifting north-east at 45,000-55,000 ft, at a speed of around 10 mph. For the first two hours, little fallout reached the ground; then an area of some 300 sq miles became coated with fairly high radioactivity (3.3 roentgens per hour). Traces of gamma radiation were detected in Santa Fe, in Las Vegas and in Trinidad, Colorado, 260 miles from the Trinity site. From New Mexico the cloud drifted across Kansas, Iowa, upstate New York, New England, and out to sea.

Evacuation plans for the immediate vicinity had been laid but were never executed. Radiation levels were constantly checked by stationary monitors and mobile monitoring teams, but all readings seemed within acceptable levels, except at a site that soon became known as 'Hot Canyon', about 20 miles from ground zero — the point directly below the explosion. Levels in parts of the canyon were too high to measure — an estimated 212-230 roentgens — with an average reading of 20 roentgens.

On 17 July, two members of the Trinity team investigated Hot Canyon and, to their dismay, discovered two elderly people, Mr and Mrs Raitliffe, living in an adobe house less than a mile away. The house was not marked on any of the monitors' maps and no one knew it was there. The observers 'noted that the Raitliffes' tin roof was used to collect rainwater for drinking. Since it rained the night after the shot, this must have washed the radioactivity into their cistern.' (Szasz, 1984). Nearby ranch families were all exposed to radioactivity. Szasz also reports that one rancher, Ted Coker, 'actually stood under the cloud for some time and had the particles fall on him. "It smelled funny," he told the scientists later...Yet no efforts were made to follow up on the health of these people.'

The National Association of Atomic Veterans (NAAV) has a Trinity co-ordinator to assist soldiers who claim they are suffering from adverse health effects as a result of the test but no Trinity-related radiation cases have been upheld by the US government up to 1984.

Such was the naïvety about radiation in the 1940s that, for years after the test, local residents sneaked into the test site to collect samples of 'trinitite' — sand that had melted into a glass-like substance due to the heat from the blast (100 million °F) — to sell to tourists.

About 30 miles north of ground zero was a limestone mesa which formed the principal grazing range for cattle in the area; it was also the region of heaviest contamination outside the restricted area. The cattle provided the first evidence of the effects of fallout. Hereford cows began losing their hair about a month after the explosion. It soon grew back, but instead of being reddish brown it was white. These 'atomic calves' became famous and were exhibited in El Paso and Alamogordo. After tests proved that Trinity fallout was responsible for the discoloration, 75 of the most badly damaged were purchased with Manhattan Project funds and shipped to the laboratories at Los Alamos and Oak Ridge for further study. By 1947, there had been no noticeable genetic changes or mutations; many lived to produce healthy offspring. Others were slaughtered and eaten.

One final trace of the fallout was revealed in May 1946. The Eastman Kodak Corporation discovered that several batches of their sensitive industrial X-ray film were flecked with imperfections. Months of detective work by the company's technical expert on such matters confirmed that the strawboard dividers that separated the films in their cartons had fallout particles enmeshed in the fibres. The strawboard was produced in two paper mills — in Vincennes, Indiana, on the Wabash River, and at Tama, Iowa, on the Iowa River. It was the river water that had carried the contamination, leached from surrounding soils. Both mills were over a thousand miles from New Mexico.

When unexplained radioactivity was also discovered in Maryland, the *New York Times* (23.5.46) reported: 'The single bomb exploded in New Mexico contaminated the air over an area as large as Australia.'

21 May 1946
Louis Slotin

Barbara Moon originally wrote this article for the October 1961 issue of the Canadian Magazine *Maclean's*. It won the President's Medal as the top magazine piece in Canada that year. It remains as significant and timely an account today as it was then. This is an edited version of the original piece, which has never been published before outside Canada.

THE NUCLEAR DEATH OF A NUCLEAR SCIENTIST

In one sense Louis Slotin may be considered interchangeable with all the other bright, disciplined, idealistic young scientists who helped the army make a bomb.

He was born to prosperous and gentle Russian-Jewish parents living in the polyglot north end of Winnipeg. He grew up and went to school there and it was early obvious that he was not cut out for the traditional eldest-son's succession to the family business, which was a livestock commission agency. He was a studious, self-possessed, bespectacled little boy and at the University of Manitoba he grew into a brilliant student of chemistry, with a particular knack for designing the swift, imaginative experiment that would test a theory and for improvising the necessary apparatus. He also grew into a seemly youth, reserved and quiet, but with a quizzical air that lent him poise, and [had] what a friend later called 'a romantic and elaborate view of himself and the world.'

[Slotin] earned his doctorate at the University of London and at the same time turned himself into a crack bantamweight boxer.

On his return from England he applied unsuccessfully for a job with the National Research Council. [He became fascinated by] a pioneer atom-smashing cyclotron [which was being developed] at the University of Chicago. With others of a small ardent group he helped build it, begging copper wire from business firms and blowing glass components himself, and from 1937 to 1940 he worked there for nothing.

By now Slotin was nearly thirty. In the laboratory with his colleagues he was a leader. At lunch with them he would neglect his food while he talked, reaching among the [plates] with his finely shaped, expressive hands, smoothing out a paper napkin, covering it with diagrams to illustrate a point, glancing up at his companions through his glasses with what one of them called 'a certain shy, eager expression'.

In 1942, when the crash program to invent an A-bomb was launched and the Manhattan Engineer District of the US Army began casting its dragnet for qualified people, Slotin was recruited from Chicago.

In 1944 he went to Los Alamos, the bomb assembly-point hidden away on the five-fingered mesa in the ancient pine-clad uplands of New Mexico. After a time there he became, in effect, chief armorer of the United States.

Slotin's job, along with some others, was to run final tests on the active core of each precious A-bomb to make sure it would produce the explicit nuclear burst it was supposed to. The way it was done was dangerous, but in wartime one takes shortcuts and believes them justified.

At the time of his death, of course, the war was over. The core he was testing was part of a one-sided arms race, and was destined for Bikini Atoll. Slotin himself intended to go along to the Bikini tests as an observer but then, as many of his fellows had already done, he was going to divorce himself from Los Alamos and return to his real work. In the fall he would go back to the University of Chicago and in fact he had already packed and shipped ahead eleven crates of books and belongings.

There is one other thing that it is necessary to know about Slotin. Just at the end of the war a young technician named Harry Daghlian had gone back to a Los Alamos laboratory one night, against all regulations, to try an experiment with fissionable materials. A moment of clumsiness had condemned him, and he was actually the first North American to die of acute radiation sickness. Slotin, as a physicist, had helped the doctors estimate Daghlian's radiation dose and as a friend sat with him for many hours during the twenty-four days it took him to die.

It was an unique seminar — for at Hiroshima and Nagasaki survivors had not understood what people around them were dying of and, besides, were too busy to analyze the process, or be able precisely to recall it later.

So it came about that on May 21, 1946, Slotin was a man specially and singularly aware of what happens to the human body when ionizing radiation deranges its fragile and miraculous chemistry.

That day, he attended a group leader's meeting held at [his] own home base, Pajarito Site, on the floor of nearby Pajarito Canyon, and, when it was over, the visiting leaders were conducted around the premises. The tour included the outlying laboratory, south of the main building, where Slotin's group, and a group led by a Dr Raemer Schreiber, did their experiments. It was a bare, white-painted room, forty feet by twenty-six, unfurnished except for a metal table near the centre of the room, a counter against the east wall near an exit ramp, and the sparse, unimposing equipment of critical assembly tests.

The visiting group-leaders finished their inspection and moved along. But one of them, Dr Alvin Graves — a sandy, thick-set physicist from Washington DC — stayed behind with Slotin. Since Slotin was leaving, Graves was being transferred to take over his work. They chatted about one experimental configuration in particular which Graves had never seen tested. Slotin said, 'Why don't I run through it for you now?'

Schreiber, the group-leader with whom Slotin shared the lab, had also stayed behind, together with his twenty-three-year-old assistant, Theodore Perlman. They set to work at the counter against the east wall, completing observations on an experiment they had done that morning. It was around three o'clock.

Three men from the regular lab staff — Marion Cieslicki, Allan Kline and Dwight Young — as well as the customary security guard, Patrick Cleary, were also in the room and they watched as Slotin set up the experiment on the centre table.

It involved a nickel-plated core of plutonium, weighing about thirteen pounds, in the form of two hemispheres which, when put together, rather resembled a gray metallic curling stone. They were the active guts of one of the three A-bombs due to be shipped to Bikini for Operation Crossroads. It is perhaps worth noting that it was this identical core that had killed Harry Daghlian nine months earlier.

The plutonium rested in a half-shell of beryllium, a metal that can bounce escaping neutrons back into the mass of an active metal so they are conserved for the fission process.

There was a matching upper half-shell of beryllium, with a hole in it, through which Slotin thrust his left thumb so that he was holding the shell rather as a south-paw holds a bowling ball.

The technique of the experiment consisted of lowering this upper shell until it almost met the lower shell. As the reflectors enclosed the core they would bounce back more and more leaking neutrons. And so at a certain point the total of neutrons available inside the core for fission would slightly exceed the total neutron loss. A slow, controlled chain reaction would start, like a car's motor idling. This could be nudged to higher speeds — but there was a danger point. If the two half-shells came to within an eighth of an inch of each other — thus making a critical surplus of neutrons available simultaneously — a fast, uncontrolled reaction called a 'prompt burst' would ensue. There would be no explosion, for in the heat generated, the components would expand, become less dense and therefore subcritical again. (And for the core to become a bomb, its components would have somehow to be grappled together from outside long enough to explode rather than expand.) But for one millisecond there would be a flare of free neutrons, an excursion of gamma rays and beta particles, and a wave of heat.

It was one of the wartime, makeshift experiments, and some months earlier Enrico Fermi, the Nobel-prize-winning physicist, had told Slotin, 'If you keep on doing that you'll be dead in a year.' Because he was leaving, Slotin expected that this would be his last time.

He proceeded to put the assembly through its paces for Graves, taking it from idling speed almost up to the critical point, much as a test pilot strains the outside limits of his plane's performance. The stages were both audible and visible to the watchers, for an instrument similar to a Geiger counter clicked ever faster as the assembly approached the critical point and a neutron monitor recorded the increasing radiation in a red-ink graph on a roll of paper.

What Slotin did next has been called by one colleague, 'something *different* — not extraordinary but not routine'. Another colleague said recently, 'This assembly had been run many times before and its characteristics were known. But this time it didn't perform according to Hoyle. So Slotin improvised.' Still other colleagues insist it was normal procedure. And there is no real arbiter among these points of view.

What Slotin did was to remove two tiny safety devices — spacers — that served to block the upper beryllium hemisphere from closing absolutely on the lower one. Then, still holding the upper shell with his cocked thumb and spreading

A copy of Louis Slotin's original ID photo, presumably taken when he first arrived at Los Alamos in 1944. (Credit: Los Alamos National Laboratory).

fingers, he lowered one side of it onto the blade of a screwdriver held in his right hand, and allowed the other side to rest on the bottom shell. The screwdriver blade still held the two parts more than the crucial eighth of an inch apart. As the Geiger counter chattered more hectically, he waggled the handle of the screwdriver back and forth to work the shell a little farther down on the bevel and a little closer to the bottom shell.

Graves, standing just behind Slotin, shifted his weight slightly and leaned to get a better view. Cieslicki of the lab staff had paused behind Graves and to his left. Clustered seven or eight feet out from the other side of the table were Cleary, the security guard, and the two other lab staffers, Kline and Young. Schreiber, busy at his own work, happened to be facing outward into the room but Perlman, his assistant, was still bent over the counter.

At exactly three-twenty Graves heard a click as the screwdriver blade escaped the crack and the beryllium shell came down on the rest of the assembly.

In the same millisecond a blue glow surrounded the assembly; the Geiger counter needle hovered shuddering; the red-ink neutron monitor line leapt off the graph; those in the room felt a quick flux of heat. That was all.

In the next milliseconds Slotin moved his left hand and shook the beryllium shell from his thumb onto the floor. It was still three-twenty and he had just been killed.

It is now known by physicists that the reaction was over, quenched by thermal expansion, before Slotin's — any human's — reflexes could work. But his action remains a disciplined, instinctive move to do what was then believed necessary: to dismantle the assembly and stop the burst. And his body, just by being there, shielded Graves and undoubtedly saved him. The others, though no one knew how to tell at the time, proved beyond lethal range.

The official report of the accident is still classified material, but some unclassified excerpts from it are available. One of them describes the actions of those in the lab immediately after the accident. It reads: 'Kline, Cleary and Young and probably Cieslicki ran out the east door of the laboratory as soon as they could react after the accident. Kline, Cleary and Cieslicki went to the MP guard at the gate, had the gate opened, the other MPs collected and the group ran up the road a short distance. Young stopped behind an earth barricade and, not seeing Slotin, went back to the end of the ramp and looked into the laboratory after perhaps a minute. He saw no one and walked around to the main laboratory building at the north end of the corridor leading from the north-east corner of the assembly room. Perlman had run up this corridor immediately after the accident. Slotin, Graves and Schreiber had followed him to the main laboratory. Slotin immediately called an ambulance, called back those who had run up the road and prepared a sketch showing the approximate positions of everyone present at the moment of the accident.'

Slotin made one other phone call — to his friend and colleague Philip Morrison, a brilliant young theoretical physicist. Morrison recalls, 'Lou said, "We've had an accident. It went prompt critical and you'd better come down here. I've called the hospital." Then either he said or I asked, "There was a blue glow." We both knew that was very bad.'

While they waited silently for the ambulance Schreiber, at Slotin's suggestion, went back into the lab with a radiation meter. It went full scale near the assembly, he hurried back, pausing only to pick up his and Slotin's jackets.

Inside an hour the eight men were bedded in three adjoining wards of the Los Alamos hospital, a sprawling, green-painted barracks up on the central mesa. Slotin and Graves shared the third ward.

Even while they were being settled in, radiation biologists and physicists from the project, including Philip Morrison, were moving into the lab to make what measurements were possible so they could attempt the complex academic reconstruction of the accident that might give a clue to how much of which kind of rays and subatomic particles had hit the men where, and in what strength, and for how long. They would not indeed be much the wiser, for there was no set of rules yet for translating their exotic new kinds of dose into biological effects.

For a while at the hospital there was a bustle of nurses taking temperatures and blood samples and collecting little piles of coins from the men's pockets, tie clips, belt buckles, rings, watches — any metal things whose induced radioactivity could be measured to give another clue to the dosage.

Slotin had vomited once, on the way up in the ambulance. Now Graves found himself waiting for symptoms and wondering, 'Did it really go critical?' and 'How bad was it?' and 'Am I beginning to be nauseated?' The first moment they were left alone Slotin said, 'Al, I'm sorry I got you into this. I'm afraid I have less than a fifty-fifty chance of living. I hope you have better than that.' Graves privately agreed with him.

Around six p.m. one of the radiation biologists, Dr Wright Langham, came into the wards to pick up all the little piles of metallic belongings collected by the nurses. Nine months earlier he had collected a similar miscellany from Harry Daghlian and Slotin later had helped him with some of his mathematical calculations. Now Slotin cocked an eyebrow at Langham and said wryly, 'I know why you're here.'

Shortly afterward, Morrison dropped in on his way back from Pajarito Site. They talked about the dosage. In a sense it was the only thing there was to talk about, for there was no antidote — nor is there now — for acute radiation sickness. There was only the faint hope that Slotin had not got enough to kill him. Before Morrison left he asked if there was anything Slotin would like, and Slotin requested something to read. That night Morrison, who had seen the aftermath of Hiroshima, consulted workmen in the special machine shop attached to the lab and together they began to invent a contrivance with a book-rack to stretch across a hospital bed, strings to clip to every page of a book, a ratchet system to turn the pages and a switch to invoke the ratchet. The switch was placed so it could be operated by the reader's elbow. It was a reading machine — for someone who was not going to have hands to use.

By six-thirty that first night — only three hours after the accident — Slotin's left hand was fat and reddened; the thumb that had crooked through the beryllium shell was numb and tingling and its nailbed was black.

By Wednesday afternoon — twenty-four hours after the accident — the hand was distended till the skin looked as though it would burst: the right hand, too, was swollen. Both were increasingly painful so the doctors ordered ice packs and morphine. Slotin's lower abdomen, which had been directly opposite the assembly, was also beginning to redden. Otherwise he felt well, appeared cheerful and had stopped vomiting. So, nine months earlier, had Daghlian. For the cells of an organism are doughty; will rally even after such an insult, and will endure until they must try to reproduce themselves. For a short space, while the major conformations of cells work through their particular cycles to this fatal point, the body may seem to carry on.

By Wednesday night the first of the huge, tightly swollen blisters had formed on Slotin's left thumb. On Thursday more arose, as big as cookies, on the palm and between all the fingers; the left arm, too, was swollen, and the right hand and part of the forearm. For that one millisecond both had been in the play of the blue glow and they were, quite literally, cooked.

Several other things happened on Thursday.

For example, there was a meeting that night of the chemists, physicists and biologists all of whom, with different techniques, had been trying feverishly to work out dosage assessments. Dr Wright Langham, who had taken away Slotin's small change and ring and watch, had already made quick calculations based on their induced radioactivity. They indicated that Slotin's dosage had been roughly four times that of Daghlian. 'Being a relatively simple person,' he recalled recently, 'I didn't see how he had much of a chance. But the physics boys were still calculating. I walked in and told them what I thought. Phil Morrison picked up my data and pitched it the length of the desk. He said, "Hell. It can't be." The physics boys kept on for three more days trying to save Slotin with their pencils.'

Thursday was also the day when the Army — for Los Alamos was still an army post — decided some sort of press release about the accident would have to be made. Their automatic tack was to protect the public from radiation hysteria. So they were shaping the same sort of noncommittal statement (...accident...laboratory...technical personnel involved...satisfactory condition...) as had cloaked Daghlian's death, when Morrison got wind of it and threatened to go to the newspapers himself if the release failed specifically to say the victims had been exposed to radiation. The actual release said almost exactly that, though it said very little else.

On Thursday, too, since a public statement had been made, Slotin was allowed to get in touch with his family. Just at nightfall he dictated a wire to his father which read, 'My trip to Pacific indefinitely postponed, will write details love Louis.' Then, late that evening, with a nurse holding the receiver, he telephoned. He sounded calm and what he said was that he'd had a little accident and he'd be in the hospital a while and since that meant he couldn't visit home as he'd planned to after the Bikini trip perhaps his parents would come down to see *him*. The Army would arrange about plane priorities for them.

Mr and Mrs Slotin left the next day, Friday; they would arrive in Los Alamos on Saturday at noon.

By now morphine and ice packs could no longer control the pain in Slotin's dying hands. So the doctors completely encased the right hand and the left hand and arm in ice, which would have the same effect as amputation but without the attendant shock. Otherwise Slotin still seemed comfortable and alert. He was getting daily blood transfusions — friends lined up from the clinic door to the street to give blood — but his appetite was good and he still grimaced ruefully as he asked each visiting physicist, 'Well, what's the dose?' Morrison was coming whenever he could to read aloud to Slotin from technical books. Wives of colleagues brought sheaves of gladioli from their gardens, for there was no florist in Los Alamos. A technical photographer was sent over from the central pool, set up his lights and took color pictures of Slotin's arms and hands and abdomen: since the case was such a rarity, records would be invaluable.

On Saturday, when Slotin's parents arrived, he was still in the phase of apparent latency and was sitting up to greet them. 'How are you, Louis?' [his father asked.]

'Why don't you speak Yiddish, father?' Slotin replied gently. 'There's nothing to be ashamed of.' They talked for a while and Slotin made light of his condition — 'Just a bit of a burn' — but Mrs Slotin, who touched his dark hair, exclaimed, 'It's stiff and dry, like wire,' and after they left Mr Slotin sought out Morrison and asked hesitantly if there was somewhere in Los Alamos it would be possible to get a bottle of whisky.

There had been two other arrivals in town. One was a doctor from Chicago who had been experimenting with lethally irradiated animals; he had found that in dogs the terminal stage was intricate massive hemorrhage and had had some success in treating it with a dye called toluidine blue. If Slotin's illness followed the same course, the medical staff might wish to try toluidine blue on him.

The second arrival, also from Chicago, was a Dr Hermann Lisco. He was a pathologist, and had prudently been alerted against the need of an autopsy.

Sunday was the fifth day after the accident and it was on Sunday that it became clear that, whatever the dosage was, it was too much. Annamae Dickie, the nurse in charge of the blood studies, did her routine count of white cells in the blood and burst into tears. The count had plummeted. The white cells — the lifesavers in the blood — had stopped reproducing themselves and were dying. Graves, on the pretext that Slotin's parents were entitled to visit him in privacy, was moved to another room that day.

Slotin was still coherent and alert. He noticed that his tongue was ulcerating opposite a gold-capped tooth and himself suggested that gold foil would effectively baffle the cap's obvious radioactivity. Everyone of course realized he knew that radioactivity strong enough to cause ulceration was itself an ominous sign. Morrison, in a letter describing the course of Slotin's illness to their colleagues in the field, reported with careful brusquerie, 'The fifth and sixth days were evidently very hard ones.'

After this Slotin passed quickly into a toxic state: his temperature and pulse rate rose rapidly; his abdomen became stiff and distended; his gastro-intestinal system broke down completely and had to be drained continuously by gastric suction through a nasal tube; all his skin turned to a deep angry puce. His body was dissolving into protoplasmic debris.

On Tuesday, the platelets in the blood, which govern its healthy clotting, suffered a fateful drop. 'This was a sure sign of the onset of the hemorrhagic phase,' wrote Morrison later, in his report to their friends. 'Both Louis and I knew enough about this to be unhappy about its coming. It is likely the next four or five days would have been very unpleasant.'

But Slotin was already having periods of mental confusion and by Wednesday he was in delirium. His lips turned blue and he was placed in an oxygen tent. By nightfall he had passed into a coma and at 11 a.m. on the morning of Thursday, May 30 — the ninth day after the accident — he died.

The newspapers and the US Army and many well-meaning acquaintances managed to find decency in Slotin's death because at the critical moment he acted like a hero. And of course it all happened long ago. So it is interesting to find that the scientists at Los Alamos chose, and still choose, to avoid thinking of him.

This summer at Los Alamos one of them said, 'I don't wish to talk about him at all.' Others spoke of the accident 'reluctantly' and held the event at arm's length. Philip Morrison, now at Cornell University, said levelly, 'It was the most painful time of my life and I don't like to go back to it.'

They do not say why they are unwilling. But it may be that if they must remember Louis Slotin they must also feel again what they felt in those first days after mankind lost its innocence.

Barbara Moon
October 1961

Hiroshima
Date: 6 August 1945. Time: 08:15.
Weapon: A primitive uranium bomb, nicknamed *Little Boy*, with an explosive power of 13kt (equivalent to 13,000 tons of TNT).
Effects: Total casualties - 136,000. Of these, 45,000 people died on the first day, and 19,000 in the following four months. 72,000 casualties were still alive in 1951. 119,000 inhabitants of Hiroshima were uninjured.
(Credit: Popperfoto)

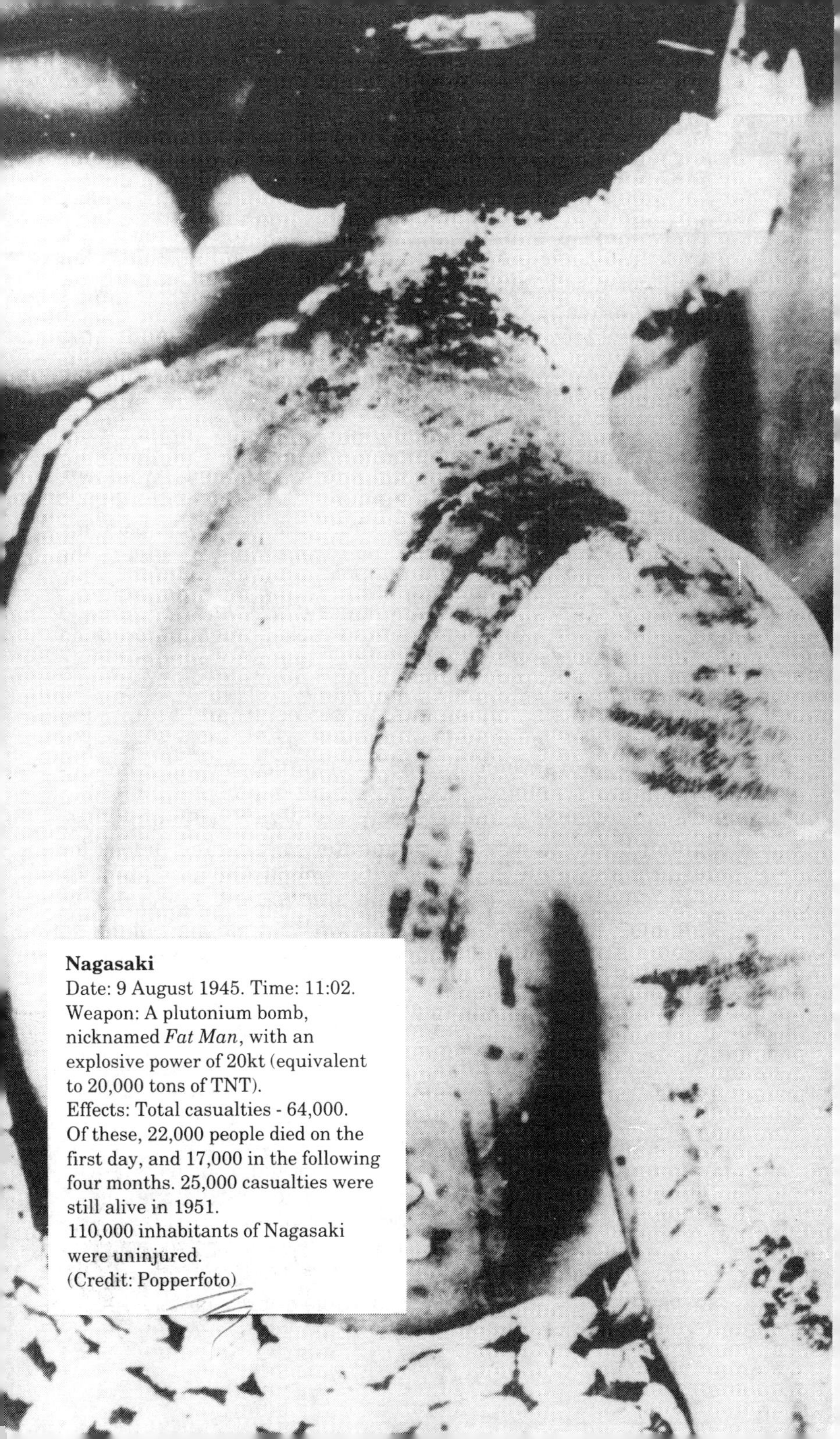

Nagasaki
Date: 9 August 1945. Time: 11:02.
Weapon: A plutonium bomb, nicknamed *Fat Man*, with an explosive power of 20kt (equivalent to 20,000 tons of TNT).
Effects: Total casualties - 64,000. Of these, 22,000 people died on the first day, and 17,000 in the following four months. 25,000 casualties were still alive in 1951.
110,000 inhabitants of Nagasaki were uninjured.
(Credit: Popperfoto)

1946-1948
US Nuclear Tests, Part 1

Micronesia — 2,100 islands and atolls scattered across the Pacific between Hawaii and the Philippines — has been colonised for nearly four centuries by four nations: Spain, Germany, Japan and the USA.

The US took Micronesia from the Japanese in 1944 after more than two years of some of the bloodiest battles of World War II. Some 6,288 US soldiers, 70,000 Japanese and 5,000 Micronesians were killed. Thousands more were wounded.

After the US takeover, Micronesians helped build US bases on Peleliu, Anguar, Saipan, Tinian and Kwajalein. Tinian became the world's largest airfield, with 20,000 military personnel stationed there. 29,000 B-29 bombing missions were launched on Japan from Tinian, as were the nuclear attacks on Hiroshima and Nagasaki.

In July 1947, the US and the Security Council of the UN signed a trusteeship agreement which brought Micronesia under the wing of the US. The Trust allowed the US to 'fortify' the islands. In return, the US promised to 'protect the health' of the inhabitants; to 'protect [them] against the loss of their land and resources'; and to 'promote the economic advancement and self-sufficiency of the inhabitants...' (Dibblin, 1988.)

Five weeks after the end of World War II, US joint chiefs of staff began to plan a series of atomic tests and to look for 'a suitable site which will permit accomplishment of the tests with acceptable risk and minimum hazard' (Dibblin.) In January 1946, Bikini (36 islands with a total area of 2.3 sq miles) in the Marshalls was selected and the islanders were relocated to Rongerik. (The US Navy took over Enewetak — formerly Eniwetok — for another series in December 1947.)

The test series was codenamed Operation Crossroads and one of the best accounts of it can be found in the 1985 General Accounting Office (GAO) report (see sources, p. 356):

'Operation Crossroads consisted of two 23-kt nuclear bombs detonated within the Bikini Island lagoon in the Pacific during the summer of 1946. The first detonation — termed Test Able — involved dropping the nuclear bomb from a plane and exploding it at an altitude of 520 feet. The second detonation — termed Test Baker — involved suspending the nuclear bomb by cable approximately 90 feet underneath a medium-sized landing ship and exploding it by remote control. Each detonation used, as its target, an array of about 80 unmanned naval ships. After each detonation a joint task force of Army and Navy personnel and civilian scientists — stationed aboard support ships more than 10 nautical miles from the center of the blast — reentered the Bikini Island lagoon and examined the damage inflicted and the radiation intensities existing on the target ships. Approximately 42,000 personnel, 240 ships (target and support), and 160 aircraft participated in Operation Crossroads. About 200 goats, 200 pigs, and 5,000 rats were also distributed throughout the target fleet so that the effects of each of the 2 nuclear bombs on animals could be studied.'

TEST ABLE

'The Test Able detonation occurred on July 1, 1946. When the bomb exploded, according to a radiological defense manual published after Operation Crossroads, a brilliant flash of light occurred, lasting a few millionths of a second, followed by a seething mass of gases, heated to a glow, which grew rapidly into a large ball of fire. A shock wave traveled from the center of the burst, visible on the water, looking like a tremendous shimmer traveling in all directions. As the glow of the ball of fire died out, a great white mushroom cloud of smoke, fission products, unfissioned particles, and dust developed and rose to a height of 30,000 to 40,000 feet.

'As the cloud began to move downwind and dissipate, drone aircraft were directed into the cloud and drone ships were directed into the target area to take radiological samples. Approximately 2 hours after the detonation, manned Navy gunboats — which were used because of their speed and maneuverability — also reentered the lagoon to measure radiation around the target ships. Once radiation levels were determined, boarding teams and salvage units began boarding ships — approximately 4 hours after the detonation — and assessing damage where the radiation levels were low. Before the day's end the lagoon was declared radiologically safe and the entire task force had reentered and anchored in the southern part of the lagoon.

'Damage inflicted by Test Able consisted of five ships sunk and six ships seriously damaged. In general, according to one of the official Crossroads reports prepared after that operation, the bomb would have been lethal for any crewmen in the open within a range of 1,300 yards and for those behind armor at about half that distance. Instantaneous exposure to the bomb's heat, blast, and initial radiation would have been the only cause of personnel deaths because the mushroom cloud created by the bomb carried most of the residual radioactive fallout downwind.'

TEST BAKER

'Following completion of the Test Able damage assessment, the Test Baker detonation occurred on July 25, 1946. Unlike the first test shot, part of the lagoon water erupted in the Test Baker blast. After an initial flash, according to a radiological defense manual published after Operation Crossroads, a huge column of water nearly a half a mile across rose 5,000 to 6,000 feet in the air. At its zenith a mushroom cloud of gas and spray developed. As the column of water collapsed back into the lagoon, a swelling wave of water and mist roughly 1,000 feet high spread out in all directions, immersing the target ships.

'Reentry into the lagoon proceeded as it had with Test Able by first sending in drone aircraft and boats to take radiological samples. From then on, however, post-Test Baker operations proceeded quite differently. Commencing about 2 hours after the detonation, low radiation readings allowed teams to board a few of the ships on the outer rim of the target array. However, intense radiation was discovered closer to the target center so further boarding and salvage efforts were abandoned that day. By day's end, almost the entire lagoon remained off limits.

'Damage inflicted by Test Baker consisted of eight ships sunk and eight immobilized or seriously damaged. According to official Crossroads reports, had crews been aboard the target ships, they probably would have fared worse during Test Baker than Test Able. Unlike the first detonation, Test Baker threw large masses of highly radioactive water onto the decks and into the hulls of the target ships, making them highly radioactive. In all, one of the official Crossroads reports estimated that topside personnel within 700 yards would have received lethal radiation doses within 30 seconds to a minute; personnel within 1,700 yards would have received lethal doses within 7 minutes; and those within 2,500 yards would have received lethal doses within 3 hours.

'With intense radiation persisting in the water and on the target ships, the joint task force devoted most of its efforts during the first week after Test Baker to the retrieval of animals and the resurfacing of target submarines submerged for the test. Commencing August 1, the joint task force began full-scale decontamination of the target ships. After first hosing down the ships with foam and salt water, nearly 2,000 Navy personnel began boarding the target ships on a daily basis to scrape, scrub, and wash the ships down to acceptable radiological dose levels. These decontamination efforts continued through August 10, 1946, when, on that day, the Crossroads Radiological Safety Officer received evidence of the probable widespread presence of plutonium on the target ships. If deposited in the body, a microscopic amount of plutonium could prove lethal. Upon learning of the probable presence of plutonium, the joint task force commander immediately halted the decontamination efforts.

'With that turn of events, Operation Crossroads came to an abrupt end. Support ships that had spent more than one day in Bikini lagoon after Test Baker were ordered to undergo decontamination, as necessary, by sandblasting hulls and by flushing salt-water systems from September 1946 to May 1947, to meet radiological clearance standards. Conversely, the target ships used during the operation met various fates. In addition to the eight ships sunk during Test Baker, six were sunk at Bikini lagoon after Test Baker because of extensive structural damage. Forty-two others, from August to September 1946, were towed to Kwajalein Island — largest of the Marshall Islands — to off-load ammunition and were subsequently sunk while 22 other target ships, from September 1946 to June 1948, were either sailed or towed back to the United States for decontamination experiments. Most of these 22 target ships were later sunk because they were not considered fit for continued use or decontamination proved unsuccessful.'

Subsequently, many thousands of US servicemen have made claims for compensation and health care due to radiation exposure received at the tests.

In May 1983, congressional hearings were held into new evidence presented by an independent body, the US National Association of Atomic Veterans (NAAV), in pursuance of these claims. The evidence consisted of an analysis of information once belonging to the late Radiological Safety Officer for Operation Crossroads, Colonel Stafford L. Warren. That analysis (Makhijani and Albright, 1983) concluded that 'every aspect of Operation Crossroads was fraught with danger for the 42,000 people present.'

Warren reported that many of the participating personnel received radiation doses hundreds of thousands of times above doses considered safe today. He also noted that General Leslie Groves, director of the Manhattan Project, 'is very much afraid of claims being instituted by men who participated in the Bikini tests.'

According to Makhijani and Albright, in the months leading up to the Baker test Warren's report had predicted that: 'If the radioactive column from the explosion did not rise more than 10,000 ft — as was in fact the case — radiological conditions would be "extremely serious".' He also noted that the heavily contaminated water in the lagoon was subsequently ingested by all the military personnel in the area, who used it for drinking, bathing and cooking — completely unaware of its potential hazard. Because Test Baker was such a radiological catastrophe, the planned third test, Charlie, was cancelled on Warren's advice.

Warren's analysis challenged preliminary statements by the Defense Nuclear Agency (DNA), who, in 1977, had been appointed administrators of a wide-ranging programme designed to investigate the correlation between atmospheric nuclear testing and subsequent health problems among test participants.

Consequently, in August 1984, the Chairman of the House Committee on Veterans' Affairs asked the GAO to review four major issues regarding Operation Crossroads: 1. the reliability of the radiation film badges used; 2. the adequacy of the personnel decontamination procedures; 3. the appropriateness of the military response to the safety recommendations made by Colonel Warren; and 4. the accuracy of the DNA's radiation dose reconstructions.

In October 1984 the DNA issued their official report on Operation Crossroads. This concluded that personnel had not been overexposed to radiation. Their position was based on radiation data recorded on film badges worn by about 6,300 of the 42,000 Crossroads participants and reconstructed internal radiation dose estimates for the participants.

The results of the GAO investigation, published in November 1985, found that: 'Crossroads personnel were exposed to four specific radiation types — internal alpha, internal and external beta, and external gamma. GAO found that DNA's calculation of exposure estimates for each radiation type may need adjustment because: 1. Film badges were not reliable for measuring both external gamma and beta radiation, as intended, and were not worn by all Crossroads participants; 2. Personnel decontamination procedures did not provide adequate protection for Crossroads personnel throughout the operation; and 3. DNA's dose reconstruction analysis for internal alpha and beta radiation has not properly estimated the possible personnel exposure from three potential pathways — inhalation, ingestion, and open wounds.'

2 December 1949

Hanford, Part 1
The Green Run

Hanford was one of three atomic cities built at high speed as part of the Manhattan Project. Beginning in 1943, a remote site, contained within a bend of the Columbia River in the desolate south-east of Washington State, was transformed by 45,000 workers into a 570-sq-mile nuclear city and work was begun on the three reactors that were to produce plutonium for the first atomic test explosion and the Nagasaki bomb.

A further five reactors were built on the site in the following decade, and these continued to operate until all eight were closed down between 1964 and 1971. By that time they had produced 50 tons of weapons-grade plutonium, enough to build 1,000 Nagasaki-equivalent bombs, which has been used for over half the weapons in the US arsenal.

In their zeal to produce plutonium for the war effort, Hanford's managers ignored health risks. Huge clouds of radioactive iodine, ruthenium, caesium and other elements were released into the atmosphere, contaminating people, animals, crops and water for hundreds of miles around. Between 1944 and 1956, 530,000 curies of radioactive iodine poured from the plant, discharge levels that today would be considered a major nuclear accident. 340,000 curies were emitted in 1945 alone. Scientists and officials were aware of the danger this large quantity of radiation posed to workers at the plant, but were chiefly concerned about people inhaling the radiation. It was not until the 1950s that scientists discovered that the main pathway by which radioactive iodine entered the body was through contaminated milk.

Much of the contamination resulted from experiments in the use of short-cooled, or 'green', reactor fuel to make plutonium in unfiltered plants. (Irradiated uranium is normally cooled in water tanks for 83 to 101 days before reprocessing. This allows the many radioactive elements that are present in dangerous quantities to decay to lower levels.

If the fuel is not cooled before it is placed in acid to dissolve the metal cladding around the fuel rods, large quantities of radioactive elements are released.)

These experiments at Hanford were carefully documented, but kept secret from the public until 27 February 1986 when 19,000 pages of documentation (*Hanford Historical Documents* 1943-57, US DOE) were released to an environmental group based in Spokane, the Hanford Education Action League (HEAL), who had filed a request under the Freedom of Information Act (FOIA).

On 2 December 1949 the most notorious of these experiments took place. Known as the Green Run, it involved one ton of irradiated uranium which was reprocessed after just 16 days instead of the normal 83-101 days. According to fresh revelations in a new document that was finally obtained under the FOIA in May 1989 ('Dissolving of twenty day metal at Hanford, 1 May 1950') the experiment was central to an elaborate military plan to try and identify the location of plutonium plants inside the Soviet Union by simulating the conditions they expected the Soviets to create in their rush to build bombs. Full details of this military mission have yet to emerge; they were deleted from the released report for national security reasons.

Karen Wheeless, a public information officer for the DOE at Hanford, told a local paper (*Tri-City Herald*, 4.5.89): 'It confirms that the release was done to develop monitoring methodology, including both instruments and technology...it was to develop a monitoring capability so that intelligence agencies could detect when other countries are involved in nuclear activities.'

The experiment began on the night of 2 December 1949 and continued until 05:00 the following morning, despite the unpredictable weather. The radiation that was released — two or three times the expected levels, containing 20,000 curies of xenon-133 and 7,780 curies of iodine-131 — emerged during a period of dead calm and as a result formed a radioactive plume measuring 200 by 40 miles. This deposited high concentrations of radiation all over the Hanford plant and in communities as far as 70 miles away, creating 'hot spots' in several neighbouring cities. There was no public health warning and no follow-up studies were conducted of the health of residents in the area. By comparison, the accident at Three Mile Island in 1979 (see p. 213) released 15-24 curies of radioactive iodine.

Radioactive iodine is particularly dangerous to children, who are up to 100 times more susceptible to radiation than adults. The US Federal Center for Disease Control began a five-year study of the incidence of thyroid diseases in three areas of Washington State nearest to Hanford in 1988, two years after initial details of the Green Run were first released publicly. They have estimated that 20,000 children in eastern Washington State could have been exposed to dangerous levels of radioactive iodine in milk produced by cows grazing in contaminated pastures; the doses may rival those received by children in the Marshall Islands as a result of the Bravo nuclear test (see p. 104).

During Hanford's first two decades, the Columbia River also became grossly contaminated by radiation from the plant. In 1954, with six reactors on-line, Hanford was dumping an average of 8,000 curies of radioactive materials into the river each day. By 1957, with eight reactors churning out plutonium at high power levels, the radioactive emissions into the river averaged 50,000 curies per day.

According to a March 1961 report 'Evaluation of pollutional effects of effluents from Hanford works', produced by the Columbia River Advisory Group, who were set up by the AEC to relieve growing public concern about radiation levels in the river: '...the capacity of the Columbia River to receive further radioactive pollution appears to be nearly, if not fully, exhausted for practical purposes, at least in the vicinity of Hanford Works.' (Thomas, 1989). (See also Hanford, Part 2, p. 204, and US Weapons Production Complex, p. 301).

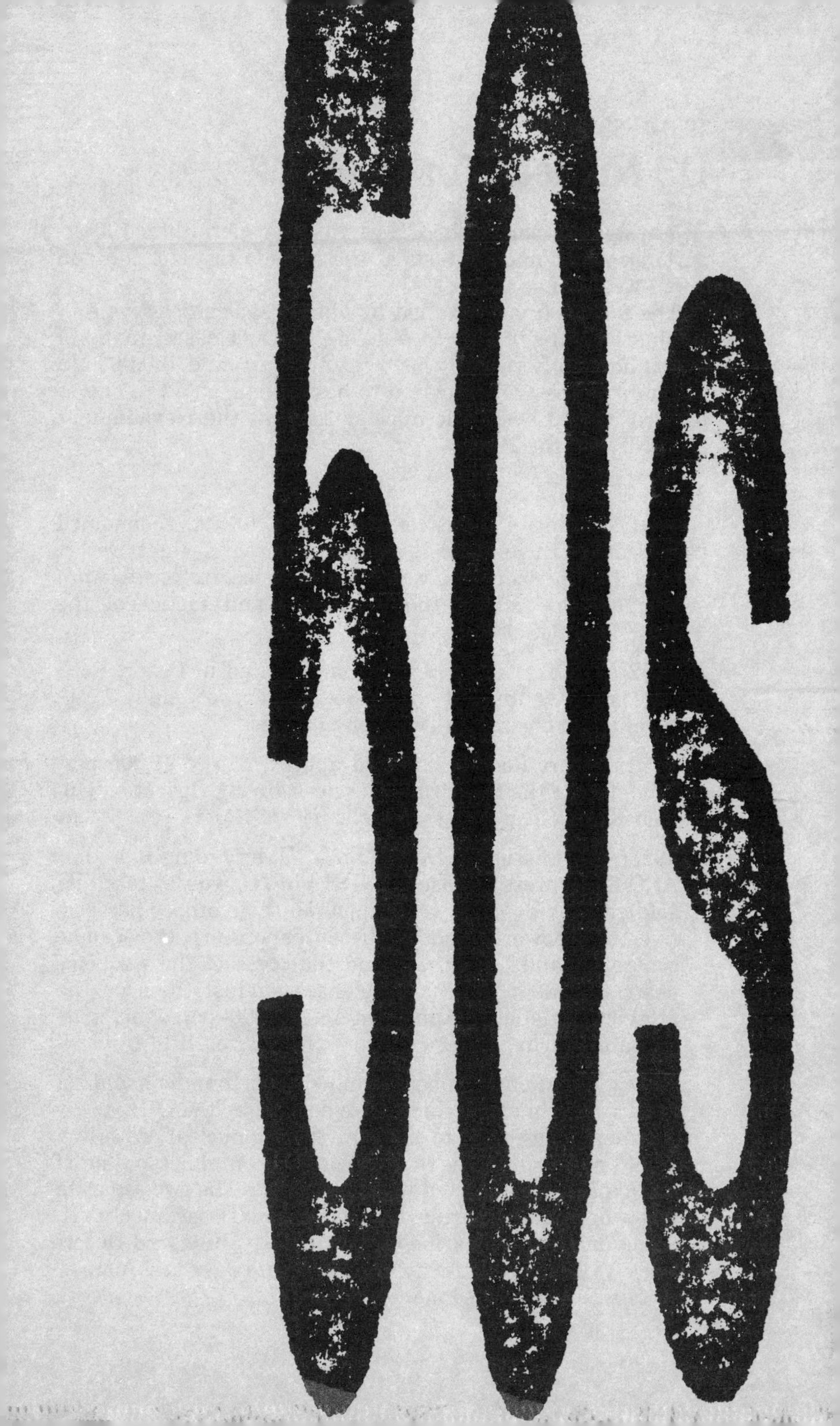
50S

1951-1958
US Nuclear Tests, Part 2

During this period the US carried out more than 100 atmospheric nuclear tests at the Nevada test site, 120 km north-west of Las Vegas.

The fallout from these led to widespread radioactive contamination off-site and to a string of court cases, many of which have only recently been resolved. Around 150,000 US military troops are estimated to have participated in one or more of the atmospheric nuclear tests at the Nevada Test Site (NTS) in the 1950s:

1951: During a series of seven test blasts, codenamed Operation Buster-Jangle, the first three of eight Desert Rock troop exercises were held. These exercises were designed to explore the conditions and tactics of the atomic battlefield.

1952: 10,600 DOD personnel participated in Desert Rock IV exercises during the 1952 series of eight tests codenamed Operation Tumbler-Snapper.

1953: Desert Rock V involved approximately 21,000 personnel from the four armed services during Operation Upshot-Knothole, a series of 11 blasts.

1955: The mission of Desert Rock VI, according to a joint AEC-DOD press release (**NWDB vol II**), was 'to teach its soldiers to view nuclear weapons in their proper perspective...that powerful though these weapons are, they can be controlled and harnessed...and that despite the weapons' destructiveness there are defenses against them on the atomic battlefield.' Approximately 8,000 personnel participated during the Operation Teapot series of 14 tests.

1957: Operation Plumbob included 24 detonations and six safety experiments, which were defined by the DOE as experiments designed to confirm that a nuclear explosion would not occur in the event of an accidental detonation of the explosive associated with the device. The two separate troop operations during Plumbob were Desert Rock VII, involving the US Marine Corps, in early July, and Desert Rock VIII, with US Army troops, in late July and August. In total, 16,000 personnel participated.

SMOKY

The most notorious of the Operation Plumbob tests was named Smoky, after a range of hills next to the test site. Here, on 31 August 1957, a 44 kt bomb was detonated atop a 210-m steel tower. The troops watched the test from a point approximately 29 km from ground zero and, three days later, carried out combat tasks and the retrieval of contaminated equipment.

Many troops were not only present at Smoky but also observed at least one other blast and were involved in manoeuvres in areas where there may have been residual contamination from previous blasts. The troops were deliberately drawn from US military units from all over the world in order to minimise fraternisation between them. The period of their assignment was officially described as Temporary Duty. This meant that, on their service record, it would be virtually impossible to prove their presence at a nuclear test. For many years this proved to be a major stumbling block to efforts to gain compensation for these men.

In late 1976 the Federal Center for Disease Control (CDC) in Atlanta, Georgia, received details of a patient, Paul Cooper, who associated his recently diagnosed leukaemia with his presence at Smoky. The case came to their attention after he filed a compensation claim with the Veterans Administration. He had been one of 200 paratroopers flown from Fort Bragg, North Carolina, to the NTS to participate in physical and psychological tests after observing the detonation. (He died, aged 44, on 9 February 1978.)

The preliminary report of the CDC's investigation indicated that nine cases of leukaemia had been found among the 3,224 men who witnessed Smoky. The expected incidence would be three and a half cases.

Two days after Smoky, on 2 September 1957, many of the same soldiers took part in manoeuvres on open ground just three miles from an 11-kt blast named Galileo. According to a report filed with the Army in March 1958 by a private research contractor, Human Resources Research Organisation (HUMRO), the paratroopers were to take apart their M-1 rifles immediately after the shock wave moved past them.

The report said the 'dust level was quite high, as expected, with the resultant impairment of vision.' (*International Herald Tribune*, 18.1.78.) These new details were revealed for the first time in 1978, at hearings by a US Congress

Health and Environment subcommittee into the after-effects of testing in Nevada and the Pacific from 1946 until 1958.

Dr Karl Morgan, the 'father' of health physics and a world-renowned authority on radiation safety, provided expert testimony. He told the House subcommittee that Smoky was 'the dirtiest of the tests we covered...There is no question there was a great deal of fall-out from the weapon. I was frightened and appalled to hear there were troops in the trenches.' (*International Herald Tribune*, 26.1.78) He said he had 'no doubt whatever' that radiation from Smoky caused the leukaemia now found in veterans of the shot.

In June 1985 the US National Academy of Sciences published their study, 'Mortality of nuclear weapons test participants', which surveyed the causes of death among 49,000 veterans. They discovered that these veterans were not dying at an unusual rate from diseases associated with radiation. But at least 10 of the 3,554 participants at Smoky had died from leukaemia before the age of 45, two and a half times the expected number. The CDC commented there was an eight in 1,000 probability of this increase in incidence happening by chance.

THE 'DOWNWINDERS'

Shot Harry took place on 19 May 1953. It was originally scheduled for 2 May, but had been postponed when the previous blast, Simon, had not only contaminated the firing area, but had also deposited fallout as far away as upstate New York. When the 32-kt Harry blast was finally staged, it produced an immense cloud of fallout which was carried downwind over ranchland and small towns in Nevada, Utah, and Arizona — areas inhabited predominantly by Mormons.

Every attempt was made by the AEC to deliberately deceive these people and no attempts were made to inform or evacuate them, even though the fact that most of them were dependent on food they raised themselves made them particularly vulnerable to fallout.

The original caption of this dramatic photo, taken at Yucca Flat, Nevada, on 28 April 1952, reads: 'Military observers watch the mushroom cloud form after the explosion of the atomic bomb in *Operation Bigshot* at the AEC proving grounds. The men are so close to the blast that they have to bend their heads back to watch the smoke'. (Credit: Popperfoto).

When the tests began at Nevada in 1951 the AEC assured the American public there was no cause for concern because the 'downwind' area was virtually uninhabited. An editorial in the *San Francisco Chronicle* commented: 'This creates an interesting new class of citizen: "virtual uninhabitants". Their voting density is slight — they are politically safe — and virtually expendable.'

A 1955 AEC public relations pamphlet — quoted in the June 1980 issue of *Life* — addressed the 'downwinders': 'You are in a very real sense active participants in the Nation's atomic test program. You have been close observers of tests which have contributed greatly to building the defenses of our own country and of the free world...At times some of you have been exposed to potential risk from flash, blast, or fall-out. You have accepted the inconvenience or the risk without fuss, without alarm, and without panic. Your cooperation has helped achieve an unusual record of safety.'

The first symptoms of radiation sickness — headaches, fever, hair loss, dizziness and vomiting — were experienced by hundreds of people in the area. For many, these symptoms were passing afflictions; for a few, those first signs disappeared, only to surface again, many years later, as cancers and other fatal conditions.

It was news of Paul Cooper's case in 1978 that first made Vonda McKinney suspect that her husband's fatal leukaemia was linked to fallout from the test site. She told a lawyer that there were many other women in the area who had family members that had died of leukaemia. His investigations led him to file an initial 100 claims for compensation on behalf of the downwinders. This figure was to grow to more than 1,000 plaintiffs, suing on behalf of 375 cancer victims.

The massive trial of Irene Allen v. The USA generated almost 9,000 pages of transcript containing the testimony of 98 witnesses and the arguments of 11 lawyers. On 10 May 1984, the Federal Circuit Judge Bruce Jenkins found against the government and awarded $2,660,000 damages to eight leukaemia victims, four of whom were children. Numerous other cases were unsuccessful because they did not qualify under the primary and secondary factors used by the judge to determine causation.

An appeal court overturned this ruling in April 1987, on the grounds that the administration enjoyed constitutional powers 'to do no wrong', and so could not be held responsible for injuries suffered as a result of powers granted to the AEC. In January 1988 this new ruling was upheld by the US

Supreme Court, which agreed that 'the sometimes harsh principle of sovereign immunity' applied. The court concluded: 'However erroneous or misguided [the government's] deliberations may seem today, it is not the place of the judicial branch to now question them.' (*New Scientist*, 21.1.88.)

1954: *THE CONQUEROR* AND THE DEATH OF JOHN WAYNE

The last film personally produced by Howard Hughes, *The Conqueror*, starred John Wayne as Genghis Khan in a strange, oriental western that regularly makes the worst-film lists. Directed by Dick Powell and co-starring Susan Hayward and Agnes Moorehead, it was an expensive production with elaborate sets.

To find locations for the Gobi Desert sequences, Powell and his assistants scoured eight states before finding the 'perfect location' near St George in Utah. (A nearby American Indian reservation provided numerous extras to act as the Mongolian hordes.) However, St George was just 137 miles from the Nevada test site, where there had been 11 atmospheric tests the year before. Most of the filming took place in Snow Canyon, which turned out to be a radioactive 'hot spot'. After filming there from June to August 1954, cast and crew returned to the studio, accompanied by 60 tons of sand from the area, to enable retakes to be completed in the studio.

Between then and 1984, 91 of the cast and crew of 220 have developed various carcinomas; and over half of them — including Wayne, Powell, Hayward and Moorehead — have died. Another of the movie's stars, Pedro Armandirez, committed suicide in 1963, at the age of 51, when he developed terminal cancer of the lymphatic system. He had survived cancer of the kidney in 1958.

Dr Robert C. Pendleton, Professor of Biology at the University of Utah, comments: 'With these numbers, this case could qualify as an epidemic. The connection between fall-out radiation and cancer in individual cases has been practically impossible to prove conclusively. But in a group this size you'd expect only 30-some cancers to develop...I think the tie-in to their exposure on the set of *The Conqueror* would hold up in a court of law.'

The children of Wayne and Hayward accompanied their parents on location. Michael Wayne developed skin cancer in 1975; his brother Patrick had a benign breast tumour removed in 1969.

According to *People* magazine (10.11.80), at least one top Defense Nuclear Agency (DNA) scientist is racked by doubt over the incident. 'Please, God,' he said, 'don't let us have killed John Wayne.'

1952-1958
British Nuclear Tests

In the late 1940s, against a background of mounting concern about the Soviet Union's nuclear intentions and the USA's nuclear monopoly, a handful of British politicians took the decision to develop their own deterrent.

'A very secret decision to make atomic bombs,' says official British nuclear historian Lorna Arnold (1987), 'was taken by Ministers in January 1947 — it was not disclosed to Parliament until May 1948 — and from this time an independent nuclear deterrent became a cornerstone in British government policy.' Thus began an enterprise ungoverned by either public scrutiny or parliamentary control.

The decision to stage the tests in Australia was welcomed by that extreme anglophile, Premier Robert Menzies, who viewed it as constituting his country's contribution to a wider defence effort. There was little parliamentary objection or public opposition at the time. In all, there were 21 British nuclear tests in the region, staged from 1952 to 1958; 12 of these were on or near the Australian mainland, and nine were at a test site on Christmas Island (see p. 97).

In 1984, due to a growing clamour of protest by Aborigines and ex-servicemen, who insisted that their health had been adversely affected by the tests, the Australian government set up a Royal Commission to investigate their claims, forcing the British government to release hundreds of previously secret documents. Thus only very recently has a far fuller picture of the test series emerged.

The Commission concluded that: 'the tests had probably caused an increase in the level of cancer among the Australian population in general, and among Aborigines living near the test sites, and thousands of servicemen and civilians directly involved with the tests, in particular.' (1985.)

Their report analysed the major test series in detail:

OPERATION HURRICANE (3 OCTOBER 1952)

A 25 kt bomb was loaded into an old frigate and blown up in a lagoon in the Monte Bello Islands, just off the north-western coast of Western Australia. The report comments: 'The Monte Bello Islands were not an appropriate place for atomic tests owing to the prevailing weather patterns and the limited opportunities for safe firing. There was fall-out on the mainland [but it is] unlikely that the fall-out exceeded the no-risk level...it is impossible to determine whether Aborigines were exposed to any significant short or long-term hazards.'

OPERATION TOTEM (14 AND 26 OCTOBER 1953)

Two test bombs, of 10 kt and eight kt respectively, were exploded on a steel tower at the Emu Field test site in South Australia. The tests were rushed and weather conditions were unsuitable.

The fallout from Totem I was twice that of Totem II. Aboriginal stories of an encampment near the site being enveloped in a 'black mist' after the test were first recorded by an anthropologist around 1970 and became a news story in Australia in 1980. Many authorities dismissed the story as myth but the Royal Commission gave credence to the Aborigines' account. The incident was later investigated by the Australian Ionising Radiation Advisory Council who concluded that the probability of long-term damage to the health of the Aborigines in the region at that time was low.

In Operation Hot Box, a three-man air crew in a specially-prepared Canberra aircraft flew through the mushroom cloud from Totem I six minutes after detonation and travelled 2,000 yd in it. The men were exposed to radiation doses of 10-15 rem. The cloud was tracked for two and a half days; it was still visible as a cloud some 24 hours after its formation.

OPERATION MOSAIC (16 MAY AND 19 JUNE 1956)

Two tests, of 15 kt and 98 kt respectively, were staged on two of the Monte Bello Islands. After the second of these tests — a spectacular explosion by all accounts — a radioactive cloud skirted the mainland of northern Australia depositing low levels of radiation on areas of the coast.

OPERATION BUFFALO (SEPTEMBER-OCTOBER 1956)

This series of four tests — two tower-mounted blasts of 15 kt and 10 kt, an airburst (three kt), and a surface detonation (1.5 kt) — were the first tests held at Maralinga, a remote area of south Australia designed as a permanent test range. (Maralinga is an Aboriginal word meaning 'Fields of Thunder'.) Part of the main A-bomb development programme, the tests were linked to what Arnold (1987) calls 'a substantial programme of biological studies and target response experiments. The military built a large village of prefabricated huts, and an army of dummy men and live domestic and farm animals were placed close to ground zero and exposed to the blast. Some 250 servicemen from Commonwealth countries, named the Indoctrinee Force, observed the surface blast from a distance of four and a half miles and then made a two hour inspection of the aftermath, wearing protective clothing and respirators.'

The Commission's report criticises the handling of all four tests in this series. 'Overall,' the report says, 'the attempts to ensure Aboriginal safety during the "Buffalo" series demonstrate ignorance, incompetence and cynicism on the part of those responsible for that safety. The inescapable conclusion is that if Aborigines were not injured or killed as a result of the explosions, this was a matter of luck rather than adequate organization, management and resources allocated to ensuring safety.'

OPERATION ANTLER (SEPTEMBER-OCTOBER 1957)

This series of two-tower blasts (one kt and six kt), and an air-burst (25 kt), was (the report says) 'better planned, organized and documented than any of the previous test series, but inadequate attention was paid to Aboriginal safety...People continued to inhabit the Prohibited Zone as close to the test sites as 130 km for six years after the tests.'

At Emu Field in 1953 and, following the major tests, at Maralinga from 1958 until 1963, a number of secret 'minor trials', with various codenames (Kittens, Rats, Tims and Vixens), were staged to test components and materials for the A-bombs and, in the case of Vixens, 'to study the dispersion of plutonium, uranium and beryllium when released in particulate form by explosions or fire, and to obtain information which will permit estimates of hazards downwind in case of accident release.' (Arnold, 1987.)

The report comments: 'In view of the long half-life of plutonium (24,000 years) the Vixens series...should never have been conducted at Maralinga.' One result of the Vixens tests was that an estimated 10,000 tiny fragments of plutonium were scattered over three 'hotspots' on the site.

The powerful force of the 98kt Mosaic blast on 19 June 1956. This was the largest explosion of the British test series in Australia; it dusted the continent's northern coast with fallout. (Credit: Popperfoto)

When Maralinga was no longer required by the British, a team from the UK Atomic Weapons Research Establishment undertook Operation Brumby (April-July 1967), an attempt to clean up the site by ploughing plutonium-contaminated soil into the ground and covering it with clean soil. The report says this operation 'was planned in haste to meet political deadlines and, in some cases, the tasks undertaken made the ultimate clean-up of the Range more difficult.'

According to Dr John Symonds, Australian government historian of the atomic tests, the clean-up 'complicated matters both physically and politically. I go along with estimates that there was about 20 kg of plutonium buried in shallow graves. Now perhaps one kilo has come to the surface — both as dust and in rather lumpy pellets — to be blown about by the desert winds, perhaps scattered by as much as a mile in several directions.' (*The Times*, 24.2.87.)

In 1967 the range became the responsibility of the Australian federal government and was extensively surveyed in 1974 and 1985. Six of the seven major trial sites were found to have radiation counts, and four other areas, principally where the Vixens experiments had been conducted, were heavily contaminated. In 1984 a Land Rights Act passed by the Australian Parliament gave Maralinga to the Tjarutja tribe. Arnold comments: 'but what use would they be able to make of it?'

The Commission concluded that Britain should bear the cost of cleaning up all the sites so as to render them 'fit for unrestricted habitation by the traditional Aborigine owners as soon as practicable'. The Australian government should accept responsibility for compensating the Aborigines affected by the tests and for the denial of the use of their tribal lands for more than 30 years. It should also give compensation to the estimated 15,000 people exposed to the tests who suffered as a result. Applicants claiming compensation must, however, prove that their disease has been caused 'on the balance of probabilities' by contamination.

In August 1986, the Australian government announced that they would make an initial compensation payment of $Aus 500,000 to Aborigines with links to the Maralinga area. This money was to be used for roads, communications and a water supply for around 450 Aborigines who were among those displaced by the tests. Further compensation would follow, the government said.

In October 1986, Britain agreed to pay half the $Aus 3 million needed to fund a two-year research programme at the

Maralinga and Emu test ranges, without committing itself to paying the cost of complete decontamination, estimated to be as high as $Aus 250 million. An air-to-ground survey was to be made to determine the extent of the contamination, samples were to be taken of soil, fauna, and flora, and a study would investigate the potential risks to local Aborigines.

The agreed details of additional joint research were announced in February 1987. Helicopters carrying sophisticated Geiger counters would fly in a grid pattern over the contaminated area of some 600 to 1,000 sq km. They would be able to detect nuclear waste on land surfaces but not locate buried concentrations of plutonium.

By June 1987, scientists involved in the Maralinga survey concluded that only 10 per cent of an estimated 22 kg of plutonium was contained in burial pits at the site. The rest, in the form of an estimated 10 million small fragments of metal and other contaminated debris, was lying in the vicinity of the test site. (This was a 100-fold increase on previous estimates.) Some plutonium was found in an Aborigine camp almost 80 miles from the test site. The helicopter surveys identified plutonium extending north from the site in three distinct plumes and in concentrations several hundred times higher than the level recommended by the US Environmental Protection Agency (EPA) as suitable for unrestricted land use.

In 1988 an assessment of the findings of the year-long survey by Australian, British and American scientists was given. It revealed higher-than-expected levels of plutonium that would delay the proposed clean-up and prevent the Aborigines from returning to their tribal lands.

On 22 December 1988, Rick Johnstone (54), a former Australian airman, was awarded $Aus 679,500 damages for radiation sickness contracted when he drove scientists and military personnel into the fallout area after four tests at Maralinga in 1956. Johnstone, president of the Australian Nuclear Veterans' Association, was awarded damages by a four-man Supreme Court jury after a hearing that lasted 62 days. His award opened the way for claims by 43 other workers at atomic sites, and for the widows of other victims.

CHRISTMAS ISLAND (MAY 1957-SEPTEMBER 1958)

With the major powers moving towards a moratorium on nuclear testing, the pressure was on Britain to develop and test an H-bomb in order to demonstrate post-war prestige.

The result was the Christmas Island 'Grapple' series of seven atmospheric tests (plus two A-bomb blasts) staged over a 15-month period. Details of the size of these tests, amongst the largest ever detonated in the atmosphere, remain classified information to this day. Many veterans have since claimed they were damaged by resultant radiation exposure.

Some knew they were exposing themselves to high doses. The men of 76 Squadron, for example, who flew modified B6 Canberras into the H-bomb clouds to take radiation samples. Prime Minister Margaret Thatcher revealed, in a written answer to Parliament on 3 February 1983, that two of these men received 30 rem.

Blakeway and Lloyd Roberts, in *Fields of Thunder* (1985), state: 'the authorities' confidence in the safety of the Christmas Island tests does not take into account the fact that...a number of the safety precautions concerning radiation were either inadequate or ill-observed. There is also a possibility that a number of men thought to have been in no danger were inadvertently put at risk.'

The first test case came in the British High Court in November 1986, when Lance Corporal Melvyn Bruce Pearce, a serviceman suffering from lymphoma, claimed compensation. He had been an engineer on Christmas Island from 1957 to 1958. His case was seen as opening the door to claims from about 1,000 members of the British Nuclear Test Veterans Association who claim to have developed cancer through exposure to radiation during the 1950s British nuclear tests.

The government opposed his right to bring a claim under Section 10 of the Crown Proceedings Act, which states that no one can sue the government for injuries incurred during time in the armed forces as a result of action by another member of the armed forces or a government servant.

The defence claimed that the tests were carried out by the UK Atomic Energy Authority (UKAEA), which has never been a part of the Ministry of Defence (MOD), although, between 1971 and 1973, its nuclear materials fabrication and weapons groups were hived off to them. In December 1986 Lance Corporal Pearce won the right to sue the government for negligence. The Crown immediately applied for, and was granted, leave to appeal, but in August 1987 the Appeal Court backed Pearce. The government was given leave to test the ruling in the House of Lords.

In January 1988, a long-awaited report by the National Radiological Protection Board (NRPB) was finally published.

(The report had been commissioned by the MOD in October 1983 to investigate media allegations that veterans of the atomic tests suffered a high incidence of certain diseases associated with exposure to radiation.)

The study found that out of 22,347 men who took part in the tests, more than 400 had died of various forms of cancer: 22 from leukaemia and six from multiple myeloma — two types of cancer associated with radiation. These findings were compared with a control group of 22,326 servicemen who were in the tropics at that time, but who did not take part in the tests. Of these, though 434 had died of cancer, considerably fewer had died from the radiation-related types: only six from leukaemia and none from multiple myeloma. Although the difference between the two groups is statistically significant, scientists who analysed the data suggested that this was due to the unusually low incidence of these two forms of cancer in the control group.

The NRPB survey concluded that: '...small hazards of leukaemia and multiple myeloma may well have been associated with participation in the nuclear weapons programme, but that such participation has not otherwise had a detectable effect on the participants' expectation of life or on their total risk of developing cancer.' (The British Nuclear Tests Veterans Association had claimed from the onset that the report would be a whitewash, as the NRPB is a government-supported body and cannot be considered totally independent.)

One good result of this report was that the UK Department of Health and Social Security (DHSS) paid out pensions to an ex-serviceman and a war widow. A spokesman said that, since the NRPB report, the DHSS was giving such people 'the benefit of the doubt'. (*Mail on Sunday*, 17.7.88.)

12 December 1952

Chalk River Reactor Ontario, Canada

The world's first major nuclear reactor disaster occurred at an experimental light-water cooled, heavy-water moderated research reactor, the NRX, situated at Chalk River, some 100 miles west-north-west of Ottawa.

The following account of the accident is from the **Bertini** report:

'The nuclear characteristics of the NRX reactor were such that a loss of light-water coolant would make the reactor more critical, whereas a loss of heavy-water would make the reactor less critical.

'Groups of control rods (banks) could be withdrawn by pressing various numbered buttons on the control panel. Red lights above these numbers indicated that the banks were in their fully withdrawn positions. The banks were inserted by increasing the air pressure above them, which was accomplished by pushing button 4. Button 3 activated a solenoid which ensured that the seal for this increased air pressure would not leak. Therefore, to drive the control rods into the reactors, the operator had to press buttons 3 and 4 at the same time; in order to facilitate this action, the buttons were located near each other on the control panel. The withdrawal buttons for the control rod banks were spaced an arm's length away.

'On 12 December 1952, the reactor was undergoing tests at low power. The circulation flow of the light-water coolant was reduced in many of the rods, since there was not much heat being generated in the fuel. The supervisor noted that several of the red lights suddenly came on. He went to the basement and found that an operator was opening valves that caused the control rod banks to rise to their fully withdrawn positions. Horrified, he immediately closed all of the incorrectly opened valves, after which the rods should have dropped back in. Some of them did, but, for unexplained reasons, others dropped in only enough to cause the red lights to turn off. These latter rods were almost completely withdrawn.

'From the basement, he phoned his assistant in the control room intending to tell him to start the test over and to insert all the control rods by pushing buttons 3 and 4. A slip of the tongue caused him to say: 'Push 4 and 1.' (Button 1 was a control rod withdrawal button.) The assistant laid down the phone, because it required out-stretched arms to push 4 and 1 simultaneously; hence, he could not be recalled immediately to rectify the error. Since button 3 had not been pushed, the air seal was not secured; thus, the air rushing into the chambers, where it should have compressed and forced the control rods in, rushed out through the seal instead. Control rod bank No. 1 was withdrawn.

'The operator in the control room soon realised that the reactor power was rising rapidly, and he pressed the scram button. Even without compressed air, the rods should have dropped in by gravity, but again for unexplained reasons many of them did not, and the power continued to climb. He phoned his supervisor in the basement and asked him to do something to increase the air pressure. After a hurried consultation with physicists and the assistant superintendent, who were present in the control room, it was decided to dump the heavy-water moderator. This succeeded in shutting down the reactor but not instantaneously, since it took some time to drain. The reactor power had peaked between 60 and 90 MW(thermal).

'More was to come. Water began to pour into the basement. It was light-water coolant. Radiation alarms sounded, both inside and outside the building, and a plant evacuation procedure was ordered. Conversations in gas masks, donned by the staff who were inside the building, proved to be too difficult to carry on, so they moved into an adjoining building. However, except for sealing the vents, there really was not much more that could be done.

'The metal sheaths containing the cooling-water annuli for about 25 fuel rods ruptured, there was some fuel melting, and the heavy-water tank (calandria) was punctured in several places. The initial power surge caused the cooling water around the rods to boil, which increased the internal pressure in the rods and ruptured metal sheaths. This boiling and loss of water increased the reactivity of the reactor, which enhanced the power surge. About one million gal of water containing about 10,000 curies of radioactive fission products had been dumped into the basement of the building.

'The core and the calandria, which were damaged beyond repair, were removed and buried, and the site was decontaminated. An improved calandria and core were installed about 14 months after the incident. There were no injuries, but there was some release of radioactivity.'

Less than six years after this, Chalk River was the scene of another serious accident. The successor to the ill-fated NRX at Chalk River was the 200-MW NRU reactor. The NRU is a heavy-water cooled and heavy-water-moderated engineering and research reactor. The **Bertini** report continues:

'The decay heat from fission products in the fuel of a reactor is sufficient to require that the fuel be cooled for some time after shutdown. It was lack of such cooling that caused the accident at the NRU reactor.

'Problems had developed in some of the fuel elements that were being used. It was found that leaks had developed in the cladding that surrounds the fuel; this allowed some of the radioactive fission products to escape the cladding and enter the reactor tank. These leaky fuel assemblies were replaced when their condition was discovered, but the tank had become somewhat contaminated, and this background radioactivity obscured the presence of other leaking fuel elements.

'For the week prior to 23 May 1958, the reactor had been in steady operation, but it shut down automatically that day on a signal indicating that the power was rising too fast. No reason could be found for the shutdown, so the operator attempted to restart the reactor four times. Each time spurious signals kept him from doing so.

'On the fifth try, he was successful, and the power was brought up by setting a switch which governed the rate of rise of power. Five minutes later, the reactor was shut down on an excessive-rate-of-rise signal again. This time the shutdown was accompanied by alarm signals indicating that a fuel element had ruptured, resulting in contamination of the coolant.

'It is postulated that the switch which set the rate of power increase was faulty, allowing the power to rise faster than desired. This, in turn, had caused the violent failure of an undetected leaky fuel element. The pressure shock from this transient had, in turn, caused a spurious signal to be generated, which caused the control rods to be withdrawn, thereby creating a rapid power transient. This transient shut down the reactor.

'Two fuel elements had been damaged, one severely and one moderately. The moderately damaged element was withdrawn from the reactor without incident.

'When a fuel element that has been in use is withdrawn from this reactor, a flask containing circulating cooling water is attached to the reactor, and the fuel element is drawn up into the flask, after which the flask is moved to a storage tank by a large crane. The flask has its own pump to circulate the water and keep the enclosed fuel element cool from the time it is withdrawn from the reactor until it is deposited in the storage tank.

'It was decided to remove a guide tube along with the more severely damaged fuel element. When the fuel element was drawn up into the flask, it was found that the guide tube had prevented cooling water from entering the flask. It was further found that the fuel element could not be reinserted into the reactor where it could be cooled by the heavy water. Attempts to do so had caused it to become stuck in an improper position in the flask. The fuel element began to get hot, and time became important.

'There was an emergency water hose available at the side of the reactor tank, so attempts were made to move the flask to this position with the crane. Since the fuel element was now stuck in an improper position within the flask and the flask was not in its normal state, a series of safety interlocks came into play that were designed to prevent the movement of the flask unless a specific operational routine was followed. Each interlock had to be overcome by time-consuming efforts. In the meantime, the fuel element began to disintegrate and burn. A piece of it fell on top of the reactor, and a much larger piece fell into a maintenance pit. Sand was dumped on the pieces to quench the burning uranium.

'The building was severely contaminated, and 600 men participated in its clean-up. The highest dose received by anyone involved in the incident was 19 rem. The area of detectable contamination outside the building was in the 100 acres adjacent to the building.

'There were no apparent injuries, but there was some release of radioactivity, which appeared to be confined to the area of the building.'

Both Chalk River accidents were shrouded in secrecy at the time, and the official word was that there would be no adverse health effects from the radiation doses received. No follow-up studies on the health of the men involved were carried out. Only in recent years has an intensive investigation been launched by the Canadian Veterans' Pensions Board.

1954-1958
US Nuclear Tests, Part 3

BRAVO

'It should be noted that no test is done without a specific purpose in mind, and at no time was the testing out of control.' So said Admiral Lewis Strauss, former director of the AEC, speaking about 'Bravo' at a press conference in the Washington Press Club, 30 March 1954. (Dibblin, 1988.)

At 06:45 on 1 March 1954, a 15-Mt hydrogen bomb, codenamed Bravo, was detonated close to the ground on the island of Bikini. It was expected to produce a six-Mt yield but actually produced 15 Mt. The explosion produced a crater 240 ft deep and 6,000 ft across and the fireball melted huge quantities of the coral atoll, sucking it up and scattering it for miles as lethal fallout. (The bomb was specifically designed to generate the maximum amount of fallout; its designers were Edward Teller and Ernest Lawrence, working at the Livermore National Laboratory in California.)

That morning, the wind was blowing in the direction of two inhabited atolls, Rongelap and Utirik, some 100 and 300 miles from Bikini respectively. During previous tests their populations had been evacuated; no attempt was made to move them before Bravo.

The islands were bathed in fallout, which coated Rongelap in a pale powder to a depth of one and half inches and wrapped Utirik in a radioactive mist. It was three days before US Navy ships arrived to evacuate the 236 islanders and 28 American service personnel to the navy base on the nearby island of Kwajalein. By then they were experiencing a range of painful symptoms: nausea, burnt skin, diarrhoea, headaches, eye pains, numbness, skin discolouration and general fatigue. They also suffered a lowering of the blood cell levels, especially the white and T-cells, which form a major part of the body's immune system. Their fingernails came off, their fingers were bleeding and their hair was falling out. The exact dose of radiation that the 236 islanders received was never measured, but it is estimated at 11 rem per person on Utirik and 190 rem on Rongelap.

Nor were they the only victims. A Japanese tuna fishing boat was 100 miles east of Bikini when Bravo exploded and was caught in the path of the fallout. By the time they reached Japan two weeks later, all 23 crew were suffering from radiation sickness. One of them, Aikichi Kuboyama, died of liver and blood damage on 23 September 1954. There was an international outcry over the incident. Two years later, the US handed over $2 million compensation to the Japanese government.

The question is, did the US deliberately use the people of Rongelap and Utirik as nuclear guinea pigs?

This is certainly the view of the islanders. According to John Anjain, a magistrate on Rongelap at the time of Bravo (Alcalay 1981): 'From the beginning of the testing programme in our islands the United States has treated us like animals in a scientific experiment for their studies. They come and study us like animals and think of us as "guinea pigs".'

Evidence for this view is also provided by a 1958 report issued by the federal facility, Brookhaven National Laboratories — who performed the radiation surveys for the AEC. It asserts that 'greater knowledge of [radiation] effects on human beings is badly needed...Even though the radioactive contamination of Rongelap Island is considered perfectly safe for human habitation, the levels of activity are higher than those found in other inhabited locations in the world. The habitation of these people on the island will afford most valuable ecological radiation data on human beings.'

Jane Dibblin interviewed Katherine Jilej, midwife and grandmother, on Mejato in 1986. The latter told her: 'We are very angry at the US and I'll tell you why. Have you ever seen a...baby born looking like a bunch of grapes, so the only reason we knew it was a baby was because we could see the brain? We've had these babies — they died soon after they were born.'

The Rongelapese now suffer from high rates of malignancies (thyroid cancer and leukaemia) and reproductive problems, all directly associated with radioactive fallout, together with a range of secondary symptoms associated with the breakdown of the body's immune system. The Marshall Islanders have one of the highest rates of diabetes in the world — probably caused either by radical diet changes or microwaves (non-ionising radiation), or a combination of both. They are also suffering from psychological phobias as the result of uncertainties about their future.

The Brookhaven scientists allowed the Rongelapese, including some who not been exposed to the effects of Bravo originally, back to their islands in July 1957, declaring the atoll to be safe, despite 'slight lingering radiation'. There had been no radiological clean-up.

After just one year, the islanders' body levels of radioactive strontium-90, caesium-137 and zinc-65 had risen rapidly. In 1966 some Brookhaven scientists themselves consumed, under controlled laboratory conditions, pandanus and coconuts grown on the islands. They recorded their strontium and caesium intake over a seven-day period, and found that these were, respectively, 20 and 60 times higher than normal. Nevertheless, the local people were told it was safe to continue living on their islands.

In 1978, the DOE completed an aerial radiation survey of the northern Marshall Islands and, in early 1979, DOE scientists informed the Rongelapese that the northern islands in their atoll — the ones the people had been using to grow copra and gather food for the last 20 years — were too radioactive to be visited. The survey also revealed that, in addition to Rongelap, Utirik, Enewetak and Bikini, 10 other atolls or islands had 'received intermediate range fall-out from one or more of the megaton range tests'. No medical programme exists for the people of these other islands and there have been no detailed health reviews of the entire Marshallese population of 43,000.

In 1984, the Rongelapese announced they wanted to leave their atoll and solicited US congressional aid to carry out a resettlement. When none was forthcoming, they contacted Greenpeace who evacuated them on the *Rainbow Warrior* to Mejato.

The Bikinians themselves were evacuated from their island after Operation Crossroads in 1946 and taken to Rongerik. In 1948 they were moved again to Kwajalein navy base for a short time, after which they were relocated on Ejit Island in Majuro Atoll, as well as being dispersed throughout the Marshalls. This sociological dislocation has had disastrous consequences.

In 1968, President Johnson announced that Bikini would be returned to its people. There had been 23 tests at Bikini, but according to a 1968 AEC report: 'The exposures to radiation that would result from the repatriation of the Bikini people do not offer a significant threat to their health and safety.' The Bikinians, unimpressed by the partial clean-up of their atoll, voted in 1972 not to return there as a group. However, the US was committed to the resettlement so it offered government employees, Bikinians and Marshallese, free food and housing if they would move there.

In 1975 the 100 people living on Bikini were found to have low levels of plutonium in their urine. By 1977, tests showed that body levels of caesium-137 had increased 11-fold. Instead of evacuating the people, an elaborate food importation system was set up, to cut their consumption of local foods.

In 1977 the DOE were quoted in the *Washington Post* (3.4.78) as describing the Bikini inhabitants as 'the best available source of data for evaluating the transfer of plutonium across the gut wall after being incorporated into biological systems.' The people were finally evacuated in September 1978.

The US Congress has since allocated $90 million over the period 1988-92 to decontaminate the island; the final choice of which method is used to do this rests with the Bikinians.

Scientists from the environmental division of the Lawrence Livermore National Laboratory in California have been visiting the atoll each year since 1972, to bring back samples of soil and food for analysis. Their conclusion is that application of potassium chloride, or fertiliser containing potassium, will block the uptake of caesium-137 by the food crops. (The plants draw caesium from the soil because its atomic stucture closely resembles potassium.)

These findings may mean that the 1,400 Bikinians, nuclear nomads for more than 40 years, may at last be able to return to their homeland.

In recent years, a new political relationship between the US and the Marshall Islands has developed. The new Compact of Free Association, which the Marshallese voted narrowly to accept in September 1983, was sold to them as a 15-year bridging agreement until the islands are ready for complete independence. Ratified by the US Senate in 1985, it gave the US an automatic 15-year extension option, well into the twenty-first century, on the use of Kwajalein (see below).

Under Section 177 of the Compact, the radiation survivors of Rongelap, Enewetak, Utirik and Bikini are compensated. The US has paid $150 million to the Marshall Islands government, in the form of a trust fund, which will yield about $270 million over 15 years — this to be shared out amongst the various islanders affected.

The compensation comes with strings attached. Under Section 177 — the so-called 'espousal' clause — all rights of radiation survivors to sue in the US courts are cut off. There will be no recourse for people yet to be identified as victims of the test's effects.

The ruling nullified lawsuits claiming $7 billion in damages that were still pending in the US courts; the courts ruled that the espousal clause was constitutional and that the Section 177 trust fund for $150 million was adequate and final. As a result, the Marshallese will never have their day in a US court.

KWAJALEIN

The fight for Kwajalein in World War II almost destroyed the island. Almost 5,000 Japanese soldiers were killed in the battle to capture the island, which had been the base from whence the attack on Pearl Harbour was launched. The island was pounded by 100 lb per sq ft of shells and bombs, the physical equivalent of 20,000 tons of TNT. One army observer who surveyed the aftermath commented in an official history of the battle: 'The entire island looked as if it had been picked up to 20,000 feet and then dropped.' (Dibblin, 1988.)

The US Navy established a base on Kwajalein after the war and moved the local population to the tiny neighbouring island of Ebeye. There are now 10,000 people living on the one-tenth of a sq mile of treeless, disease-ridden dirt that is Ebeye — a three-mile ferry ride from the spacious air-conditioned luxury enjoyed by the approximately 3,000 US Army and technical personnel at the US base on Kwajalein; an island 10 times larger.

Kwajalein is now one of the most strategic sites in the world for the US. Since 1959 it has been used as a target zone to test US Intercontinental Ballistic Missiles (ICBMs) for accuracy and speed. These unarmed missiles are launched 4,200 miles away — from Vandenberg Air Force Base (AFB) in central California — and for 40 years, virtually all long-range missiles have been tested here. Kwajalein has proved vital for the development of the Star Wars programme. According to a Pentagon spokesman (*Washington Times*, 26.6.85): 'All of our high-tech strategic systems have some relationship to the Kwajalein facility. If we didn't have Kwajalein we wouldn't be able to test such long-range stuff over open, largely uninhabited areas of the earth's surface.'

In the sheltered lagoon of Kwajalein, surrounded by a ring of the 93 islands that make up the atoll, the test missiles shatter as they hit the water at 6,000 mph. Many of these missiles are tipped with 'spent' uranium-238, which is used to weight the missile in order to simulate a nuclear warhead on radar screens; when the missile is destroyed, the uranium is scattered in the water.

Army Public Affairs Officer Ed Vaughn, stationed at Huntsville, Alabama, was quoted in the *Marshall Islands Journal* (24.8.82): 'We do have a clean bottom policy at Kwajalein. We send a small submersible and divers out to recover the debris from the re-entry vehicles that break up...We don't claim that we get it all, by any means, but we attempt to get as much of it as we can locate.'

But, according to the *Christian Science Monitor* (8.6.89), the lagoon cleaners may have missed one. 'After a launch in July 1987, a missile nose cone containing sensitive flight data couldn't be found. There's speculation that it was snatched by a Soviet mini-sub before US divers could recover it. But, Pentagon officials say, it also could have been buried in the sandy sea bed or lost when it was shipped back to the US.'

27 July 1956

Broken Arrow 1 Lakenheath AFB, UK

A US B-47 bomber — part of the 307th Bombardment Wing based at Lincoln AFB, Nebraska — on a routine training mission called Operation Reflex, skidded and slid off the runway at Lakenheath RAF Base, 20 miles north-east of Cambridge, England. It burst into flames and crashed into a nuclear bomb storage igloo in which there were three Mark 6 nuclear bombs, each containing 8,000 lb of TNT as part of the trigger mechanism designed to explode the nuclear cores. There was no danger of a nuclear explosion because the bombs had not been armed, but if the TNT had gone up, plutonium would have been released into the atmosphere and scattered over a wide area.

According to a retired USAF general (unnamed), who was a B-47 pilot based at Lakenheath at that time, if the fire had ignited the TNT 'it is possible that a part of eastern England would have become a desert'. (*Observer*, 9.8.81.) Another US officer who was there admitted that a disaster was only prevented by 'a combination of tremendous heroism, good fortune and the will of God'.

It appears that a major incident was only averted because the base fire chief at the time, Master-Sergeant L.H. Dunn, ordered his crew to pour flame-suppressing foam on the weapons store and to ignore the burning bomber with its four fliers trapped inside. There was apparently little chance of rescuing them anyway.

The incident only rated three paragraphs in the newspapers at the time. It occurred on the same day that Nasser announced his intention to nationalise the Suez Canal and two days after the Italian liner *Andrea Doria* had sunk off the coast of the USA with the loss of 51 lives.

Aware that the incident could fuel a growing anti-nuclear movement in Britain, Prime Minister Anthony Eden and US President Eisenhower ordered that any mention of nuclear weapons should be excised from the accident reports.

The first full picture of the incident was published in the *Omaha World-Herald* in November 1979. When asked about the incident the then UK Defence Secretary, Francis Pym, told Parliament: 'The United States authorities have already stated no nuclear materials were involved either within the crashed aircraft or in any building affected by the resulting fire.' (*Observer*, 9.8.81) The MOD has consistently refused to confirm or deny the incident and maintains the same policy in relation to nuclear bombs held at RAF and USAF bases in Britain, although officials claim that all munitions are now protected against air crashes.

22 May 1957

Broken Arrow 2
Kirtland AFB, New Mexico

A B-36 bomber with a 13-man crew was ferrying a weapon from Biggs AFB in Texas to Kirtland, an air force base on the southern edge of Albuquerque, when, at 11:50 and at an altitude of 1,700 ft, the plane hit turbulent air just as an officer was removing a locking pin that secured the bomb. (It was standard procedure at that time that the locking pin be removed during take-off and landing to allow for emergency jettison of the weapon, if necessary.) According to George Houston, radio operator on the flight, the officer lost his balance and, in reaching out to steady himself, he 'grabbed the mechanism the bombardier uses to release the unit. The bomb-bay doors were closed at the time but [the bomb] took them with it.'

Houston likened the accident to the closing scenes of the 1964 movie *Dr Strangelove*. (*International Herald Tribune*, 30/31.8.86.)

The 42,000 lb, 10-Mt bomb (which was one of the largest hydrogen bombs ever made, and was known as Mark 17) fell 1,700 ft — too short a drop for its parachutes to slow its descent — and landed four and a half miles south of the Kirtland control tower. Conventional explosives in the bomb detonated when it hit the ground, killing a cow and creating a crater 12 ft deep and 25 ft in diameter in empty land owned by the University of New Mexico. Recovery and clean-up operations conducted by Field Command, a division of Armed Forces Special Weapons Project, revealed no radioactivity beyond the lip of the crater. Pilot Richard Meyer later stated: 'We weren't even carrying all the essential materials needed to arm it for dropping as a weapon.'

First official confirmation of the incident did not come until 1981, but no details were released until the *Albuquerque Journal* published an account based on a military document it had obtained under the FOIA in August 1986. (Ben McCarthy of the Department of Energy press office at Albuquerque told the *Daily Telegraph* (1.9.86) that the story *had* been released in the US — in 1957, 1959, 1971 and 1984. He said: 'We can't understand why everyone is so excited when the story has been released four times previously.')

Ironically, Kirtland is the home of the National Atomic Museum where a version of the Mark 17 is on display alongside bombs of the same type as Fat Man and Little Boy, the two dropped on Japan.

11 September 1957

Rocky Flats, Colorado

Built secretly in 1952 under the auspices of the AEC, the Rocky Flats plant is unique in that, unlike other such facilities which are situated in remote areas, it is just 16 miles upwind and north-west of downtown Denver. Almost two million people live within 30 miles of the plant.

Here plutonium is drilled and shaped into triggers for nuclear warheads, and plutonium from old warheads is extracted and recycled to make new bombs. Both activities are extremely hazardous. Plutonium is notoriously dangerous to handle and oxidises readily on contact with oxygen — particularly if it is the form of fine shavings — burning to produce plutonium oxide. This compound can form extraordinarily fine particles, sometimes no more than a molecule in size. Plutonium is most damaging to human health when inhaled and these fine particles are the most dangerous and the most difficult to filter out of the air. Potentially fatal lung doses are measured in billionths of a gram.

On the evening of 11 September 1957, a fire began in Building 771 when plutonium shavings spontaneously ignited in a glove-box, a work station at which workers use lead-lined gloves to handle the plutonium. The highly-flammable Plexiglas glove-boxes caught fire and, as workers attempted to extinguish the blaze, an explosion blew out all 620 industrial ventilation filters in the building, sending thick plumes of black smoke into the air, which contained radioactive particles from between 30 and 44 lb of burning plutonium.

A secret AEC sampling report, dated the following March, found radiation levels at up to 8,000 times background level at a nearby ranch and at two elementary schools. The report was not made public at the time and the official position is still that little or no radioactivity escaped (*Washington Post*, 12.12.88). Dr Carl Johnson, former director of the Jefferson County Health Department, which covers an area adjacent to Denver, recalls that: 'Plutonium-contaminated smoke escaped from the plant for half a day, and high levels of radioactivity were found in both the schools that were checked downwind.' Despite this, he asserts, no warning was given at the time to schools, the county commissioners, the state health departments or local city authorities.

A second major fire at the plant occurred more than 10 years later, on 11 May 1969. Few safety improvements had been implemented since 1957 and the fire had a similar cause. Chips of plutonium on a conveyor belt inside a glove-box spontaneously ignited and the effects were, once again, worsened by the flammable glove-box material. Still regarded as the worst accident in the plant's history, the fire caused $50 million worth of damage and shut down all new weapons production for six months.

Some residents doubted the plant manager's assurances that the fire had been contained. At the request of a citizen's group, scientists from the National Center for Atmospheric Research tested soil near the plant and found it contaminated with plutonium and americium (one of the products resulting from the radioactive decay of plutonium) at up to 210 times background levels.

Faced with this evidence, the Rocky Flats management owned up to a new problem. Hundreds of drums containing plutonium-contaminated oil that had been stacked along the plant's eastern fence had been leaking for more than a decade. Dow Chemical, operators of the plant under AEC contract, estimated that 86 g of plutonium had been lost from the drums.

In 1974, the government purchased thousands of acres of contaminated land to the east and south-east of Rocky Flats, and thousands more acres were added following a 1985 lawsuit, when the government agreed to pay $9 million for further plutonium-impregnated land close to the plant. (See US Weapons Production Complex, p. 301.)

10 October 1957

Windscale, Sellafield, UK, Part 1

In early 1946 Britain's nuclear scientists had returned from participating in the Manhattan Project at Los Alamos and from working on Canada's Chalk River civil reactor programme, and the US had reorganised its atomic research programme, effectively cutting off the supply of both information and fissile material to the UK. Britain was determined to push ahead with its own independent research into nuclear weaponry and therefore a supply of plutonium had to be secured.

To meet this need, two plutonium production units were hurriedly built at Sellafield, an old munitions factory on the Cumbrian coast. The installation, named Windscale, consisted of two simple, air-cooled atomic piles in which natural uranium fuel, held in a matrix of graphite, was bombarded with neutrons, transmuting it into plutonium-239.

The graphite acted as a moderator, slowing the neutrons down and increasing the chances of a successful collision with the uranium nuclei, but this moderator exhibited a strange property. The neutrons knocked carbon atoms out of their normal positions in the graphite molecules, causing the graphite to change shape and to store energy. The need to release this energy (called Wigner energy after the scientist who first explained it) in a controlled manner was recognised when, in September 1952, the energy was spontaneously released. Fortunately, the reactor was shut down at the time. Controlled releases were subsequently carried out in a process that involved heating the graphite up by starting the nuclear chain reaction in the pile while shutting off the fans that normally provided cooling air.

On the evening of Monday 7 October 1957, when the Wigner energy had been allowed to build up for rather longer than usual, this procedure was begun. The fans were turned off, the pile was made critical and allowed to heat up overnight, and was then shut down the following day. However, temperature sensors in the core appeared to indicate that the full release of Wigner energy had not taken place, and so, unusually, the pile was heated up again. This time the sensors showed an abnormally rapid rise in temperature and so the power of the pile was reduced. During Wednesday 9 October all seemed normal, with the exception of one part of the core in which the temperature was steadily rising. On the Thursday morning cooling fans were turned on, and the temperature throughout the core dropped, apart from this hot spot in which the temperature continued to increase. At about the same time, monitoring equipment in the filters of the plant's 150-m chimney registered a rise in radioactivity. By midday there was a further release of radioactivity, which also registered on monitors around the Windscale site, and air samples revealed 10 times the normal levels of activity.

By now it was clear that something had gone badly wrong and a burst fuel cartridge was suspected. Attempts to see inside the core using a scanner failed because the scanner had jammed, and, in the end, two members of staff wearing protective clothing removed a charge plug and looked inside the pile. The fuel channels that they could see were ablaze. The temperature at one point in the core had risen too far, a fuel cartridge had split, and the uranium had oxidised, releasing enough heat to ignite the graphite. The fans that had been turned on to bring the temperature down had had the effect of fanning the blaze. At the height of the fire, three tonnes of uranium were alight.

Workers attempted to push out the fuel elements and restrict the fire but progress was too slow and, in the early hours of Friday 11 October, the decision was taken to flood the pile with water. With the fire brigade in position and the

Some of the 2,000 workers who, in the 1950s, passed daily through the tiny station of Sellafield on their way to work at the nuclear plant. According to the original caption, some of the workers lived in a nearby government hostel called Greengarth Hall. It reads: 'The 2,500 men and women who work at the factory are forbidden to talk of or about their job.' (Credit: Popperfoto) police on standby (for no one knew whether the water might not cause a hydrogen/oxygen explosion), hoses were inserted into the core and the water was turned on. There was no explosion, and by 11:00 the fire was under control, though water was pumped in for a further 24 hours just to be sure.

Only now did the UKAEA make news of the accident available to the press. An official spokesman was quoted (*Manchester Guardian*, 12.10.57) as saying: 'There was not a large amount of radiation released. The amount was not hazardous and in fact it was carried out to sea by the wind.' None of this was true.

On site, construction workers had been exposed to up to 150 times the maximum permissible level of radioactivity. They were told to go indoors but were not told what was happening to them. Local farmers and villagers received 10 times the maximum permitted lifetime radioactive dose. The UKAEA and the government of the day knew this but decided not to evacuate anyone.

Two days after the fire the government did take some action, as it was clear that local milk supplies had been contaminated by the radio-isotope iodine-131, which affects human thyroid glands. Some two million litres of milk from cows grazing in an area of more than 500 sq km around the plant were poured away into the sea and rivers. Local waterways gave off a sour stench for weeks afterwards.

Few contingency plans had been made for such an accident. At the time when the piles were originally being built, Sir John Cockcroft, the leading nuclear physicist of the time, had insisted that filters be installed in the chimneys as a safety measure. Known as 'Cockcroft's Folly', these filters prevented a major accident from becoming a catastrophe.

Pile No. 1 never operated again and Pile No. 2 was shut down shortly afterwards. Now entombed in concrete, they stand today as 'monuments to our ignorance' in the words of Sir Christopher Hinton, the man responsible for their design and construction (*New Scientist*, 14.10.82). In Pile No. 1 there are still around 22 tonnes of melted and partly-burned fuel. The decommissioning of both piles began in 1987 and it will take decades to dismantle them completely.

The full report of the inquiry into the fire, written by the father of the British bomb, Sir William Penney, was not published at the time; it only became a public document in 1988 under the Thirty Year Rule (whereby British government papers are withheld from public scrutiny for a 30-year interval). The UKAEA had backed the view of its then chairman,

Sir William Plowden, that the report should be published, as did the MOD. But Prime Minister Harold Macmillan kept it secret on the basis that a recently forged agreement with US President Eisenhower, for joint research into nuclear defence, could be jeopardised by the news.

It was not until 25 years after the fire that official estimates of the total population dose resulting from the release of radiation were made public. In 1982, the NRPB, a regulatory body advising the government on nuclear safety and radiation limits, published a report ('An assessment of the radiological impact of the Windscale fire, October 1957') that claimed to be a 'complete description of the radiological impact of the fire'. In this they considered the effects of 41 isotopes released at the time, and estimated that the fallout from the accident had caused 260 cases of cancer, 13 of them fatal. But these investigations ignored the effect of another more dangerous isotope which was released in significant quantities, has a high take-up rate by the body, and has a half-life of 140 days — polonium-210.

Polonium, a strong alpha-emitter, is a little-known element that formed a vital component in early atomic bombs — the 'initiator' at the heart of the triggering mechanism. At the time of the fire, polonium was being produced by irradiating bismuth in a side channel at the Windscale pile.

The main radioactive cloud from the Windscale fire travelled south-east across most of England and on over Europe. According to John Urquhart, a statistician at Newcastle-upon-Tyne University, this cloud contained 370 curies of polonium, which translates into a dose of some 850,000 man-rems. Urquhart says, 'We are therefore talking about more than a thousand deaths from the Windscale accident.' The NRPB subsequently revised their figures, claiming there would be 32 deaths, half of them due to polonium. Many people affected by the fire are now seeking compensation.

A research paper by P. M. E. Sheehan and I. B. Hillary in the *British Medical Journal* (November 1983) revealed that in a study of 47 married fertile women, who had been students at an Irish boarding school across the Irish Sea from Windscale in 1957, *six* had given birth to Down's Syndrome children. The high incidence of Down's Syndrome is especially significant given that the average age of the mothers at the time of birth was 26.8 years; it is unusual for mothers of this age to give birth to Down's Syndrome babies.

In January 1989, under the Thirty Year Rule, it was revealed that the fire at Windscale was not the first incident at the plant to be officially covered up on the direct instructions of Prime Minister Harold Macmillan. He also personally sanctioned a news black-out on an incident earlier in 1957.

In spring that year there was a leak of strontium-90 from Windscale, which contaminated milk from more than 800 Cumbrian farms. In a secret memorandum John Hare, the Minister for Agriculture, told Macmillan: 'During the summer, the readings on some farms have been many times higher than the national average and up to ten times the highest recorded figures for weapon fallout in the wet hill areas.' (*Independent*, 2.1.89.) He added that no action was being taken 'to prevent milk being consumed or produced on farms in the area', and warned that 'The readings in the area cannot be concealed indefinitely.'

On 24 October Hare wrote again to Macmillan to inform him that the Medical Research Council had set new standards for the 'permissible daily intake of strontium-90'. The standards were considerably relaxed, a development that Hare described as 'very satisfactory'. Macmillan's response was short and to the point: 'Nothing must be published without my seeing and approving.'

Hare developed a plan for a carefully orchestrated and gradual release of the news following a softening-up operation which culminated in a planted written Commons question. No information about the accident was released for more than 18 months.

Furthermore, it now appears that the incident was just one of a number of unpublicised releases of strontium-90 in the mid-1950s caused by the bursting of spent fuel elements being prepared for reprocessing.

RADIATION DISCHARGES

In 1956, levels of radioactive discharge from the Windscale plant into the Irish Sea were deliberately raised for two years, 'partly to dispose of unwanted wastes but principally to yield better experimental data'. (*Lancet*, 25.8.84)

Dr John Dunster, chief health physicist with the UKAEA at the time and later director of the regulatory NRPB, told a UN conference on the peaceful uses of atomic energy in Geneva in 1958: 'The intention has been to discharge fairly substantial amounts of radioactivity as part of an organised and deliberate scientific experiment...the aims of this experiment would have been defeated if the level of radioactivity discharged had been kept to a minimum.' (Cutler and Edwards, 1988) He said that the policy had been necessary in order to ensure that radioactivity at higher levels behaved in the way that laboratory data predicted.

This came to light almost 30 years later, when James Cutler of Yorkshire TV suggested that the results of this policy were now being seen in the raised incidence of leukaemia (almost 10 times the national average) in the nearby village of Seascale. This suggestion was rejected by Dr Dunster, saying that discharge levels were *far higher* in later years. (*The Times*, 25.8.84) And indeed, the figures supplied by the plant's owners, British Nuclear Fuels Limited, to the 1983 Black inquiry — an independent inquiry set up by the British government to investigate cancer levels in the vicinity of the Windscale plant (see p. 271) — show this to be the case. They indicate that from 1956 to 1960, between 60,000 and 100,000 curies of radioactivity were released annually, whereas in 1975 discharges reached a peak of 250,000 curies.

Then, in 1986, doubt was cast over the validity of the BNFL figures when it was revealed that releases of uranium from Windscale into the atmosphere between 1952 and 1955 were not, as BNFL had told the Black inquiry, 400 g but 20 kg - 50 times as much. Doses of radiation received by the population over this period may have been five times as large as previously thought.

1957/1958

Chelyabinsk-40, USSR

An accident that contaminated thousands of sq miles in the Central Ural Mountains of Russia may have caused hundreds of human casualties. Yet all news of it was suppressed and, but for some ingenious scientific detective work, it might still be a secret.

In the late 1940s, the Soviets began hasty construction of Cheylabinsk-40, a plutonium-production facility, about 10 miles east of the industrial town of Kyshtym, on the southern shore of Lake Kyzyltash. The nuclear scientists were under intense pressure to produce enough plutonium to carry out the first Soviet bomb test before Stalin's 70th birthday in December 1949. (The bomb was, in fact, exploded in August that year.)

The first hint of problems at Chelyabinsk came in the late 1950s, when sketchy press accounts appeared in western Europe describing a catastrophic accident in the Soviet Union that had generated high levels of chemical and radioactive fallout.

It was November 1976 when an *émigré* Soviet biochemist, Zhores Medvedev, casually referred to a 'major disaster in the Urals' in a *New Scientist* (4.11.76) article on Soviet dissident scientists. Such was the level of interest that Medvedev and other researchers began digging deeper and a mass of Soviet ecological research papers was uncovered, indicating that lakes, soil and more than 200 animal and plant species in an unidentified area covering several thousand miles had been contaminated by radiation.

What was especially significant was that the radioactive 'fingerprint' provided by these studies indicated an accident involving waste storage operations associated with the production of weapons-grade plutonium. (The five radioisotopes that were reported — strontium-90, ruthenium-106, caesium-137, cerium-144 and zirconium-95 — become dominant in high-level liquid wastes after one to two years of decay following removal from a reactor.)

The location of all this Soviet research was finally confirmed by an acknowledgement in one of the research papers which had been missed by the Soviet censors, indicating that the samples had come from the Chelyabinsk region.

Bit by bit, the details fell into place. The accident almost certainly occurred in late December 1957 or early January 1958, because on 9 January, Radio Moscow devoted a large segment of its airtime to radiation sickness and a detailed list of preventative measures.

This was confirmed by Soviet exiles who had been in the area at the time and who witnessed the after-effects. The government had apparently ordered the hasty evacuation of surrounding towns, and rest homes and hotels had been commandeered as hospitals. Huge quantities of food had been destroyed and fresh supplies brought in from outside. Wartime rationing had been imposed and the area sealed off. The main north-south road was closed for nine months.

Professor Leo Tumerman, in a letter to the *Jerusalem Post* (*New Scientist*, 30.6.77), said he had travelled through the devastated area in 1960 on his way to visit the construction site of the first major Soviet atomic power plant at Beloyarsk (see, p. 228). One hundred km from Sverdlovsk, a road sign warned drivers not to stop for the next 30 km and to drive through at maximum speed with the windows closed. 'On both sides of the road for as far as one could see, the land was "dead": no villages, no towns, only the chimneys of destroyed houses, no cultivated fields or pastures, no herds, no people... nothing.'

These anecdotal accounts were later backed up in a December 1979 report — an analysis of Soviet maps of the area, undertaken by scientists at the Oak Ridge National Laboratory in the US. By comparing maps made before 1958 with those made during the 1970s, the investigators discovered that the names of 30 small communities with populations under 2,000, as well as some larger towns, had been removed. In one L-shaped 60-mile-long sector, all the towns marked on earlier maps had disappeared, indicating a mass movement of population. The water drainage system of the area had also been modified. They estimated that most of the nearby industrial town of Kasli was contaminated, and 14 lakes and some 625 sq miles of land were poisoned.

The exact cause of the disaster remains open to speculation, but Medvedev believes that a conventional chemical explosion caused by the accumulation of gases around hot nuclear waste is the most likely explanation. Technological procedures for handling nuclear waste in the 1950s were crude and many short cuts were taken to save time.

Medvedev's evidence was dismissed at the time by many nuclear scientists including the chairman of the UKAEA, Sir John Hill, who, according to **Pringle and Spigelman**, described the idea as 'pure science fiction'. The whole question of the disposal of nuclear wastes was especially controversial in 1977 because a major inquiry into plans to expand the reprocessing plant at Windscale was being held.

The Oak Ridge analysis suggested another possible cause for the disaster. They believe that the caesium-137 was separated from the nuclear waste products before they were stored, using a process which would, in turn, produce waste containing large quantities of ammonium nitrate. Failure of a cooling system on a high level waste storage tank could have triggered the accident, creating a powerful explosion.

In a discussion on the US television programme *60 Minutes* (broadcast on CBS, 9.11.80), the chief of scientific intelligence for the CIA from 1955 until 1963, Herbert Scoville, confirmed that both the CIA and the AEC knew about the disaster at the time. Ralph Nader, whose Critical Mass organisation obtained this information under the FOIA, believes there was a government cover-up in the US too. In the *60 Minutes* discussion, in which Dr Medvedev also took part, he said: 'In the last eight years, there has been mounting public debate on nuclear power. That debate *would* have occurred in the late 1950s and early 1960s — in that critical period when the governments and the utilities were deciding to go full blast with nuclear power plants around the country — if our government had disclosed evidence about this radioactive waste catastrophe in the Soviet Union, made it public, and informed the people about the real risks of nuclear power.'

The anchor man of the programme, Dan Rather, concluded the discussion: '...disposal of nuclear waste... is still the biggest problem dogging nuclear energy programs today. As the US Nuclear Regulatory Commission has said, 'a nuclear accident anywhere in the world is a nuclear accident everywhere in the world.'

A three-year investigation carried out by the Los Alamos National Laboratory for the US Department of Energy and made public in January 1982 denied the possibility of any major accident involving nuclear waste. It suggested that three factors were responsible for slowly but completely contaminating the environment around Kyshtym: 1. Water used to cool the reactors from 1950 onwards, heavily contaminated by leaky fuel rods, was drained into the Techa River, thus spreading the contamination; 2. Acid rain devastated vegetation for 20 km around the plant. Nitric acid was used to extract the plutonium-239 from irradiated fuel elements, and the nitrogen oxides, iodine and xenon produced were vented from a stack to form severe (and possibly radioactive) acid rain in the atmosphere; 3. The Soviets stored their nuclear waste in open ponds rather than in sealed tanks. As the liquid evaporated, the lake bed would have been covered with highly radioactive dust that high winds would disperse.

They speculated that widespread radiation forced the evacuation of the local people and that houses in the region were burnt to prevent them returning. In the 1960s, the report claims that a 'death squad' of prisoners was brought in to dump three ft of sand and soil over contaminated areas. The wasteland, fenced off and renamed the 'Chelyabinsk All-Union Radiological Manoeuvre and Exercise Range', was then used as a training ground for tank platoons.

Medvedev still believes there was a major accident there and says his analysis of events was confirmed by a member of the Soviet Academy of Sciences (see note, p. 348).

More news on Kyshtym surfaced in December 1988. The head of the Swedish-based Space Media network claimed that their analysis of satellite photos taken by Spot and Landsat satellites in 1987 and 1988 show that the Kyshtym complex is back in operation and a new facility is being expanded three miles to the north-east of the original complex.

The Swedish company also quoted Dr Medvedev as saying he had heard from Soviet nuclear safety officials that a second major accident occurred in the area in 1967.

Also in December 1988 came the first official Soviet acknowledgement of the Kyshtym accident. Yevgeny Velikhov, vice-president of the USSR Academy of Sciences, was in Japan to lecture at a meeting sponsored by the Japanese nuclear industry and government. He told his audience: 'We do not have enough information because it happened before glasnost. But details should be made public, and I promise I will try hard in that direction. I think they will come out.' (*Nucleonics Week*, 8.12.88.)

11 March 1958

Broken Arrow 3
Florence, South Carolina

A squadron of B-47s from the 308th Bomb Wing of the Strategic Air Command (SAC), stationed at Hunter AFB just outside Savannah, Georgia, was primed for another field manoeuvre. Codenamed Operation Snow Flurry, the planes were to fly to one of four US bases in North Africa, in a simulation of combat conditions.

Like the other planes in the squadron, No. 876, *The City of Savannah*, had a real atomic bomb on board, known as a 'pig' to the airmen who flew the plane. The plane also carried a pluglike detonator in a locked safety container, which could be used to arm the weapon if a special code signal was received and acknowledged. Shortly after take-off, at a height of some 14,000 ft, the bomb came loose from its shackles and smashed through the bomb bay doors.

The bomb landed in the garden of the home of Walter Gregg in the small town of Mars Bluff, near Florence, South Carolina, making a crater 35 ft deep and 75 ft across. 'It blew out the side and top of the garage just as my boy ran inside with me,' Gregg later recalled. 'The timbers were falling around us. There was a green foggy haze, then a cloud of black smoke. It lasted about 30 seconds. When it cleared up, I looked at the house. The top was blown in and a side almost blown off.' (**Burleson**.)

The chemical 'trigger', designed to set off the nuclear warhead, had exploded with the power of several hundred lb of TNT. The intense heat had vaporised the nuclear material, throwing a ring of plutonium contamination around ground zero — the point directly below the blast. The shock wave from the blast knocked cars out of control, knocked trees flat, and damaged houses within a half mile radius. Walter Gregg's house was a total write-off, though he and his family escaped with only minor injuries.

A special disaster crew under the direction of Major-General Charles B. Dougher, commander of the 38th Air Division at Hunter, moved into operation. Within 72 hours the radioactive materials were scoured from the site, and the area was declared contamination-free, although the health of the local population was monitored for several months to ensure they had not been exposed to any radiation.

After being assured by the government that he would receive full compensation, Gregg began to look on the bright side. 'I always wanted a swimming pool,' he said, 'and now I've got the hole for one at no cost. I may open it to the public. Charge them for swimming in uranium-enriched waters.' Five months later, he finally received $54,000 in compensation from the Air Force, a fraction of the $300,000 he had asked for. The bomb had come within 50 yd of wiping out his whole family.

There were wider implications. The US Air Force, having determined that the fault was mechanical, ordered all of their planes carrying nuclear bombs to manually and positively 'lock in' their weapons while on practice combat runs. This reduced the possibility of accidental drops, but increased the hazards to the crew if the plane crashed.

Additional Stories

1950 February 13 A B-36 bomber, en route from Alaska to Texas on a simulated combat mission, developed serious mechanical difficulties and crashed on Vancouver Island, off Canada's British Columbia coast. The 16 crewmembers, and one passenger on board, parachuted to safety and were rescued. The nuclear weapon on board, which contained a dummy capsule with no nuclear material in it, was jettisoned from 8,000 ft over the Pacific Ocean. The weapon's high explosive detonated on impact. No attempt has been made to recover it. (**DOD/CDI**)

1950 April 11 Three minutes after take-off from Kirtland AFB, New Mexico, a B-29 crashed into a mountain on Manzano Base, a 'dead storage' site where outmoded weapons were stored. The plane burst into flames killing the crew of 13. The one nuclear weapon on board had its detonators installed and some of its high-explosive material burned in the gasoline fire; other pieces of unburnt high explosive were scattered throughout the wreckage. A capsule containing nuclear material was on board the aircraft but had not been inserted into the weapon 'for safety reasons'. (**DOD/CDI**)

1950 July 13 A B-50 on a training mission from Biggs AFB, El Paso, Texas, went into a nose-dive and crashed near Lebanon, Ohio, killing the 16-man crew. High explosive in the weapon on board detonated on impact, creating a crater 25 ft deep and 200 sq ft in area. There was no nuclear capsule on board. (**DOD/CDI**)

1950 August 5 Near midnight — within minutes of the fifth anniversary of Hiroshima — a B-29 bomber that had taken off from Fairfield-Suisun AFB crashed near the end of the runway after the pilot had told the control tower that two engines were malfunctioning and that the plane's landing gear was stuck. The plane was carrying at least one nuclear weapon (a fact still officially denied), about a dozen 500-lb demolition bombs, 8,000 gal of gasoline and a crew of 20. Within minutes the plane was engulfed in flames, and 19 people died, including the co-pilot and base Commander, General Robert F. Travis. Sixty others were wounded. (**DOD/CDI**; D.E. Kaplan, 'Where the bombs are', *New West*, April 1981)

1950 November 10 Because of an in-flight emergency, a weapon, which did not contain a capsule of nuclear material, was jettisoned over water (no precise location given except 'outside United States') from an altitude of 10,500 ft. A high-explosive detonation was observed. There is no record of recovery of this nuclear weapon. (**DOD/CDI**)

1955 January 17 Launched on this date the *Nautilus*, was the world's first nuclear submarine and travelled 62,559 miles powered by a single golf-ball-sized lump of uranium-235. It suffered its first accident before it even put to sea. A steam pipe in the reactor compartment burst, revealing that ordinary piping instead of a seamless type had been used in its construction. All the suspect piping had to replaced. In December 1957, the *Washington Post* reported that the submarine's reactor compartment had flooded when a valve malfunctioned. The following April one of the steam condensers sprang a leak, and the reactor was shut down. The captain ordered 140 pints of car radiator sealant to be poured into the cooling system. This cured the leak and the reactor was restarted. (**Neptune**; E. Gray, *Sea Classic International*, Autumn 1986)

1955 November 29 The EBR-1 was an experimental U-235 fast-breeder reactor located at the US National Reactor Testing Station in Idaho. It was the world's second 'power-reactor' to produce electricity, (the first was the Soviet APS-1 reactor at Obninsk). During tests of EBR-1's behavioural characteristics, power was to be increased to a point at which reactivity doubled every tenth of a second. A staff scientist stood by to give a verbal command to scram the reactor - thus ending the experiment instantaneously.

The test proceeded, but when the command was given, the technician at the controls pressed the button for a slow shutdown instead of a scram. In the few seconds he took to hit the correct button, almost half of the fuel rods melted into a lump in the bottom of the containment vessel, which fortunately did not reach critical mass.

Not even Lewis Strauss, the chairman of the AEC, which owned and ran the reactor, knew of this accident until six months later, when he was questioned by the *Wall Street Journal* and had to acknowledge his ignorance of it. Only then did the AEC finally admit that the accident had occurred. (See: SL-1 Reactor, Idaho Falls, p. 136.) (John G. Fuller, *We Almost Lost Detroit*, Berkley Books, 1984; **Patterson**; **Bertini**.)

1956 March 10 A B-47, one of four aircraft on a scheduled non-stop flight from MacDill AFB at Tampa, Florida, to 'an overseas base', failed to make contact at its second refuelling point over the Mediterranean. An extensive search failed to locate the aircraft, its crew, or the two capsules of nuclear material in carrying cases on board. This disappearance was treated as an accident. (**DOD/CDI**)

1956 August 19 The experimental sodium-cooled nuclear reactor aboard the submarine USS *Seawolf* leaked during a full-power test run while the new ship was at Groton, Connecticut. After makeshift repairs, the vessel completed its initial sea trials on reduced power in February 1957, but the problems associated with its sodium-cooled reactor led the navy to replace it with a water-cooled type and only these were used in subsequent nuclear submarines. (**Neptune**)

1957 July 28 A C-124 transport plane en route from Dover AFB, Delaware, to Europe with three nuclear weapons and one nuclear capsule (not installed in a weapon) on board, experienced a loss of power in two of its engines. Unable to maintain level flight, the decision was made to jettison the cargo to save the plane and the crew. Two of the nuclear weapons were jettisoned over the Atlantic Ocean — one at 4,500 ft, one at 2,500 ft. Both were presumed to have been damaged on impact, but no detonation occurred and neither the weapons nor any debris were ever found. They are still in the ocean. (**DOD/CDI**)

1957 October 11 A B-47 carrying one nuclear weapon and one nuclear capsule in a carrying case crashed shortly after take-off from Homestead AFB in Florida, USA, after one of its outrigger tyres burst. The plane came down in an uninhabited area approximately 3,800 ft from the end of the runway. The nuclear weapon was enveloped in flames and burned and smouldered for four hours, during which time there were two high explosive detonations. The nuclear capsule and its carrying-case were recovered intact. Four crewmen were killed. (**DOD/CDI**)

1958 January 31 A B-47 aircraft at an 'overseas base', possibly a US AFB at Sidi Slimane in French Morocco, crashed while making a simulated take-off during an exercise alert. It was carrying one nuclear weapon. The aircraft burned for seven hours. The high explosive did not detonate but radioactive contamination meant that the wreckage and the asphalt beneath it had to be removed. Following the crash, exercise alerts were temporarily suspended and B-47 wheels were checked for defects. (**DOD/CDI**)

1958 February 5 A B-47 from Hunter AFB in Savannah, Georgia, on a simulated combat mission with a nuclear weapon on board, collided in mid-air with an F-86 aircraft. The aircraft attempted three times to land at Hunter but could not reduce airspeed. To avoid exposing the base to the risk of a high explosive detonation (the nuclear capsule was not on board), the nuclear weapon was jettisoned from 7,200 ft into the ocean several miles from the mouth of the Savannah River in Georgia. The weapon was never found; the aircraft landed safely. (**DOD/CDI**)

1958 November 4 A B-47 carrying one nuclear weapon caught fire on take-off from Dyess AFB, Texas. Three of the four crew ejected safely, but one was killed when the aircraft crashed from an altitude of 1,500 ft. The weapon's high explosive detonated, blasting a crater 35 ft in diameter and six ft deep. 'Nuclear materials were recovered near the crash site.' (**DOD**)

1958 November 18 The Heat Transfer Reactor Experiment Facility at the National Reactor Testing Station in Idaho was designed to test high-temperature reactor cores. On this occasion, an automatic servo-mechanism, responding to a reactor power indicator, was to be used to raise the power of a reactor core from 90 per cent to 100 per cent. The reactor was brought up to the 90 per cent level manually and the automatic mechanism was switched on. The power began to rise as planned, but then the power indicator incorrectly signalled a decrease in power. The servo-mechanism responded by pulling out the control rods, rapidly increasing the power beyond the 100 per cent level. The reactor scrammed automatically but the core had already suffered extensive damage. A quantity of radiation was released within the facility and some was detected downwind. The length of time from the switching on of the servo-mechanism to the automatic scram was about 20 seconds. (**Bertini**)

1958 November 26 A B-47 caught fire on the ground at Chennault AFB, Lake Charles, Louisiana, and one nuclear weapon was destroyed in the fire, contaminating the aircraft wreckage. (**DOD**)

1959 January 18 An F-100 Super Sabre interceptor aircraft, designed to carry nuclear-capable air-to-air missiles, exploded in flames on the runway of a Pacific base when its external fuel tanks inadvertently jettisoned. (No precise location is given but US bases in the area at that time were in Okinawa, Taiwan, South Korea and Thailand.) The nuclear capsule was not in the vicinity of the aircraft and was not involved in the accident. (**DOD/CDI; Hanson**)

1959 July 6 A C-124 on a nuclear logistics movement mission crashed on take-off from Barksdale AFB in Louisiana and caught fire, destroying the plane and nuclear weapon.

'Limited contamination was present over a very small area immediately below the destroyed weapon.' (**DOD**)

1959 July 26 At the AEC's Sodium Reactor Experiment reactor at Santa Barbara, California, a series of test runs revealed that tetralin sealant had leaked into the sodium coolant, where it had decomposed and coated the fuel elements, thus reducing the transfer of heat. Intermittent attempts were made to purge the coolant and clean up the fuel elements.

During the final run there were 10 scrams and four forced shutdowns. When the operators finally shut the reactor down to investigate the cause of these problems, they found that 10 of the 43 fuel assemblies were severely damaged, and that some radioactivity had been released. (**Bertini**)

1959 August 18 A helicopter engine that exploded while being tested in its hangar aboard the USS *Wasp* caused a series of fires which took over two hours to extinguish. The ship was carrying nuclear weapons. The ammunition magazines were flooded as a precaution and preliminary steps were taken to flood the nuclear weapons magazine, though this proved unnecessary. However, water from the fire-fighting efforts eventually leaked into the magazine around electrical cables. (**Neptune**)

1959 October 15 A B-52 carrying two unarmed nuclear weapons collided with a KC-135 tanker plane which was refuelling it, at 32,000 ft above Hardinsberg, Kentucky. Four of the eight-man B-52 crew ejected safely; all four crewmembers of the tanker plane died. The two unarmed nuclear weapons were recovered intact; one had been partially burned but 'this did not result in the dispersion of any nuclear material or other contamination'. (**DOD**)

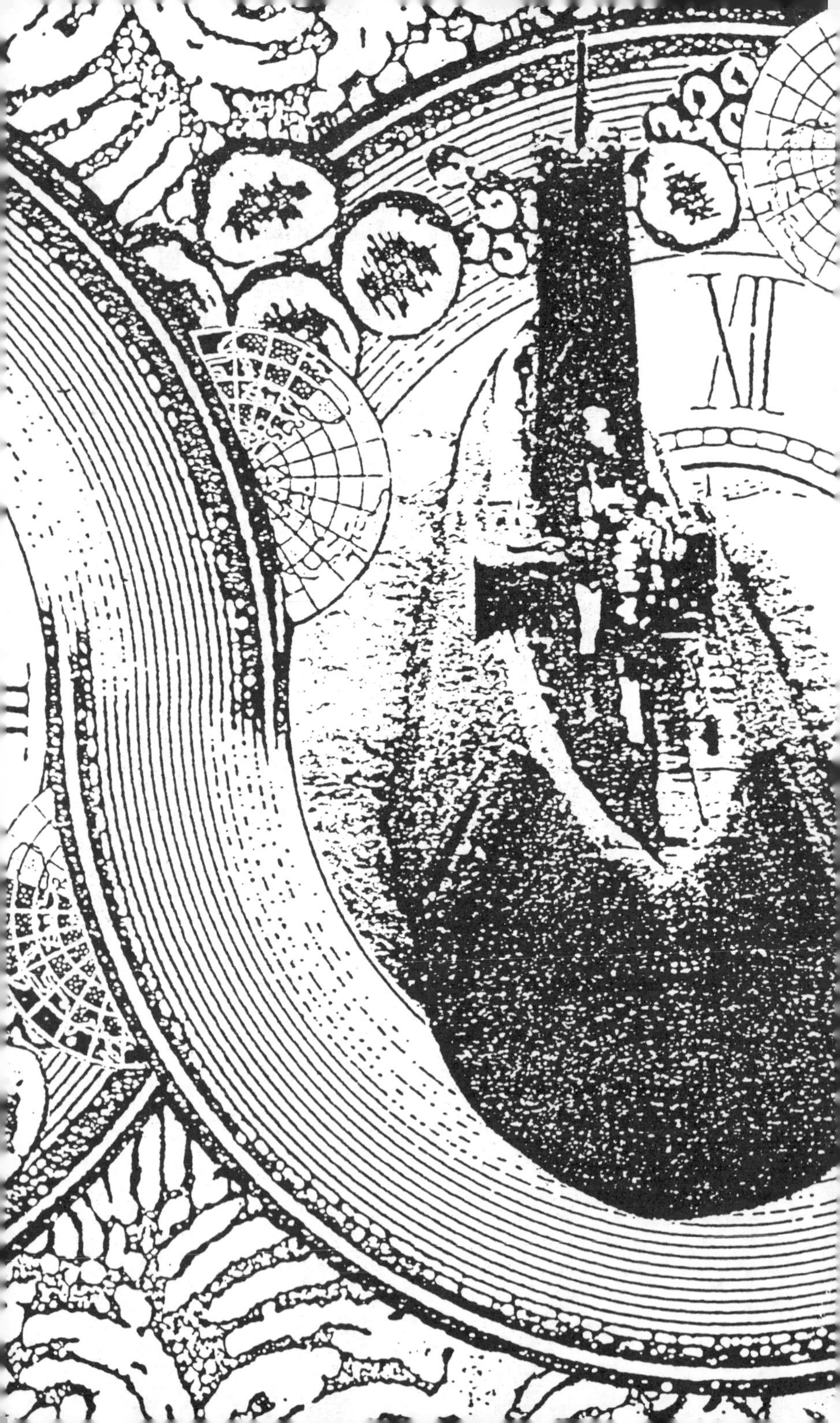

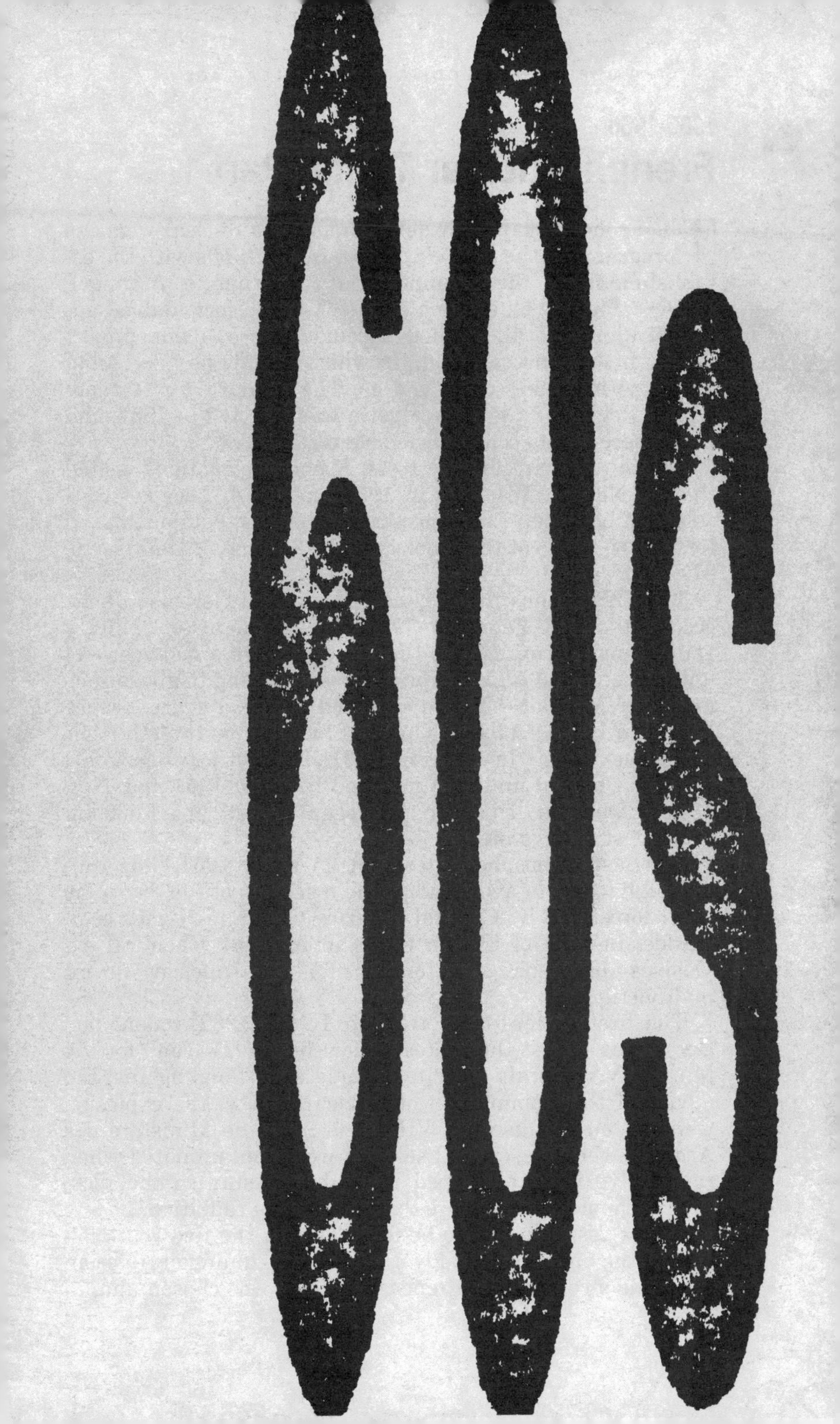

1960-1968

French Nuclear Tests, Part 1

The French began their development of a nuclear weapons programme, the *force de frappe*, in the 1950s with the establishment of the Commissariat à l'Energie Atomique (CEA). Pierre Guillaumat, the CEA's formidable administrator-general, pushed through the weapons project during his seven year reign, in which he saw no fewer than 11 prime ministers come and go. The search for a test site began, Reggane in French Algeria was selected in 1957, and plans were laid to conduct the first test in 1960.

To help prepare for the tests, French delegations visited the US Nevada Test Site in 1957 and 1958. They tested a selection of underground personnel shelters, equipment and test instruments at the Smoky test on 31 August 1957 (see p. 87).

The first French nuclear test, codenamed Gerboise Bleue, occurred on 13 February 1960. A vast array of military equipment was exposed to the blast, as were a menagerie of animals and 150 Algerian prisoners (according to allegations made on Algerian TV and reported by the Agence France Presse in Paris, 11.5.85). This was followed by three further atmospheric tests in 1960 and 1961. The testing programme was then moved underground, and between 1961 and 1966 13 tests were carried out in the Hoggar Massif, about 560 km south-east of Reggane.

The last atmospheric test, on 25 April 1961, was conducted hastily, to avoid the possible seizure of the device by rebel forces led by General Maurice Challe (a former commander-in-chief of French forces in Algeria). Chemical explosives used as a detonator for the nuclear device malfunctioned.

The first underground test, on 1 May 1962, codenamed Beryl, was to test the prototype for the AN 11 bomb for the Mirage IVA aircraft. Despite adverse winds, and against the advice of the Commission of Nuclear Safety, the explosion went ahead because two VIPs, one from the Ministère des Armées, were present. 12 soldiers were contaminated when radioactive vapour escaped through a fissure in the rock; nine of them received more than 100 rem of radiation.

A new test site had to be found within the five years following the end of colonial war and the beginning of Algerian independence in 1962. French Polynesia was chosen and, in

1963, legionnaires occupied the then uninhabited atolls of Moruroa and Fangataufa and began building the infrastructure for a new test programme.

Moruroa, a long and narrow reef (30 by 10 km) surrounding a lagoon, is made of porous coral resting on a bed of hard, brittle, and permeable basalt rock, the remains of an old volcano. (In Tahitian, the name means 'a place of a great secret'. The name was corrupted to 'M*u*ruroa' by French Navy cartographers.)

The first nuclear test in French Polynesia was carried out on 2 July 1966. Between 1966 and 1974, France conducted 44 atmospheric tests in the Pacific, 39 at Moruroa and five at Fangataufa.

From the very beginning these tests were surrounded by controversy. There were official assurances from the French that 'Not a single particle of radioactive fall-out would ever reach an inhabited island.' (Danielsson, 1986.) Nobel laureate Dr Albert Schweitzer remained unconvinced. A letter written on 17 April 1964 to John Teariki, the Deputy of the Tahitian Territorial Assembly (TA), read: 'Long before I received your letter I was worried about the fate of the Polynesian people. I have been fighting against all atomic weapons and nuclear tests since 1955. It is sad to learn that they have been forced on the inhabitants of your islands. Yet I knew that the French Parliament would not come to your assistance. The deputies do not have the courage to resist and they do not dare to oppose the military brass who are determined to undertake nuclear tests in your country. Those who claim that these tests are harmless are liars. Who could have imagined that France would be willing to deliver its own citizens to the military in this manner?'

According to the Danielssons (1986): 'Up until June 1963 statistics, listing among other things the number of deaths and their causes, were published every month in the *Journal Officiel*, and detailed information about diseases and epidemics was readily available. From the date when the CEA established its headquarters in Polynesia, however, and down to the present time, public health statistics have no longer been published. What is more, if anyone is bold enough to make an enquiry about this taboo subject, he is immediately reported to the secret police.'

In September 1966, De Gaulle visited Moruroa aboard the cruiser *De Grasse* to observe an atmospheric test. Already running late on his busy schedule, he ordered the test — the highest-yield atmospheric explosion to date — to proceed, despite the fact that unfavourable winds threatened to blow the fallout towards inhabited islands.

The explosion of this 120 kt bomb on 11 September resulted in widespread radioactive contamination over an estimated 3,000 km. It reached all the islands west of Moruroa in a matter of hours or days, depending on their distance, including Western Samoa, 2,000 miles downwind; Fiji; and the Cook Islands. The National Radiation Laboratory of New Zealand estimated that four days later in Apia, the capital of Western Samoa, the radioactive content of the rainwater in catchment tanks was 135,000 picocuries per litre.

Rumours that fallout shelters were being built on some of the islands were confirmed by inhabitants who fled to Tahiti. At the first session of the Territorial Assembly in June 1967, a resolution was adopted asking the French government to find out the exact nature of the radioactive contamination and to invite three foreign specialists plus three French scientists to carry out on-the-spot research. The request was ignored.

That month and the one following, three small nuclear devices were exploded in the atmosphere, one at sea level. As a result two French meteorologists on the island of Tureia (126 km north of Moruroa) were evacuated and hospitalised for a week for decontamination and tests. The 60 *inhabitants* of Tureia were not moved and no precautions were taken for their safety.

By May 1968 there were 5,936 French military and 2,265 civilian technicians on Moruroa, Hao (a permanent support base 200 nautical miles to the north-west), Fangataufa and Tahiti. An aircraft carrier and three cruisers with a further 7,018 personnel on board were in the area for that year's season of tests.

The fourth of the five explosions was France's first time thermonuclear explosion which produced a 2.6-Mt blast that so heavily contaminated Fangataufa that it was declared off-limits to humans for the next six years. As a result, the second thermonuclear bomb was exploded (on 8 September) over Moruroa.

The terrible beauty of an atmospheric test over the French test site at Moruroa in the South Pacific.

7-8 June 1960

BOMARC Missile Fire New Jersey

The BOMARC missile was a 47-ft-long, winged air-defence weapon designed to intercept high-flying supersonic aircraft at distances up to 450 miles and destroy them with its nuclear warhead. About 45 launch bases were planned. Ten were operational at various times from 1959 until 1972, and 238 missiles at peak were stationed in them.

On 7 June a BOMARC surface-to-air missile in 'ready storage condition', one of 28 housed at the 46th Air Defense Missile Squadron in Jackson Township, New Jersey, 10 miles east of McGuire AFB, was destroyed by an explosion and fire after a high-pressure helium tank exploded and ruptured the missile's fuel tanks. The warhead was destroyed by the fire, but the high explosive did not detonate. An unknown amount of plutonium was released into the atmosphere.

The *New York Times* reported that 'The atomic warhead apparently dropped into the molten mass that was left of the missile, which burned for 45 minutes.' Radiation was caused when the 'thoriated magnesium metal which forms part of the weapon, caught fire...The metal, already radioactive, becomes highly radioactive when it is burned.' (**CDI**)

In 1985, US Air Force officials agreed to give all unclassified information about the accident to the New Jersey Department of Environmental Protection (DEP) so that state officials could determine a) whether plutonium particles released in the fire could have landed in a populated area; b) whether the missile site, now abandoned and partially covered with reinforced concrete, was safe. DEP officials remain concerned that contamination, which the Pentagon claimed was limited to within 100 ft of the wreckage, may now have been leached out of the soil into groundwater.

3 January 1961

SL-1 Reactor, Idaho Falls, Idaho

The Stationary Low-Power Reactor (SL-1) was a three-MW prototype military nuclear-power plant, one of 17 reactors at the AEC's National Reactor Testing Station (NRTS), a remote site with no fire, medical or other emergency services, which covered 892 sq miles at Idaho Falls. The SL-1 facility had been used to gain operating experience, develop plant performance tests, obtain core burn-up data, train military personnel in operations and maintenance, and test components that might subsequently be used in improved versions. Part of the army's programme to develop simple and compact power plants to be transported to remote Arctic sites by air, it was the smallest known power reactor when it was made operational in August 1958. By the beginning of 1961 it had logged 1,100 hours of operation. On 3 January 1961 the first operating reactor incident to cause direct fatalities occurred at Idaho Falls.

According to the **Bertini** report the reactor 'was shut down for maintenance purposes in December of 1960 with the intention of starting up again on 4 January 1961. The additional instrumentation that had been installed prior to the planned start-up required the disconnecting of the control rods from their drives. The installation of the instruments had been completed during the day-shift on 3 January, and it was the job of the crew of the 4:00 to 12:00 p.m. shift to reconnect the control rods.'

At 16:00 on the day in question, three young servicemen — Richard McKinley, John Byrnes and Richard Legg — were reassembling the control-rod drives to prepare for the reactor's start-up. Two of them were experienced and qualified reactor operators, and the third was a trainee.

Bertini describes what happened:

'When disconnected, the control rods could be lifted completely out of the reactor manually. The justification for this was that maintenance in remote areas should be as simple as possible. However, lifting the central control rod about 16 in. was sufficient to make the reactor critical.

'The first indication of an accident was the sounding of alarms at 9:01 p.m. in the Fire Stations and Security Headquarters for the NRTS located some distance from the SL-1 facility. Since the alarms could have been set off either by fire, radiation, or pressure surges in the facility, members of the fire department and the plant security force, as well as a health physicist, responded.

'They searched for the three men in the building adjacent to the reactor building and also the ground floor of the reactor building, but the radiation levels were greater than the limits of their meters (25 roentgens/hr), so they withdrew.

There were no indications of smoke or fire. Calls were placed to other facilities at the NRTS, but the missing men were not there, so it was concluded that they must still be in the reactor building. Other health physicists and support and military personnel began to arrive.

'Wearing protective clothing, two men went up the stairs of the reactor building, but when they encountered radiation levels of 200 roentgens/hr, they withdrew. They were followed by another pair who reached the top of the stairs and looked into the basement of the reactor building and in doing so encountered fields of 500 roentgens/hr, so they quickly withdrew also. They could see no one, but they did see some evidence of damage. It was then about 10:30 p.m.

'Two others reached the basement and saw two men, one of whom was moving. Five others then went in, placed the man who was moving on a stretcher, and raced out. They had ascertained that the second man inside [Byrnes] was dead. The man on the stretcher [Legg] was placed in an ambulance, but he died before it travelled very far. The ambulance then returned to the SL-1 area.

'Four more men entered the basement in search of the third missing man [McKinley]. Upon looking up, they found him pinioned to the ceiling by a control rod. Assuming that he was dead, they did not try to remove the body. Since both men in the building were assumed to be dead, rescue operations were suspended temporarily.

'About 6:00 a.m. the following morning the dead man in the ambulance was taken out of it for decontamination purposes; lead shielding had to be used in removing his clothing. The radiation levels of his body measured about 300 roentgens/hr (upon subsequent removal, the bodies of the other two men measured about the same).

'At about 7:30 p.m. the second body was recovered by men working in teams; one team carried it part way out of the high-radiation area, and other teams completed the removal. It took six such teams to remove the body of the third man. The recovery operation was completed on 9 January.

'The recording instruments had been turned off while the control rods were being re-attached to their drives, and there were no survivors; thus, the cause of the accident is conjecture only. Based on a careful examination of the remnants of the core and the vessel during the clean-up phase, it is generally concluded that the central control rod was withdrawn manually and withdrawn quickly. Examination revealed that it had been withdrawn about 20 in. at the time of the excursion, sufficient for a large increase in reactivity...It is believed that the resulting short power surge, which reached a peak of (about) 20,000 MW...created a sudden volume of steam in the core, causing the water above it to rise with such force that when it hit the lid of the pressure vessel, the vessel itself rose nine feet in the air and then dropped back to its approximate original position.

'Monitoring of the area for radioactivity began shortly after the accident. An aerial survey early the next day revealed no increase except in the immediate vicinity of SL-1. Four flights were made in the next nine days, and some air samples taken revealed a radioactivity level about 50 per cent above background. Sagebrush samples downwind indicated maximum levels about 40 times greater than background. Even though the radioactivity was high, it was apparent that almost all (99.99 per cent) of the radioactive fission products were contained in the reactor building in spite of the fact that the building had no air locks, airtight seals, reduced pressure, etc.

'The reactor vessel and core were removed, the building razed, the area decontaminated, and the site made suitable for other purposes by July 1962. There were three fatalities, and some radioactivity was released.'

The bodies of the three men were found to be saturated with highly contaminated water from the reactor. 'In addition,' according to Horan and Gammill (1963), 'particles of fuel had penetrated the skin, resulting in large open wounds due to blast effect.' During the recovery of the bodies of Byrnes and Legg, and then of McKinley, 14 men received radiation doses of more than five roentgens, and six of them received doses of more than 20 roentgens.

Following detailed post-mortem examinations, the bodies were placed in lead-lined coffins and flown by military aircraft to airfields closest to the cemeteries chosen by the families.

The following requirements were placed on the burial sites: burial in a perpetual-care cemetery with adequate records of grave locations; the graves would not be re-opened without the express permission of the AEC; the coffins would be surrounded by at least 12 in of poured concrete and at least three ft of packed earth.

Monitoring of radiation levels downwind from the accident detected iodine-131 in air, vegetation and milk, though, according to published figures, contamination did not exceed the maximum permissible levels for continuous exposure to the off-site population.

The rescue of the contaminated corpses — and the subsequent clean-up operation of the reactor building — contaminated vehicles and equipment, which in turn contaminated areas around the plant and the public highway. It took 18 months to decontaminate the building, with men working four-hour shifts during which they could only spend eight minutes in the reactor building.

According to an article in *Nuclear Engineering* (March 1961), at the time of the accident the reactor was in operation when it was known to have a number of problems, and any running of the reactor was hazardous. 'For example...control rods had been sticking so that even under scram operation they had to be motored in and required manual assistance when they were withdrawn...Furthermore, boron was being lost from the core and the reactivity of the reactor was greater than it should have been.' (See also Additional Stories, 1955, November 29, p. 126.)

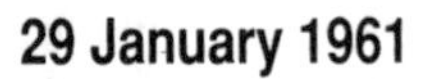

29 January 1961

Broken Arrow 4, Goldsboro North Carolina

During a routine airborne SAC-alert training mission from Seymour-Johnson AFB, the right wing of one of the B-52s began to break apart. The crew began to burn off excess fuel to lighten the plane for an emergency landing. Five of the eight crew members parachuted to safety; three others were killed in the crash.

When the wing came apart, the two multi-megaton nuclear bombs on board were automatically jettisoned. The parachute attached to one of them was deployed and was found hanging from a tree; the other bomb fell free and broke apart on impact but no explosion occurred.

Controversy surrounds the incident, which has been described by the Stockholm International Peace Research Institute (**SIPRI**) as 'perhaps the single most important example [of an atomic] accident which nearly resulted in a catastrophe.' According to Dr Ralph Lapp, former executive director of the DOD's Atomic Research and Development Board, each warhead 'was equipped with six interlocking safety mechanisms, all of which had to be triggered in sequence to explode the bomb...When Air Force experts rushed to the North Carolina farm to examine the weapon after the accident, they found that five of the six interlocks had been set off by the fall. Only a *single switch* prevented the 24-Mt bomb from detonating and spreading fire and destruction...' (*Mother Jones*, April 1981)

The Pentagon claimed at the time that *two* of the six switches remained untriggered; they reiterated this claim in 1983. In 1969 a further twist was added to the story when they claimed the bomb was 'unarmed', i.e. a crucial piece of fissionable material called 'the capsule' was not in it. Both SIPRI and Lapp questioned this story. By the 1980s the Pentagon had changed their version again: 'To trigger the bomb, the crew would have to perform a vital operation.' The truth about this story remains ambiguous to this day.

The plane crashed on the farm of Ellen and Buck Tyndall, 12 miles from the town of Faro, which was sealed off by air force personnel while firemen struggled to control the blaze. Bits of metal and debris were collected from a wide area around the crash site and men with Geiger counters searched and re-searched the area for five months.

One day, Buck Tyndall saw some of the men looking at a piece of the bomb which had cracked open. His wife was told, 'There's nothing here that could hurt you, but it could hurt your grandchildren.' (*Mother Jones*, April 1981.)

Shortly afterwards, a special engineering group from Fort Ord, California, arrived and, using tractors with pans attached to draglines, dug a huge hole which covered almost three acres and was some 50 ft deep. More than four million cu ft of soil were moved but the object of the search, a portion of one weapon, containing uranium, 'could not be recovered,' according to the official **DOD** report. How big a piece and how much uranium was never revealed. This accident occurred four days after John F. Kennedy became president but details of the accident were not released until 1969. As a result of Goldsboro, many new safety devices were placed on US nuclear weapons and the Soviets were encouraged to do the same.

7 October 1962

Nukey Poo Reactor, Antarctica

The Antarctic Treaty, which came into effect on 23 June 1961, bans nuclear explosions as well as the disposal of nuclear waste in the Antarctic, but it does not forbid the use of nuclear power in the region. As a result, the US were able to install the Portable Medium Power Plant 3A, an experimental 1.8-MW PWR at the McMurdo base to try to find a more economical way of providing heat and power there.

The reactor, soon nicknamed Nukey Poo, arrived by ship on 21 December 1961 and was sited half-way up Observation Hill near Mount Erebus, an active volcano on Ross Island. Power production began in July 1962. Four years later, the US Navy announced that it had broken the record for the longest continuous operation of a military nuclear reactor, and in 1971 the power output was increased by 10 per cent.

During this whole period, the navy claimed that Nukey Poo's only serious problem had occurred in 1962, when a faulty seal allowed hydrogen to accumulate in the top of the containment vessel and a spark from a short circuit ignited the gas, causing a fire. In reality, the reactor's 10 years of operation had been an expensive story of shutdowns, fire damage and radiation leakages. In 1972, a temporary shutdown, caused by coolant water leaking into the steam generator tank, coincided with a navy cost-effectiveness study. The study concluded that it would be too expensive to overhaul and upgrade the plant to new standards (the reactor had been built with no emergency core-cooling system). Consequently it was closed down and demolished over the next three Antarctic summers at a cost of $1 million.

The reactor and 101 large drums of radioactive earth were shipped away to the US. The reactor vessel, fuel rods, and buildings were buried at the Savannah River plant. Later, another 11,000 cu m of soil and rock were also removed and shipped to the US. It took several more years of cleaning up before the site was declared to be 'decontaminated to levels as low as reasonably achievable,' and the land was finally released for unrestricted use in May 1979.

10 April 1963
USS Thresher, North Atlantic

The USS *Thresher* was the most advanced nuclear-powered attack submarine of its time. Commissioned in August 1961 as the first of a new class of submarines designed to destroy other submarines in the deep ocean, her sudden and inexplicable sinking about 220 miles off the New England coast caused shockwaves in the US Navy.

From the beginning *Thresher* was plagued by fires, accidents and equipment breakdowns. Two thirds of the time between her commissioning and her loss was spent undergoing repairs. Yet, according to John Bentley's detailed book about the accident (1975), she was, nevertheless, sent out on sea trials with two fatal defects. These defects — one in the high-pressure piping system that carried water to her ballast tanks, one in the emergency deballasting system that emptied these tanks — had been pointed out by Admiral Hyman G. Rickover, chief architect of the nuclear navy and head of the Navy Reactor Branch of the AEC.

Shortly after 09:00 on 10 April that fateful morning the USS *Skylark*, a surface vessel working with *Thresher* on her test dive, received this message via underwater telephone: 'Experiencing minor difficulties. Have positive up-angle. Am attempting to blow. Will keep you informed.' Four minutes later her last and garbled message was heard. Only the words 'test depth' were distinguishable but it was clear *Thresher* was in serious trouble and heading for the bottom. None of the 129-man crew survived. (*National Geographic*, June 1964.)

After a painstaking two-week survey of the area the bathyscaphe *Trieste*, at that time the navy's only craft capable of operating at such depth, located the wreck at a depth of 8,500 ft and was able to photograph the compressed remains of the submarine. The *Thresher* was never recovered.

A naval Court of Inquiry could not say with certainty why *Thresher* sank but thought it was likely that a sea-water pipe broke which caused water to jet in to the submarine with terrific force, smashing the power circuits so that *Thresher* was unable to surface.

It is now known that a major hazard to submarines occurs where sharp salinity gradients in the ocean are found in conjunction with sharp temperature gradients. This combination can produce severe underwater turbulence and this effect may have been partly responsible for *Thresher*'s demise.

But questions still remain as to whether the *Thresher* accident could have been caused by a malfunction of the submarine's PWR reactor. The full proceedings of the naval and subsequent congressional inquiries into the incident remain secret, but the available record shows that they both accepted unquestioningly the evidence of Admiral Rickover, who dismissed such a possible cause out of hand. He consistently refused to give a reactor accident serious consideration, and his theory of a 'burst pipe' was adopted as 'the most probable cause' of the tragedy. (**Pringle and Spigelman**.)

Evidence submitted to the inquiry by a chemist from the Portsmouth naval shipyard appears to have been ignored. He had found that pieces of plastic from the submarine, of the type used for reactor shielding, had been scorched and had fragments of metal embedded in them, suggesting that there had been some kind of explosion in or near the reactor. His evidence was consistent with a core meltdown, but this line of inquiry was not pursued.

Admiral Ralph James (former chief of the US Navy's Bureau of Ships) and Norman Polmar (for 10 years US editor of the authoritative *Jane's Fighting Ships*) are two critics of the official, 'burst pipe' view. They believe that there *was* a burst pipe, but that this caused a stream of pressurised water to hit the nuclear reactor's control console, causing an emergency shutdown of the reactor. Without power, the *Thresher* was unable to pull out of its dive and it eventually imploded due to the water pressure.

This accident had far-reaching effects. As Edwin A. Link, pioneer explorer and inventor of the flight simulator, wrote in *National Geographic* 1964: 'The sinking of the *Thresher* sent out ever-widening ripples — ripples that will not be stilled. This disaster changed the course of our study of the ocean, greatly increased its impetus, and dramatized the need for a new science: oceanology.'

1964 onward

Chinese Nuclear Tests

When the USA, USSR and UK signed the Limited Test Ban Treaty in 1963, which prohibited atmospheric nuclear testing, Soviet Premier Krushchev had made a deal with the Chinese to supply them with nuclear assistance in return for their acceptance of a test ban. When he reneged on that agreement, China withdrew its support of the Treaty.

China became a nuclear power on 16 October 1964, when she exploded an A-bomb in the atmosphere, equivalent to 20,000 tons of TNT. On 17 June 1967, she exploded a 3-Mt H-bomb. This was the shortest development period between fission and fusion weapons for any of the nuclear powers (32 months).

A test site was established in the Lop Nur region of Xinjiang province in the remote north-west of the country. Following an atmospheric test there on 26 September 1976, the fallout from the blast was carried across the Pacific and deposited by torrential rain over the east coast of the USA on 3 October. According to the *New Scientist* (14.10.76): 'At one point, officials at the Peach Bottom nuclear power plant near Philadelphia feared that their reactor had sprung a leak, so rapidly were their radiation counts rising.'

In 1981, a United Press International report said that there were an increasing number of cases of liver, lung and skin cancer in the region of Lop Nur. Officials told Western diplomats that peaches grown there had developed 'rubber-like patches'. (*International Herald Tribune*, 23.8.81) They said, 'Many years ago, people never died of cancer, but in recent years they have been dying this way. Some people say it is because of the testing.' In 1985 *The Times* (31.12.85) reported that there had been demonstrations by 100 Muslim students in Urumqui, 500 miles north-west of Lop Nur, in protest against nuclear testing and the presence of labour camps in the region. The students claimed that the tests had spread radiation sickness and death among a relatively large percentage of the population.

China conducted 34 nuclear tests between 1964 and 1988. In March 1986 Premier Zhao Ziyang pledged that China would conduct no more nuclear tests in the atmosphere. In May 1986 a senior Chinese military official, Mr Qian Xuesen, a consultant for the national defence committee of the scientific and technological industry, stated: 'Facts are facts. A few deaths have occurred, but generally China has paid great attention to possible accidents. No large disasters have happened.'

1965-67

Operation Hat, the Himalayas

If some nuclear-accident stories read like scripts for Hollywood disaster movies, then the story of Operation Hat reads like the script for a spoof spy caper.

Operation Hat began shortly after China's first nuclear bomb test in 1964. The US CIA, with the co-operation of the Indian government, planned an expedition to the Himalayas to plant a nuclear-powered monitoring station on the summit of 26,600-ft-high Nanda Devi, from where it would eavesdrop on the Chinese nuclear test programme in over-the-border Xinjiang Province.

The Agency recruited several top US civilian climbers who, together with four of India's best climbers from the 1962 Everest expedition, formed the cadre of the ill-fated Operation Hat. Far from succeeding in eavesdropping on the top-secret Lop Nur nuclear test site, Operation Hat was destined to threaten one of the world's great rivers with plutonium contamination.

The US climbers and their Indian colleagues set out up the south face of Nanda Devi in the autumn of 1965. A squad of porters carried the disassembled monitoring station, together with its SNAP power pack, on their backs.

The SNAP — Space Nuclear Auxiliary Power — generator was a nuclear battery originally developed for the American civil and military space programmes. Shaped like a cone, SNAP was fuelled by between one and eight lb of plutonium, was small enough to be carried by one man, and would power the monitoring station until its task was completed. The CIA would then send a second expedition to retrieve the station, SNAP and all.

Operation Hat ran into the first of its many difficulties when the expedition encountered severe weather and rock conditions. 2,000 ft from Nanda Devi's summit, the climbers decided to turn back, but not before they'd cached the monitoring station which would await their return — when conditions improved.

The Operation Hat climbers ventured back up Nanda Devi in the spring of 1966, but were dumbfounded to discover that a winter avalanche had swept the spy station from the mountainside. The vital SNAP generator, and its plutonium, were now entombed under a mound of rock and snow the size of a Giza pyramid.

The CIA and its Indian government partner were in a quandary. The southern slope of Nanda Devi, where SNAP lay buried, is a major source of headwater for the Ganges, the sacred river of 500 million Hindus. A holy bathing place for pilgrims was just a few km downstream from the SNAP site. If SNAP were to break open under the weight of the avalanche, there was a real risk that the hallowed waters of the Ganges would be polluted with deadly plutonium, and both the Agency and the Indian government would face the wrath of millions of people.

Over the next two years, expeditions to locate and recover SNAP returned empty-handed.

Eventually, after water sampling of the Ganges revealed no contamination, the decision was made to abandon SNAP in the hope that it would remain intact and that Operation Hat would remain a secret.

The secret of the lost SNAP was kept until May 1978 when the US journalist Howard Kohn revealed the existence of Operation Hat in *Outside* magazine. In a masterly-worded non-statement to the Indian Parliament, Prime Minister Morarji Desai tried to defuse the danger of SNAP: 'The indirect evidence so far is that the safety precautions built into the nuclear-fuelled power pack may be as effective as has been claimed and, if so, pollution effects may not take place in the future.' (*The Times*, 18.4.78.)

In 1967, Operation Hat finally scored a success. A second SNAP-powered spy station was placed on, and eventually retrieved from, the slopes of nearby Nanda Kot mountain.

The first SNAP is still there, entombed under thousands of tons of rubble. Many nuclear experts disagree with Morarji Desai. They say that the SNAP generator will eventually corrode and disintegrate, releasing plutonium into the headwaters of one of the world's great rivers.

5 December 1965

USS Ticonderoga, North Pacific

'At Sea, Pacific: An A-4 aircraft, loaded with one nuclear weapon, rolled off the elevator of a US aircraft carrier and fell into the sea. The pilot, aircraft, and weapon were lost. The incident occurred more than 500 miles from land.' That was how a 1981 **DOD** report described the only recorded nuclear weapons accident involving a US ship. The terse entry left a great deal unsaid and falsified an important detail. The full facts did not emerge until 1989.

On 5 December 1965, the aircraft carrier USS *Ticonderoga* was some 70 miles east of the Japanese Ryukyu Islands, en route to the US Navy base at Yokosuka, south of Tokyo. It was returning from a bombing-support mission off North Vietnam. An A-4E Skyhawk attack jet, carrying a B43 nuclear bomb, was being moved from its hangar towards an elevator when, possibly because of a brake failure, it rolled off the deck and into the sea. The plane, with a 1-Mt bomb and its pilot aboard, immediately sank in 16,000 ft of water. Searchers failed to find the pilot, Lieutenant D. M. Webster, and since the navy possessed no salvage equipment that could operate at this depth at the time, no recovery was attempted.

This version of the story raises three important points, all of which the DOD was anxious not to publicise — hence the innocuous official account.

Firstly, although it would be true to say that the location of the accident was 500 miles from mainland China, it was just 70 miles from inhabited Japanese territory. The bomb had an explosive power 70 times greater than that which devastated Hiroshima in 1944. The US deliberately misled the Japanese government about the location.

Secondly, the incident provides clear confirmation that the US Navy was flying nuclear alert missions from aircraft carriers during the Vietnam War. Carrier-based aircraft have since been taken out of the US strategic nuclear war plan.

Thirdly, by continuing its journey and docking in Japan, the *Ticonderoga* violated Japan's official ban on the introduction of nuclear weapons into its territory, a ban which the US is still believed to be routinely flouting.

When full details of the accident were made public in May 1989, a Pentagon spokesman claimed that the accident had occurred off the coast of Vietnam and that the ship had been heading for Vietnam at the time. He subsequently admitted that the DOD's siting of the accident — 500 miles from land — was 'incomplete' (**Neptune**). Greenpeace researchers then released copies of the *Ticonderoga*'s deck-log, which specified the location and proved that the vessel docked in Yokosuka two days later.

The revelations led to Japanese calls for the recovery of the bomb, which may have been damaged by the pressure three miles down in the Pacific and may already be leaking radioactive material.

17 January 1966

Broken Arrow 5, Palomares, Spain

The most dramatic nuclear accident of the decade occurred when a USAF B-52 bomber, carrying four H-bombs, collided with a refuelling tanker off the coast of Spain. Plutonium from two of the bombs contaminated several hundred acres in and around Palomares, a village of some 2,000 people.

The B-52, codenamed TEA 16, was returning to base in North Carolina, USA, after flying a routine 'air alert' patrol. TEA 16 was attempting to take on fuel from a KC-135 'flying tanker' from the USAF base at Moron, south-western Spain, when the two aircraft collided 30,000 ft above the Spanish coast. Both planes broke up: the KC-135's 40,000 gal of jet fuel exploded, killing the four-man crew instantly. Four of the B-52's seven crew parachuted to safety. Wreckage from the two planes (which were worth $11 million) was strewn over an area of 100 sq miles. Although it was initially denied by the US DOD, TEA 16 was carrying four B-28 H-bombs. Bomb one was recovered, relatively intact, from a dry riverbed. The high explosives in bombs two and three detonated, scattering plutonium over the village of Palomares and its surrounding fields. Bomb four fell into the Mediterranean Sea.

For the next three months, 1,700 US servicemen and Spanish Civil Guards carried out a vast decontamination and removal operation at Palomares. Crops were buried or burned and some 1,750 tonnes of radioactive soil were removed to the United States. (The US personnel wore protective clothing, were thoroughly screened, and were rotated every two weeks. No such precautions were applied to the Civil Guards.)

Bomb four, lost in the ocean, triggered off what has been described, according to the CDI, as 'the most expensive, intensive, harrowing and feverish underwater search for a man-made object in world history.' The recovery operation lasted 81 days: 33 naval vessels sealed off the search zone while a small armada of mini-research subs, diving bells and scuba teams scoured the sea-bed, aided by sonar experts, oceanic photographers and 3,000 navy personnel.

Two weeks after the search began, the midget sub *Alvin* located the missing bomb at a depth of 2,500 ft, approximately five miles offshore.

After a series of mishaps, the bomb was finally recovered, dented but intact and with no known radiation leakage, at a cost in excess of $11 million.

Yet the story of bomb four's recovery, which obsessed the world's press for weeks, only served to deflect attention away from bombs two and three — bombs which had contaminated both the land and the people of Palomares.

According to The Palomares Summary Report, published by the US Defense Nuclear Agency in 1975, it was not until 19 January, two days after the accident, that 'first attempts were made to delineate the area and extent of contamination.' (This contradicts a DOD press release issued on 20 January — when the contamination survey was hardly underway — which stated that 'radiological surveys have established that there is no danger to public health or safety as a result of this accident.')

'Monitoring in the village of Palomares, in conjunction with the Spanish nuclear energy commission (Junta de Energia Nuclear — JEN) personnel,' continued the Summary, 'was begun on 24 January 1966. This consisted initially of monitoring houses and random crop monitoring. By 3 February 1966, however, it was established that a pattern of contamination ran through the village...'

Palomares is situated in the province of Almeria — at that time the second-poorest province in Spain. The villagers lived by farming, growing just one crop — tomatoes — for export. On the day of the crash, many people were working on their land. Antonia Flores, who later became mayoress of the community, was six years old at the time and was playing in the fields with her younger brother: 'We heard a loud explosion and it was as if the sky was on fire,' she recalls. She and other villagers watched the flaming wreckage fall to earth and later handled fragments of the debris.

Although the radiation survey found that no fewer than 650 acres of village, farms and cropland were contaminated, there was no effective large-scale monitoring of the villagers. Neither were there established criteria for 'acceptable' plutonium contamination: 'The Spanish government had not established criteria for permissible levels,' says the Summary, 'which is completely understandable because plutonium-producing facilities and nuclear weapons were non-existent in Spain. Significantly, there were no criteria in the United States for accident situations. The available criteria pertained only to plutonium processing plants and laboratories. There were, however, the broad guidelines established from the Nevada tests...'

Those guidelines, as we now know, were inadequate. It was as if a decision had been already made, in advance, that serious contamination of the villagers was a possibility so remote that it could be ignored.

Major-General Delmar Wilson (left), head of the US 16th Air Force, and Admiral William S. Guest, Commander of the US Navy Task Force 65, standing beside the H-bomb recovered from the ocean. This official US Navy photo, taken on 8 April 1966, was the first photo of a US hydrogen bomb ever to be released. (Credit: Popperfoto)

After the last barrel of contaminated debris had been shipped out of Palomares, the United States and Spain agreed a programme under which JEN would monitor land and people for several years.

In November 1971, says the Summary, Dr Wright Langham from the AEC's Los Alamos Scientific Laboratory visited Palomares. His findings paint a depressing picture of what should have been a thorough, systematic follow-on study. Apart from negative lung-counter findings and urine sampling of just 100 villagers (29 of whom tested positive, but were deemed not 'statistically significant'), Langham noted '...that no further measurements have been made on the Palomares residents.'

Furthermore, Langham found that air monitoring in and around the village continued for just 24 months after the accident: 'Positive air samples were obtained occasionally at all stations, with the highest (plutonium) values coinciding with periods of high wind velocity.' Yet, at the time of his visit, only two of the four air-monitoring stations were operational. As time passed and equipment broke down, two of the stations had been dismantled to keep the other two in operation, and these were '...about ready to go also'.

As far as crop monitoring was concerned, 'the number of samples to be processed is large, and the Spanish were given only one alpha spectrometer which has given poor service,' and, 'the soil studies program is a slow, arduous task fraught with many difficulties...'

In general, Langham found that 'Some of their [JEN's] equipment is now obsolete and their facilities still poor by US standards. The equipment we gave them is now six years old. It has not been updated, improved, or added to...

'Enthusiasm for the work [amongst JEN personnel] did not seem as high as it once was. This could be a result of their having to turn their attention during the last year to a fission-product release into a major river used for irrigation of vegetable crops for the Madrid market. It could be also that we have not maintained the interest and attention in the Palomares program manifested originally...they are understaffed technically...'

Langham concluded by suggesting that domestic American considerations might prompt more effective research: 'Current concern in this country over plutonium environmental contamination from the breeder reactor developmental program and from projected uses of plutonium-238 might justify considering revitalization of the

Palomares program...' If so, he continued, the Americans should 'update their equipment and certainly provide at least one additional alpha spectrometer...' and, 'consider...providing them with a new lung counter to re-count a number of the 100 Palomares residents examined the first year after the accident...Results on people who have lived in a contaminated area for six years after an accident might be of value even if all negative — as I am relatively sure they would be.'

The authors of the Palomares Summary endorsed Langham's findings: 'Unless the political implications of a continuing radiation monitoring program at Palomares are overriding, a program such as Dr Langham...propose[s] should probably be supported. Palomares is one of the few locations in the world that offers an on-going experimental laboratory, probably the only one offering a look at an agricultural area.'

The villagers of Palomares, the 'on-going experimental laboratory', are well aware of the 'political implications' of their plight. It wasn't until 1985 that they were able to gain access to their medical records — due to a vigorous campaign by Mayoress Antonia Flores who, as a girl, had watched the crash.

Later that year, Dr Francisco Mingot, director of JEN's Institute of Radiobiological and Environmental Protection, broke a 20-year official silence by addressing an open village meeting. He assured the villagers that there was no health hazard in the residual amount of plutonium remaining in the environment: 'We have detected plutonium in ten per cent of the population but these are well below danger levels...' (*Evening Argus*, 26.11.85.)

This is disputed by Dr Eduardo Rodriguez Farre, a radiobiologist at the state-run Board of Scientific Research in Barcelona and a member of Palomares' independent medical commission: 'Plutonium is one of the most toxic substances known to man. I find it inconceivable that [JEN] should say otherwise.' (*Evening Argus*, 26.11.85.) Farre claims that medical examinations have been insufficient because they do not include chromosome testing; that the population should have been evacuated after the accident; and that the area of contamination was greater than admitted. (His latter claim is backed by the Palomares Summary which reveals that, during the initial survey, prevailing winds churned up plutonium dust and 'the total extent of the spread will never be known'.)

PALOMARES — A PROBLEM OF DETECTION

All Palomares radiological surveys were conducted with the standard US military alpha survey meter used to check for the presence of plutonium contamination: the PAC 1S.

Alpha particles are a product of the radioactive decay of plutonium. They have a very small range; they can be stopped by a sheet of paper, a blade of grass, or even dew. The greatest danger from alpha particles results from ingestion of radioactive material. A microscopic amount of plutonium taken into the body — by, for instance, inhalation — will release a large number of alpha particles. Even though their range is short, alpha particles can cause huge amounts of damage which can eventually lead to cancer.

Bearing this in mind, one would assume that the PAC 1S — the only alpha detector available at Palomares — would be able to measure plutonium contamination accurately. But this was not the case. The Palomares Summary Report catalogues a series of shortcomings in the equipment which, when taken together, seriously question the adequacy of the USAF's to conduct a plutonium survey at Palomares: '...portable equipment for alpha detection has historically been troublesome for open terrain surveys,' says the Summary. Because alpha particles have such short ranges — only 3-4 cm in air — PAC 1Ss had to be held right next to the surface being surveyed. This proved problematic. The devices also suffered 'unusually high failure rates' and were operated largely by untrained personnel. 'Under no circumstances,' concluded the Summary, 'should PAC 1S instruments be deployed in the field again without pertinent directives and a repair capability.'

Major-General Delmar Wilson, Commander, 16th Air Force, who headed the clean-up operation, was even more forthright. The Summary quotes him as saying '...that the US Air Force was unprepared to provide adequate detection and monitoring for its personnel when an aircraft accident occurred involving plutonium weapons in a remote area of a foreign country.'

5 October 1966

Fermi Reactor, Detroit

In January 1956, the Power Reactor Development Corporation (PRDC), a consortium of some 35 utility companies and manufacturers headed by Detroit Edison, submitted a proposal for a commercial fast-breeder reactor. The plant was to be situated at Lagoona Beach on the western shore of Lake Erie, just 30 miles south-west of Detroit, a city with a population of two million.

Detroit Edison, one of the largest power utilities in the Midwest, had been working on feasibility studies for such a reactor since 1951, in conjunction with the AEC, basing their plans on the prototype EBR-1 experimental reactor at Idaho Falls. When the development of civil nuclear power was made possible by Eisenhower's Atomic Energy Act in August 1954, their plans began to take on a more concrete form.

Their proposed reactor, to be named after the atomic physicist Enrico Fermi, was to be a scaled-up version of the EBR-1, with a small, dense core in which 14,700 uranium fuel pins would be positioned very close together. Critics suggested that this configuration meant that any mishap within the core could result in a rapid meltdown and an explosion that could spread radioactive material over a wide area. The EBR-1 accident in 1955 (see 1955, November 29, p. 126) did not bode well for the future of the project.

On 6 June 1956, the Advisory Committee on Reactor Safeguards, a panel of experts established by Congress to advise the AEC on the safety of proposed reactors, submitted a report which stated that not enough was known to guarantee public safety if such a plant was operated near an urban centre.

Far from acting upon this advice, the chairman of the AEC suppressed the report, and although this cover-up was subsequently made public by a member of the commission, the AEC weathered the consequent criticism and issued the construction permit on 4 August 1956.

Walter Reuther, the head of the United Auto Workers union (UAW), which had half a million members living within a 30-mile radius of the plant, demanded a public hearing into the AEC's decision. The AEC, in turn, gave assurances that when construction was completed, no operating licence would be issued unless the safety of the plant could be guaranteed; however, the UAW were convinced that, once tens of millions of dollars had been spent, the AEC would be unable to resist pressure to grant the licence.

The hearings into the AEC decision, which took place at the AEC headquarters, began on 8 January 1957 and ran for more than two years. During this period, the WASH-740 report was released (see Reactor Risk, p. 13), which painted a horrifying picture of the potential effects of a reactor accident in the USA.

In addition, a University of Michigan study of a potential accident at the Enrico Fermi plant put figures for deaths, injuries, and damage to property even higher than the WASH-740 report. The UK Windscale reactor caught fire in October 1957, and in May 1958 there was an accident at the Chalk River NRU reactor. Public and commercial disquiet were growing, and Congress was forced to pass the Price-Anderson Act (see Reactor Risk, p. 13). Despite all this, construction of the Fermi plant continued apace, and on 26 May 1959 the hearings ended with the AEC ruling that the construction permit would stand.

On 25 July 1959, the UAW and the American Federation of Labour - Congress of Industrial Organizations (AFL-CIO), the main US trades union organisation, brought an action in the US Court of Appeals. Almost a year later, on 10 July 1960, the court delivered its ruling — the plant's construction permit was illegal and building must stop. A petition to the court by the PRDC and the AEC was denied, so they appealed to the US Supreme Court who, in the spring of 1961, ruled that the AEC had been within its rights in permitting the construction of the Fermi reactor.

At no stage had the legal wranglings interrupted the construction of the reactor, and in July 1963, once the operating licence had been granted for low-power operation, the fuel assemblies were loaded into the core. In August the first commercially-sponsored fast-breeder reactor went on-stream.

It had already shown its potential for mishaps. In August 1959, when tests on the volatile sodium coolant began in an abandoned gravel pit 20 miles from the Fermi site, there was an explosion that hospitalised six people and injured many others. Later that year, tests revealed that the fuel rods would only last for one third of the time that had been predicted, and the sodium coolant was found to erode the ribs that kept the fuel rods apart. In 1960 it was discovered that the fuel pins could swell and block the passage of the coolant, and in consequence the power output of the reactor had to be halved. The sodium also reacted with the graphite shielding and much of this had to be replaced, an operation that took 15 months and cost $2.5 million.

On 4 January 1963, sodium coolant leaked from a faulty valve and burst into flames. Since the fuel had not yet been loaded into the reactor the sodium was not radioactive, but 'what if' questions raised the public safety issue yet again, and the Department of Health made several requests for improved standards of safety and monitoring at the plant.

After the loading of the fuel, technical difficulties continued to plague the reactor and, by the summer of 1966, the plant had cost some $120 million, but had produced no more than $303,000-worth of electricity.

All this was by way of a prelude to what happened on 6 October 1966 — an accident that almost fulfilled the critics' worst fears.

The reactor was started up for a series of tests to be carried out at 67 MW power output. When the reactor reached 20 MW, an erratic signal was noticed, but it soon disappeared and the power rise was continued. At about 30 MW the signal showed again, and it was found that some of the control rods were not in their expected positions and that the temperature of the coolant above two fuel assemblies was high. Soon afterwards, radiation alarms went off in the containment building and the operator scrammed the reactor.

Tests on the sodium coolant showed the presence of highly active fission products, suggesting that part of the fuel in the core had melted. The problem then was what to do about it. Since the cause of the trouble was unknown, no one knew what to expect from the reactor. Any attempt to use remote handling gear within the core could disturb the delicate geometry of the fuel pins and produce a major disaster.

Patterson states that, at this point, all police and civil defence authorities were alerted to prepare for the evacuation of Detroit and other major population centres in the area, but that all official records of the alert were later expunged.

For several days the operators waited to see what would happen, while outlining plans to investigate the core. Cautious tests confirmed that some of the fuel had melted, but the cause remained a mystery. Then, a piece of crumpled metal was discovered at the bottom of the core which it was thought might be the key to the accident. Elaborate remote-controlled tools were devised to remove it and finally, 18 months later, the cause of the fuel melt was positively identified.

A steel cone had been built in the bottom of the vessel so that, in the event of a meltdown, melting fuel would be spread over a wider area, preventing the fuel from reaching critical mass. At the last minute, as an extra safety precaution, it had been decided to protect the steel with plates of zirconium, but this change of plan was never noted on the design blueprints. Ironically, it was one of these zirconium plates that had come loose and had been forced up into the core by the flow of sodium coolant, thus blocking the cooling of two of the fuel assemblies and causing them to melt.

It took until the end of 1968 to remove the last pieces of the zirconium plate, and a further year to gain AEC approval for the reactor to be restarted. In May 1970, as the reactor cooling system was being refilled with sodium, with AEC inspectors looking on, some 200 lb of the reactive coolant leaked from a burst pipe, mixed with water and exploded, contaminating the building. Nonetheless, preparations continued and in July the reactor was restarted.

By now, the reactor that was supposed to point the way towards a new era of cheap power had cost $132 million and had still produced negligible quantities of electricity. The AEC extended the operating licence in 1971, but suspended operations at the plant in 1972. The plant, which had been an economic disaster, was then shut down permanently.

Nuclear Powered Ships

During the winter of 1966-67, the worst known accident aboard a nuclear-powered ship occurred when one of the three reactors powering the Soviet ice-breaker *Lenin* suffered a meltdown.

The accident is believed to have rendered the ship too radioactive to use and, for a year, she was abandoned in the Arctic. According to a CIA report first released in 1987 (*Toronto Globe and Mail*, 8.6.87), between 27 and 30 people died in the accident.

The use of nuclear reactors to power marine vessels began in 1954 with the commissioning of the US Navy's submarine USS *Nautilus*, followed closely by the *Seawolf* and the *Skate*. These submarines quickly made headlines — the *Nautilus* by crossing the Arctic Ocean under the polar ice-cap and the *Skate* by actually surfacing through the ice at the North Pole. The nuclear-powered submarine clearly lent itself to patrolling the frozen waters of the far north, and the strategic implications were not lost on the Soviet Union.

The USSR not only pushed ahead with its own nuclear-submarine fleet (as did Britain and France), but also adapted nuclear power for use in a surface vessel — an ice-breaker that could patrol the northern coast of the USSR as well as opening up shipping routes that are frozen over for most of the year. The *Lenin*, commissioned in 1959, quickly succeeded in forcing a passage through the Vil'kitskogo Strait, off the northernmost coast of mainland USSR, a feat that no other ice-breaker had achieved.

The US followed suit with nuclear-powered warships (the cruiser USS *Long Beach* in 1959 and the aircraft carrier USS *Enterprise* the next year) and, in keeping with President Eisenhower's policy of 'Atoms for Peace', launched the first nuclear merchant ship. In the words of John Robb, chief of construction for the AEC and the Maritime Administration at the time, the most important job of the NS (Nuclear Ship) *Savannah* was '...to break down political, legal, and psychological barriers to the use of nuclear energy. We want her to show the world that the atom can be put to work at sea like any other source of power.' (*National Geographic*, August 1962.) The *Savannah*, a dry-cargo freighter with the capacity to carry 100,000 tons, was launched from Yorktown, Virginia, in March 1962.

Insurance for the first nuclear merchant ship, as for commercial power reactors, proved to be problematic, and before she put to sea the US government found it necessary to pass an Act of Congress to provide an indemnity fund of $500 million for the ship in the event of an accident. Even so, port authorities were unwilling to host a ship that had the potential to put their docks out of action indefinitely.

The next to be built was the West German ore carrier, *Otto Hahn*, commissioned in 1964. Foreign ports were even less welcoming to this ship and it was forced to operate on the less profitable routes. The initial promises of cheap sea freight had quickly begun to prove somewhat empty. Oil prices remained low, capital costs and running costs for the nuclear reactors were high, as were insurance premiums and wages for the crews. Government subsidies to keep the *Savannah* at sea ran to $3 million per year.

On 18 November 1967, at a meeting of the US Society of Naval Architects and Marine Engineers, it was suggested that in future, nuclear-powered ships could be made more economical by doing away with many of the secondary safety features. Within a week, the *Savannah* was back in her home port after instruments indicated that the reactor's secondary cooling system was leaking.

Two years after the reported meltdown aboard the Soviet ice-breaker *Lenin*, following which she was towed to Murmansk for an official overhaul and to have her reactors replaced, the Japanese took their first steps in the field of nuclear propulsion. Their experience was not encouraging.

In 1969, Japan's first and only nuclear-powered ship, the cargo vessel *Mutsu*, was launched from the port of Mutsu. She was a focus of popular opposition even while she was being built, and local people, fearing that radiation leaks might ruin the fishing in Mutsu Bay, prevented the ship's sea trials in 1972. Two years later, on 26 August 1974, when bad weather hampered a blockade by fishermen, she finally left port to test the nuclear reactor. Her voyage 'probably did more to set back the cause of nuclear marine propulsion than anything in the industry's worst nightmares'. (**Patterson**.)

When the *Mutsu* was 500 miles out to sea, the reactor was brought into operation and the power was gradually raised. Before it reached two per cent of full power, radiation levels outside the reactor were found to be high and the reactor was shut down. The problem proved to be a six-inch gap in the reactor shielding — the result of a technical error. Carrying out repairs at sea posed a problem, and the crew had only limited success using boiled rice as shielding cement. They finally resorted to stuffing the gap with old socks.

Unfortunately for the *Mutsu*'s owners, news of the difficulties reached the mainland. Returning to Mutsu was out of the question given the anger of the local population. Several other ports refused the crippled ship entry, and she remained at sea for more than six weeks while her owners negotiated for her return to Japan. The Japanese government had to promise to spend almost two million pounds on improving the town of Mutsu and its fisheries before the townspeople would allow her to dock. The ship's owners also had to agree to find a new port for the *Mutsu*, to leave all the nuclear fuel in the reactor, to remove all shore-based nuclear facilities from Mutsu port and to give the keys of the fuel-handling crane there to the mayor of the town.

She was finally able to dock, but redesign and repair work took more than 10 years and it is expected that the *Mutsu* will be decommissioned in 1989/1990.

Despite the serious reactor accident aboard the *Lenin*, several more ice-breakers joined the Soviet nuclear fleet during the seventies (the *Arktika*, which in August 1977 became the first surface vessel to reach the North Pole, and the *Sibir*) and eighties (the *Rossiya*, the *Leonid Brezhnev*, the *Sovietskiy Soyuz*, the *Taymyr*, the *Oktyabrskaya Revolyutsiya* and the *Vaygach*).

On 11 November 1988, according to the Soviet trade newspaper *Vodny Transport* (18.2.89), one of the two reactors aboard the *Rossiya* almost suffered a meltdown while the ship was docked in the northern port of Murmansk. During a 'potentially dangerous nuclear operation', the chief physicist apparently gave an 'erroneous command' to open a drain valve in the cooling system of the No. 2 reactor, which was in operation at the time. Had members of the crew not closed the drain valve again, 'the reactor would have remained without cooling and the fuel might have melted.' Murmansk, the largest city inside the Arctic Circle, has a population of half a million.

In December 1988, the giant 33,500-tonne nuclear-powered barge carrier *Sevmorput* made its maiden voyage, but this ship has met with public opposition even within the USSR. Reports, soon after its launch, that the $260-million vessel had developed 'fractures', and was seeking port facilities for repairs, led to protests from Soviet environmental groups, and in March 1989, after officials in four ports sided with dock workers and refused the ship entry, it remained at anchor outside Vladivostok for several days.

In the West, commercial interest in nuclear-powered ships has remained slight during a period in which there has been a recession in world trade and a growth in awareness of the risks of nuclear power. In 1972, the *Savannah* was withdrawn from service, and in 1979 the German government announced that the *Otto Hahn* was to be laid up, since the information that she could provide would not be worth the costs of further refuelling. The navies of the US, USSR, France and Britain now have many nuclear-powered submarines and warships, and these provide a means of testing for developments in nuclear propulsion without the results being subject to public scrutiny, since in most countries the regulations concerning civil nuclear power do not apply to military reactors.

21 January 1968

Broken Arrow 6, Thule Greenland

'No danger to man or animal and plant life was created by the Thule accident — that is now a well-established fact.' Hans Henrik Koch, Chairman, Danish Atomic Energy Commission.

'All the Thule workers who were in the clean-up, they're all old men now. My husband is only 49 but he's an old man now.' Sally Markussen, wife of Danish worker at Thule.

HOBO 28, a nuclear-armed USAF SAC B-52 bomber, was on 'airborne alert' patrol over the Arctic when fire broke out in the navigator's compartment. Realising that the fire could not be contained, the pilot requested authority for an emergency landing at Thule Air Base, Greenland.

Shortly after HOBO 28 began descending towards Thule, all electrical power was lost and the aircraft became uncontrollable. Just after 16:30 on 21 January 1968, the giant eight-engined plane went into a steep left bank and struck ice-covered North Star Bay, some seven miles from the Thule runway, at 500 knots. Six of the seven-man crew ejected to safety. The seventh was killed.

HOBO 28 was immediately destroyed by the impact and the simultaneous explosion of over 200,000 lb of jet fuel. High explosive within the casings of its four 1.1 Mt H-bombs also detonated, blowing the warheads into highly radioactive plutonium and tritium fragments. Most of the shattered plane drove through the three-ft-thick ice and settled, surrounded by radioactive debris, onto the sea-bed some 700 ft below. The remaining wreckage, reduced to pieces the size of a cigarette pack, formed a black, radioactive scar across the refrozen ice. Above the smoking wreckage, the fire and explosion created a cloud of burning fuel and further radioactive debris which dispersed over North Star Bay.

Alerted by Thule Air Base, the USAF Command Post deep inside the Pentagon immediately convened its Broken Arrow Control Group for the clean-up operation.

The clean-up — officially called Project Crested Ice but dubbed by Thule workers 'Dr Freezelove' — involved specialists from more than 70 US agencies. They were augmented by scientists from Denmark (Greenland is a Danish dependency) and hundreds of US servicemen and Danish civilian workers. By the middle of September, the wreckage of HOBO 28 and its bombs, together with thousands of tonnes of radioactive ice and snow, had been shipped to the United States. Project Crested Ice was over.

An official USAF report, published in 1970, was little more than an exercise in self-congratulation. Major-General Richard O. Hunziker, USAF, who headed Project Crested Ice, wrote that 'A major disaster was turned into a classic example of international co-operation...The seemingly insurmountable task of recovering and removing all traces of the accident proves again that truth may be stranger than fiction — and fully as exciting.'

He played down the problems of contamination from the dispersed plutonium and tritium, asserting that 'The scientific evidence collected by both countries confirmed that, as hitherto assumed, there were no risks involved for human beings, animal and plant life as a consequence of the crash.'

Two decades later, however, the health of some 500 Danes who worked for Project Crested Ice casts doubt on Hunziker's confident assertion. They claim to be suffering from a range of ailments, including cancer and sterility, which may be associated with radiation exposure, and their condition has aroused growing public concern in Denmark.

But the plight of the Danish civilian workers may just be the tip of an iceberg of radiological problems stemming from the crash and the subsequent clean-up operation. Like the Danes, the US personnel were given no long-term radiation checks. Unlike the Danes, however, they were servicemen, subject to military regulations, secrecy and bureaucracy. So the USAF personnel were not only exposed to higher levels of radioactive contamination than was considered safe for their Danish counterparts, but they also will never *know* the true extent of their contamination.

From the very beginning, Crested Ice faced formidable obstacles. Greenland was in perpetual darkness until February. Fierce, driving winds reduced the average temperature of -28 °F to -70 °F. And the project was working to a tight, natural schedule: the spring thaw, just weeks away, would melt and break up North Star Bay's ice cover.

Initially, USAF personnel determined the extent of the radioactive impact zone. They found that about 25,000 sq m of North Star Bay had been contaminated with plutonium and tritium debris from the four destroyed bombs. This zone, roughly rectangular, and centred on the blackened area resulting from the explosion and fuel fire, was cordoned off by a 'zero line' outside which, claimed Hunziker, no radiation was detected.

As the airmen searched, shoulder-to-shoulder, for pieces of HOBO 28 and its deadly cargo, other Crested Ice workers created, from scratch, a township on the ice of North Star Bay. Day and night, USAF transports thundered into Thule Air base, carrying prefabricated buildings, a heliport, generators, and lighting, heating and communications equipment. Engineers strung telephone and teletype lines and built three roads over the ice.

The search for aircraft wreckage and bomb parts was conducted in impossible conditions. Although the search parties wore Arctic clothing — insulated trousers known as 'iron pants', sweater, flight suit and heavy parka — it was often so cold that they were confined to quarters. They worked for between 13 and 18 hours a day, seven days a week, in darkness. Their only light came from petrol lanterns and torches. Winds of up to 85 mph blew the snow horizontally in storms which often lasted all day. Their breath froze on face masks and parka hoods. The masks restricted breathing, so many workers simply discarded them, even though this increased the risk of inhaling plutonium particles. The intense cold also took its toll on equipment: torch batteries lasted only 10 minutes. Engines had to be kept running — once stopped, they could never be re-started.

The search teams located debris with a variety of radiation monitors, most of which were inadequate to the task. Eventually, concedes the USAF report, it was accepted that, due to the conditions at Thule, 'absolute determination of the quantities of contamination...was essentially impossible.'

The search teams were replaced by a mixed American and Danish workforce. These workers, who, like the search teams, wore no radiological protection clothing, began clearing wreckage and scraping contaminated snow and ice into containers for removal to the United States. (The latter measure, explained Hunziker — again playing down the radiological problem — 'had been jointly agreed earlier as a housekeeping measure, even in the absence of concern regarding biological hazards.')

The Americans — all Air Force personnel — drove the scrapers and mechanised loaders, while the Danes — taken from more than 1,000 civilians working at the Thule Air Base — filled the containers, which were later dumped into surplus 25,000-gal fuel tanks. Spills were unavoidable: contaminated snow and ice were simply brushed or shovelled back into the heavy containers. Several of the Danish workers later claimed that contaminated material often leaked from the tanks and that radiation monitoring was infrequent and haphazard.

All workers were decontaminated at the end of their shift. They were checked with counters, their urine was sampled and their nostrils were swabbed to detect inhaled radioactive particles — procedures which caused concern to many of the Danes. 'Numerous questions were raised as to the reasons for these precautions,' noted Ole Walmod-Larsen of the Danish AEC. 'One of the most significant problems only appeared after more than a one-and-a-half-hour discussion - the problem was one of sterility and impotence. When it was explained that these fears were groundless, no further questions were asked!' (USAF report, 1970.)

In fact, the decontamination procedures were *less* thorough than they ought to have been. The nasal swabs were often used in lieu of routine urine samples in order to save time. According to Captain William K. McRaney, Health Physicist, USAF Directorate of Nuclear Safety, the swabs '...in general revealed no detectable activity. However, since most everyone's nose ran profusely in this climate, there was a reasonable doubt as to the validity of this check.' McRaney's comment is one of the few expressions of doubt in the entire Thule report.

By September 1968, a total of 237,000 cu ft of radioactive snow, ice and debris had been removed from the contaminated area, dumped into the tanks and shipped out to the United States for burial. (The remains of the four H-bombs were returned to the assembly plant in Amarillo, Texas.) The impact zone was treated with black sand to hasten the spring melt, which would take the residual radioactive material into the water of North Star Bay. The last 25,000-gal tank of melted snow was loaded aboard a US Navy cargo ship on Friday 13 September 1968.

Project Crested Ice ended with no ceremony except a scrawled 'That's all folks' on the tank's cap.

In December 1986, the Danish Prime Minister, Poul Schlueter, announced that surviving workers from the Thule base would be examined by radiological experts. This decision was seen as a belated response to growing public concern over the incident, following an investigation by Mrs

Sally Markussen, the wife of the personnel manager at Thule in 1968. She had collected names and medical details from some 800 Danes involved in the Thule clean-up. Over 500 were found to have health problems and 98 claimed to have developed cancer. Her own husband, Ole, became sick in 1979 at the age of 41 and suffered breathing difficulties, vomiting and the excretion of blood. Common symptoms found among the former base workers were weight loss, constant tiredness, loss of concentration and balance, loss of co-ordination between hand and brain, congestion in the lungs and sores on the shins and arms. Only 20 of the workers surveyed had been able to have children, and of those born, several suffered from deformities. Further studies would need to be based on more accurate knowledge of the materials to which the men had been exposed.

Consequently, in March 1987, the Danish government officially requested the US to submit a list of the radioactive and toxic materials released into the environment by the crash, and information on the health of US Crested Ice personnel.

In November 1987, an official study was released by the Danish Institute for Clinical Epidemiology, which had taken a cursory look at disease rates amongst Crested Ice personnel and compared them with 3,000 workers at Thule who had been there either before or after the crash. Although the study found that the proportion of Thule clean-up workers hospitalised with diagnosed cancer was 40 per cent higher than the general population, this was still not considered statistically high enough for conclusions to be reached.

A report by the Institute of Cancer Epidemiology was also released in November 1987. This investigation concluded that although workers involved in the clean-up showed a 50 per cent increase in cancer compared with the general population, this was the same for Thule workers who were *not* involved in the clean-up. The incidence of lung cancer at the Thule base was twice the national average and the study concluded that exposure to radiation was not the cause.

In late 1987, nearly 200 of the Danish workers filed a suit for damages in the United States under the Foreign Military Claims Act. In February 1988, the US Air Force decided against allowing this claim under the Act; if they wanted to pursue the case, the Danes were told, it would have to be done under NATO's Status of Forces Agreement, making it a much tougher legal battle to fight.

In mid-January 1988, after 10 months of silence, hundreds of secret documents were delivered from the Pentagon to the Danish Embassy in Washington. They proved to be of no help in answering the central questions because the US had refused to list the contents of the nuclear weapons. One thing was confirmed though; no long-term tests had been carried out on the hundreds of Americans who had taken part in Crested Ice.

In July 1988, the Danish National Institute of Radiation Hygiene published its report — 'Plutonium Excretion in Former Thule Workers' — which concluded that no traces of plutonium could be detected in the urine of 53 former Thule workers. On the basis of these findings, the Institute recommended to the government that there should be no further centralised investigation into the health effects of the Thule clean-up, and that any workers wishing to have a medical check should apply to their local clinics.

The joint conclusions of the official reports found no grounds on which Thule workers might claim that cancers were caused by radiation exposure in 1968. However, Sally Markussen's reports suggest that workers are suffering from various other illnesses, and none of these have been taken into account by the official studies. Nor have the effects of any other, as yet undisclosed, substances on the ice been taken into account.

END OF ALERT

A few days after the crash — the tenth accident in 11 years involving a nuclear-armed SAC bomber — the US Secretary of Defense, Robert McNamara, ordered the removal of nuclear weapons from planes on airborne alert. The alert missions were later curtailed and finally suspended altogether. The official USAF explanation was that airborne alert was too demanding in terms of wear and tear on the planes and on crew time. At the time, B-52s were being used extensively in the Vietnam War.

SAC now maintains 25-30 per cent of its bombers on a ground-alert status; armed and ready to take off in minutes in a crisis period. (Nuclear weapons in the US are still transported by air, using Military Airlift Command C-141 and C-130 aircraft.)

THULE'S ROLE

The Thule base, some 700 miles from the North Pole, forms part of the US early-warning system. The first alarm is sounded by the three Defense Support Program (DSP) satellites that directly monitor Soviet launch sites. In the event of a Soviet attack, the radar at Thule — known as the Ballistic Missile Early Warning System (BMEWS) — has a range of 3,000 miles and will pick up Soviet ICBMs 10 minutes after they leave their silos. To prove that a missile warning is read correctly, North American Aerospace Defense Command Centre (NORAD) requires both satellite and radar confirmation. However, this equipment was designed in the late 1950s and is now antiquated by today's technology standards, as are the computers and other support equipment.

On 5 October 1960, the central war room of NORAD received a top-priority warning from Thule, indicating a massive missile attack. It proved to be a radar malfunction.

ONE LOST BOMB?

In December 1987 a Danish civil engineer involved in the clean-up revealed that a secret, highly sensitive film in the Pentagon archives — shot from a miniature US submarine during the initial search — showed one of the H-bombs from the B-52 lying intact on the sea-bed off Greenland.

Greenland MPs demanded that the film be brought to Denmark for viewing by government representatives. (Greenland is a semi-independent Danish dependency with full control over its domestic affairs. Foreign and defence policies are dealt with in Copenhagen.) A Pentagon spokesman responded by quoting from a **DOD** statement issued at the time of the accident which said that fragments of all four bombs had been found and identified by serial numbers matching those in SAC files, thus making it 'impossible for one or more of the bombs to have gone through the ice.'

21 May 1968

USS Scorpion, North Atlantic

Details remain classified of another submarine accident, in which 99 men were lost. The 3,075-ton, nuclear-powered attack submarine, USS *Scorpion*, was last heard from on 21 May 1968, 250 miles south of the Azores. The nine-year-old sub was returning home to Norfolk, Virginia, after a three-month exercise with the Sixth Fleet in the Mediterranean.

Commander Francis A. Slattery, USN, commanding USS *Scorpion* in May 1968. (Credit: DAVA Still Media Repository).

The navy announced five months later that the oceanographic research ship *Mizar* — which played a key role in locating the lost bomb off Palomares — had identified the wreckage of the *Scorpion* lying at a depth of 10,000 ft on the sea-bed about 400 miles south-west of the Azores. The hull was split and part of the bow was missing.

The incident was investigated by a seven-man naval Court of Inquiry which sat for 11 weeks and heard 90 witnesses. It learnt that the *Scorpion* had reported at least five 'relatively minor' mechanical problems during its last mission and that the vessel had been only partially improved following recommendations made after the *Thresher* incident. Some five weeks before its disappearance, the *Scorpion* had collided with a barge in Naples harbour during a storm, sinking the barge. A week later the submarine had put into Naples again, and divers had had to untangle a fishing net from its propeller. They reported no damage. The Court of Inquiry concluded that 'the certain cause of loss cannot be ascertained from evidence now available' (*Washington Star*, 2.8.81), but that there was no indication of sabotage, enemy action or other such deliberate destruction.

In 1985 Norfolk's local newspaper, the *Virginian-Pilot*, obtained classified documents which revealed that the loss of the USS *Scorpion* occurred when an accidentally-armed Mk 37 torpedo exploded while the crew was attempting to disarm it. The original navy accident report did not mention an explosion, and the possibility was officially discounted.

The documents showed that investigators had discovered two 'acoustic events': the first they attributed to the explosion, the second to the implosion of the hull when the *Scorpion* fell below crash depth. (These 'events' were probably recorded by hydrophones in the US east coast submarine detection network.) The *Virginian-Pilot* article quoted Dr John Craven, the former navy scientist who headed the Naval Inquiry, as saying that a torpedo mishap 'is the one scenario that, in my opinion, fits all of the evidence'.

The nuclear weapons on board the *Scorpion* may have been either SUBROC or ASTOR, probably the latter. The SUBROC, first deployed in 1965, is a submarine-launched nuclear anti-SUBmarine ROCket with a range of 25-30 miles and high explosive power. It breaks the surface of the water, travels through the air, and re-enters the water to attack submarines. It was designed to replace ASTOR, a nuclear-armed Anti-Submarine TORpedo developed as a last-ditch defensive weapon for submarines. The short range of these torpedoes meant that the launching submarine might itself suffer the effects of a nearby nuclear explosion.

The *Scorpion* was not the only submarine casualty in the spring of 1968. On 22 January, the Israeli diesel-powered submarine *Dakar* disappeared in the Mediterranean, 250 miles off the coast of Israel, with 69 men on board. Five days later, the French diesel submarine *Minerve* sank in the Mediterranean, off Toulon, with the loss of 52 lives. On 11 April 1968, a Soviet Golf-class ballistic-missile submarine sank about 750 miles north-west of the island of Oahu, Hawaii, killing approximately 80 men (see Project Jennifer, p. 196). The loss of the *Scorpion* in May brought the number of submariners killed in just over four months to some 300.

November 1969

Operation Holystone

The nuclear-armed, nuclear-powered submarine USS *Gato* was on a surveillance patrol which had taken her within one mile of the Soviet coast, when she collided with a Soviet submarine 15-25 miles off the entrance to the White Sea, in the Barents Sea off northern USSR.

A crew-member aboard the submarine was quoted (*New York Times*, 6.7.75) as saying that the *Gato* was struck in the heavy metal plating that protects the nuclear reactor, but she sustained no damage. Nonetheless, the submarine's weapons officer immediately ran down two decks and took the first steps towards arming a SUBROC missile and three nuclear-armed torpedoes, and the submarine was 'manoeuvred in preparation for combat'. According to the *New York Times* report: 'Only one authentication — either from the ship's captain or her executive officer — was needed to prepare the torpedoes for launching.'

The USS *Gato* was patrolling as part of Operation Holystone, one of the most dangerous and provocative US submarine surveillance and intelligence-gathering programmes (also known variously by the codenames Pinnacle, Bollard and Barnacle).

First authorised during the Eisenhower administration, Holystone was placed under the direct control of the Chief of Naval Operations at Atlantic Fleet Command headquarters in Norfolk, Virginia. It involves the use of at least four SSN-637 or Sturgeon-class nuclear-armed attack submarines, specially fitted with sophisticated electronic equipment.

Holystone submarines engage in a wide range of intelligence-gathering operations including: close-up photography of the undersides of Soviet ships and subs; plugging into Soviet underwater communication cables to intercept high-level information considered too sensitive to send by less secure means; monitoring Soviet SLBM tests; recording the 'voice autographs' of Soviet submarines — the distinctive sound that distinguishes one from another. Submarines involved in Holystone surveillance missions have frequently operated within the 12-mile territorial limit claimed by the Soviet Union, and often within the three-mile limit. Any incident in such a sensitive area could trigger off a disastrous chain of events. The general orders for Holystone missions reportedly state that, if threatened, the submarines have authority to use weapons.

The collision of the USS *Gato* is the most serious reported Holystone incident, but there have been many others. In the mid-1960s a Holystone submarine collided with a Soviet Echo-class submarine inside Vladivostok harbour while photographing its underside. Some of the US submarine's equipment was knocked off. Some time before January 1968, the USS *Ronquil*, while engaged on a Holystone patrol, caught fire while below the surface off the Soviet coast. Soviet destroyers surrounded the submarine and attempted to force it to surface, but it successfully eluded the Soviet ships and made its escape.

In the late 1960s, a Holystone submarine collided with a North Vietnamese minesweeper in the Gulf of Tonkin, sinking the minesweeper. In the autumn of 1969 another US submarine ran aground off the Soviet coast, giving rise to fears of an international incident should it be discovered. In March 1971 it was reported that a Sturgeon-class submarine had collided with a Soviet submarine 17 miles off the Soviet coast while on a secret reconnaissance mission.

On 1 May 1974, while on an intelligence-gathering mission inside Soviet territorial waters, the USS *Pintado* collided almost head-on with a Soviet Yankee-class nuclear-powered submarine 200 ft below the surface of the approach to Petropavlovsk naval base on the Kamchatka Peninsula. The *Pintado* left at top speed and put in to Guam for repairs which lasted seven weeks. The collision damaged the submarine's detection sonar and a torpedo hatch was jammed shut. The same year, a Holystone submarine ran aground inside Vladivostok harbour, having struck the harbour floor while running at low power to avoid detection.

Additional Stories

1960 April 3 During power increases at the Westinghouse Test Reactor at Waltz Mills, Pennsylvania, a fuel element melted and there was some release of radioactivity to the atmosphere. It is thought that the cladding around the element separated from the fuel and prevented the transfer of heat to the coolant, allowing the fuel to heat up. (**Bertini**)

1960 May 16 This was the first occasion on which global US military forces had been put on alert status of DEFCON 3. The crisis lasted for less than 24 hours. On 14 May, the Big Four summit conference — Eisenhower, Khrushchev, Macmillan and de Gaulle — opened in Paris in an atmosphere of hope for an East-West *rapprochement.* On the evening of 15 May, President Eisenhower and Secretary of Defense Thomas Gates received information suggesting that Khrushchev might scuttle the conference in protest against the 1 May 1960 U-2 incident in which Gary Powers (see p. 359) was shot down. Although there were no signs of Soviet military mobilisation, Gates suggested that the American military alert system be tested, in order to ensure US military commanders could receive further communications if necessary. Gates issued an order for 'a quiet increase in command readiness'. Within hours the 'quiet alert' had got louder and the whole system was on DEFCON 3. The US media not only announced forces were being mobilised but also revealed the level of the alert, thus adding fuel to Khrushchev's claims of US 'provocation'. This incident

DEFCON The process by which American military forces are placed on alert is called the Defense Condition (DEFCON) system, created by the joint chiefs of staff in November 1959. There are five DEFCONS, or gradations of alert, but precise details are kept highly classified. It is known that most American forces are kept on the lowest alert status, DEFCON 5, in normal peacetime. The Strategic Air Command (SAC) is an exception, being routinely kept at DEFCON 4. Military forces close to areas of intense combat are also kept at higher levels of alert. The Pacific Fleet during the Vietnam War was kept at DEFCON 3.

There have been three occasions on which global US military forces have been put on DEFCON 3 or above during a crisis with the Soviet Union. (See Additional Stories: 1960, May 16; 1962, October 22; 1973, October 24.)

proved it was impossible to keep high level alerts secret. (Scott D. Sagan, 'Nuclear alerts and crisis management', *International Security* vol. 9, No 4, Spring 1985)

1960 November 28 While the nuclear-powered submarine USS *Nautilus* was docked in Portsmouth, New Hampshire, six men were soaked in radioactive reactor-coolant when one of them accidentally knocked a valve. Dose meters and contaminated clothing were thrown away, making radiation measurement impossible. (**Neptune**)

1961 March 14 A USAF B-52 experienced failure of the crew compartment pressurisation system, forcing the plane to descend to 10,000 ft. This increased fuel consumption and the plane ran out of fuel before it could rendezvous with a tanker aircraft. All of the crew baled out except the captain, who stayed with the plane to 4,000 ft to steer it away from populated areas. The two nuclear weapons on board, which may have been free-fall bombs or air-to-ground missiles, were torn from the plane when it crashed near Yuba City, California. There was no explosion, but one fireman died extinguishing the fire. (**DOD/CDI**)

1962 January 6 A submarine belonging to the US or another NATO power was damaged and forced to the surface by a 20-Mt underwater nuclear test blast some 100 miles from the submarine in the Barents Sea off the Soviet Union. (**Neptune**)

1962 March 26 The Royal Navy's first nuclear-powered submarine, the *Dreadnought*, suffered a fire while under construction in Barrow-in-Furness. A second minor fire occurred in the submarine's control room in December 1965, while she was undergoing repairs in Rosyth, Scotland. (**Neptune**)

1962 June 19 Test-shot Starfish, the second in a series of high altitude tests of Thor ICBMs from the US Pacific Test Range at Johnston Island, produced a 1.4 Mt explosion 400 km above the mid-Pacific. Light from the blast could be seen from Hawaii to Australia, and its electro-magnetic pulse (EMP) destroyed satellite equipment and blocked all high frequency radio communications across the Pacific for half an hour. It also modified the Van Allen Belt of charged particles around the earth. The EMP phenomenon, which is produced by all nuclear explosions, is of particular concern to military authorities, whose major means of communication could be instantly put out of action in the event of a nuclear attack.

Two more of the five Bluegill Prime test-shots failed within minutes of launch when their tracking systems malfunctioned. These Thor missiles and their warheads had to be destroyed in the atmosphere. On 25 July 1962, during test-shot Bluefish Prime, the Thor missile blew up on the launch pad, causing extensive damage.

The warhead was detonated by radio command and there was considerable alpha contamination of the launch pad. (Hayes, Zarsky and Bello, *American Lake*, Penguin, 1986; Nuclear Weapons Databook II; Daniel Ford, *The Button*, George Allen and Unwin, 1985.)

1962 October 22-November 20 SAC was placed on DEFCON 2, and other commands were placed on DEFCON 3 during the Cuban Missile Crisis. See box, 1960. (Sagan.)

1963 July 1 While the Oak Ridge Research Reactor, in Oak Ridge, Tennessee, was being brought up to full power, a gasket in the reactor tank came loose and dropped into the core, blocking the flow of coolant water around one of the fuel elements. Part of the element melted and radiation was released into the building, which had to be evacuated, and into the atmosphere. It was later estimated that some 1,000 curies in fission products were released into the primary cooling water. (**Bertini**)

1963 November 13 While three employees at the AEC storage igloo at Medina Base, San Antonio, Texas, were dismantling the high-explosive component of a nuclear bomb, it began burning spontaneously, setting off 123,000 lb of high explosive components of obsolete nuclear weapons. The workers escaped with minor injuries. (**DOD/CDI**)

1964 January 13 A B-52D en route from Massachusetts to its home base at Turner AFB, Georgia, was climbing to avoid turbulent air when it hit violent turbulence. The aircraft suffered a 'structural failure' and crashed in an isolated mountainous and wooded area near Cumberland, Maryland. According to the DOD report, 'Both weapons were in a tactical ferry configuration (no mechanical or electrical connections had been made to the aircraft and the safing switches were in the "SAFE" position)'. Both nuclear weapons remained in the aircraft until it crashed and they were recovered relatively intact. One crew member died in the crash. The four others ejected, but two subsequently died in the sub-zero conditions. (**DOD**)

1964 May A USAF Voodoo from Bentwaters in Suffolk, with a nuclear weapon on board, crashed near Argyll in Scotland. According to Malcolm Spaven, 'there is evidence to suggest that [it] was carrying a nuclear weapon.' (**Bradford**; Malcolm Spaven, *Fortress Scotland*, Pluto Press, 1984)

1964 December 5 An LGM 30B Minuteman-1 three-stage intercontinental ballistic missile was on strategic alert at Ellsworth AFB, South Dakota, when a retrorocket fired, causing the re-entry vehicle containing the arming system to fall some 75 ft to the floor of the silo. There was no detonation or radioactive contamination. (**DOD/CDI**)

1964 December 8 While taxiing behind another aircraft, a B-58 supersonic bomber containing five nuclear weapons skidded on the icy runway at Bunker Hill (now Grissom) AFB at Peru, Indiana. The plane struck a concrete electrical manhole box and caught fire. One crew member died when he ejected in his escape capsule and it landed more than 500 ft from the plane. Portions of the nuclear weapons burned, but contamination was limited to the crash site. (**DOD**)

1965 October 11 According to the US DOD: '/C-124/Wright-Patterson Air Force Base [near Dayton] Ohio. The aircraft was being refuelled in preparation for a routine logistics mission when a fire occurred at the aft end of the refueling trailer. The fuselage of the aircraft, containing only components of nuclear weapons and a dummy training unit, was destroyed by the fire. There were no casualties...Minor [radioactive] contamination was found on the aircraft, cargo and clothing of explosive ordnance disposal and fire fighting personnel, and was removed by normal cleaning.' (**DOD**)

1966 January 19 The nuclear warhead of a Terrier surface-to-air missile separated from the missile and dropped on to the deck of the destroyer USS *Luce* while it was docked at Jacksonville, Florida. Apart from a dent in the warhead, no equipment was damaged. (**Neptune**)

1966 October 26 A crewman aboard the aircraft carrier USS *Oriskany*, operating off Vietnam, panicked when a flare accidentally ignited while being moved, and threw it into a locker containing 650 more flares. The resulting fire took three hours to bring under control, killed 44 people, destroyed or damaged six aircraft and put the aircraft carrier out of action for several months. (**Neptune**)

1966 November 4 A flash fire occurred on board the aircraft carrier USS *Franklin D. Roosevelt* off Vietnam. The fire, in a storage compartment containing oil and hydraulic fluid, killed seven crew members. (**Neptune**)

1967 May There was a partial meltdown in one of the four Magnox reactors at the Chapelcross site in Annan, Scotland, which were built in order to produce weapons-plutonium to augment the output from the Windscale reactors. The No. 2 reactor had been restarted, after a shutdown for refuelling, when sleeving around some of the plutonium fuel broke up and clogged one of the cooling channels. The fuel overheated and caught fire at a temperature of 1,000 °C. The reactor was then shut down, but one fuel element fell out of the bottom of the core and melted. The incident was contained, but some radiation was released to the atmosphere. The reactor core was damaged, permanently affecting its efficiency, and the reactor was out of action for more than a year while special tools were designed and made in order to bore out the fuel from the blocked cooling channel. (**Patterson**; *Guardian*, 13.10.86/9.2.88)

1967 August 31 The Polaris submarine USS *Simon Bolivar* collided with the target ship USS *Betelgeuse* during torpedo practice, 70 miles south-east of Charlestown, South Carolina. The submarine, which was carrying 16 unarmed Polaris missiles, suffered damage estimated at $1 million. (**Neptune**)

1967 November 5 The nuclear-powered ballistic missile submarine HMS *Repulse* ran aground 30 minutes after her launch from Barrow-in-Furness in the UK. It took seven tugs to pull her free. (**Neptune**)

1968 January 30 The nuclear-powered submarine USS *Seawolf* struck the sea-bed 65 miles east of Cape Cod and damaged its rudder. (**Neptune**)

1968 February 1 The destroyer USS *Rowan*, carrying ASROC Anti-Submarine ROCkets, was in collision with the Soviet freighter *Kapitan Visiobokov* in the Sea of Japan, about 100 miles east of Pohang, South Korea. (**Neptune**)

1968 February 12 A B-52 with nuclear weapons on board crashed near Toronto, Canada. (**Bradford**)

1968 March During the scheduled shutdown of an unidentified US test reactor, modifications were made to the reactor pool cooling system. In order to prevent the shielding water that was covering spent fuel, capsules and other radioactive material from escaping while they were working, a 'regulation basketball', wrapped in rubber tape to increase its diameter by two in, was inserted into the suction line and inflated. The force of the water blew the basketball out, and 14,000 gal of radioactive water flooded into the basement. The AEC commented: 'Where the risk of fuel melting and personnel safety are involved, consultation with knowledgeable people should be made prior to questionable operation.' (Pollard, Nugget File, UCS, 1979)

1968 April 9 The Polaris submarine USS *Robert E. Lee* became snagged in the nets of the French trawler *Lorraine-Bretagne* in the Irish Sea. (**Neptune**)

1968 April 19 An RAF Shackleton aircraft crashed at the Mull of Kintyre, Scotland — the fourth of these planes to go down within six months. Five weeks later, on 25 May, a Soviet TU-16 Badger aircraft crashed in the Norwegian Sea while 'buzzing' the US aircraft-carrier *Essex*. In both cases, it is uncertain whether the planes were on photo reconnaisance missions or on anti-submarine warfare patrol, in which case they could have been carrying nuclear depth charges. (**Bradford**)

1968 June 12 The aircraft carrier USS *Wasp* collided while refuelling from the USS *Truckee* off the coast of Virginia, badly damaging both ships. (**Neptune**)

1968 June 16-17 US Navy A-4 attack jets sank a US patrol boat and attacked the cruiser USS *Boston* and the Australian destroyer *Hobart* in the Tonkin Gulf off Vietnam, when they mistook the ships for low-flying enemy helicopters. The *Boston* is capable of carrying nuclear-armed Terrier anti-aircraft missiles. (**Neptune**)

1968 June 26 A nuclear-capable RAF Scimitar aircraft crashed on the Isle of Wight, in the UK, after hitting overhead power lines. (**Bradford**)

1968 August 9 The Polaris submarine USS *Von Steuben* collided with the merchant ship *Sealady* about 40 miles off the south coast of Spain, sinking the ship. (**Neptune**)

1969 January 14 The nuclear aircraft-carrier the USS *Enterprise*, positioned 70 miles south-west of Pearl Harbour, suffered a series of fires and explosions which killed 28 crew members and injured 343 others. A tractor used for starting aircraft aboard the *Enterprise* had been unintentionally backed under the wing of an F-4 Phantom carrying Zuni rockets. Heat from the tractor's small jet engine had played on a Zuni warhead for about a minute, at which point the rocket exploded, blasting shrapnel across the deck, puncturing the fuel tanks of other planes and starting numerous fires. The fires caused other rockets and bombs, aboard planes and stacked on deck, to explode, blowing holes in the carrier's steel deck and spilling aviation fuel from storage tanks. Despite limited fire-fighting equipment on the flight deck, the fire was contained and burned itself out after three hours. (**Neptune**)

1969 January 21 The Swiss took their first steps into nuclear power by building their own design of experimental reactor, which was cooled by carbon dioxide and moderated by a tank of heavy water, in a cavern under a hill at Lucens Vaud, near Lausanne. Its underground location proved to be fortunate. A seal in the carbon-dioxide cooling system burst, causing the pressure of the gas to drop from its working level — which was 60 times atmospheric pressure — down to atmospheric pressure, with a consequent reduction in its cooling capability. The reactor was scrammed, but the fuel had heated up to such a point that one element melted, along with its cladding, releasing radiation into the cavern. The fuel channels in the moderator tank collapsed and a large amount of heavy water leaked out. Radiation levels reached several hundred rem per hour in the cavern, which was automatically sealed off, preventing any radiation from being released into the atmosphere. The reactor was severely damaged and there were so many decontamination problems that, in the early 1970s, it was decided to use the cavern as a storage place for nuclear waste. (**Bertini**; **Patterson**)

1969 April At a US nuclear reactor, unidentified by the NRC, samples of water taken from taps at a laboratory sink were found to contain radiation levels above background levels. Further checks revealed radioactivity in one of the plant's drinking fountains. Investigators discovered the cause: a hose had been connected from a drinking-water tap to a 3,000-gal radioactive waste tank. The AEC commented: 'The coupling of a contaminated system with a potable water system is considered poor practice in general...' (Pollard, Nugget File, UCS, 1979)

1969 October 17 At Saint Laurent des Eaux in France, while the unit 1 gas-cooled, graphite-moderated reactor was running at full power, a handling error led to a coolant-flow-restrictor rod being placed in the reactor core instead of a fuel rod. This caused the temperature in part of the core to rise, fusing five fuel elements together, and 50 kg of uranium were dispersed through the reactor core, producing high levels of radioactive contamination. The reactor was out of action for a year for repairs and clean-up. (**Bertini**; Le Dossier Electronucléaire, Syndicat CFDT de L'Energie Atomique, Sciences, 1980)

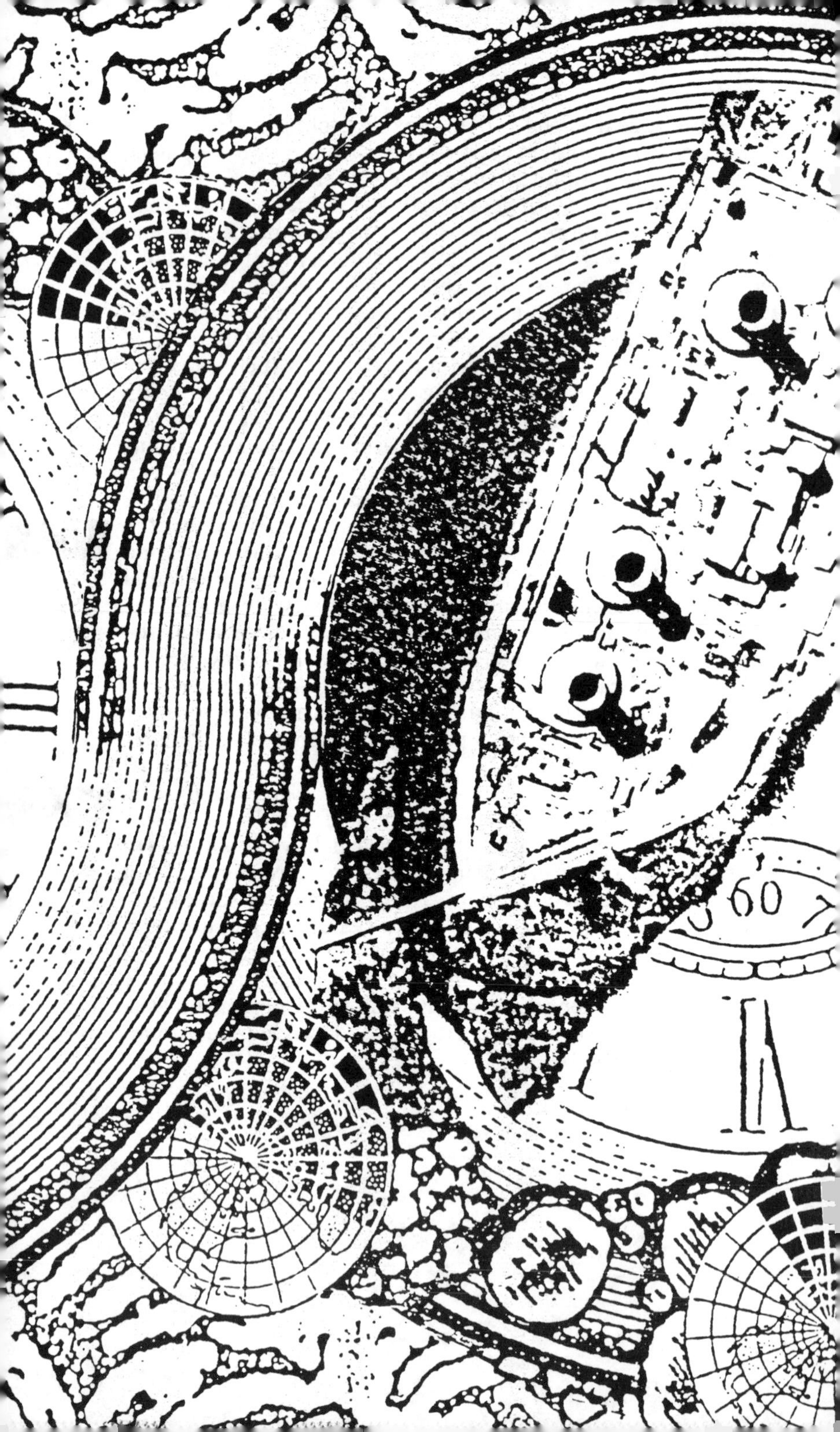

70S

1970-1979

French Nuclear Tests, Part 2

In 1969 De Gaulle resigned after a referendum did not go his way, and Georges Pompidou became the new French president. The test programme that year was cancelled due to a lack of funds so an accelerated test programme of eight detonations was carried out in 1970. This was followed by five tests in 1971 and a further three the following year.

By now the French were beginning to feel the pressure of world opinion. They claimed their annual reports to the UN Scientific Committee on the Effects of Radiation had proved the tests to be 'innocuous'. Yet at the UN Conference on the Human Environment held in Stockholm in June 1972, the French nuclear tests in the Pacific were condemned, the resolution being, according to the Danielssons (1986), 'motivated precisely by the unsatisfactory character of the French UN reports'. This condemnation by the world community was confirmed at a UN debate on 29 November that year when 106 nations voted against the testing.

On 10 June 1972 the new Minister for Defence, Michel Debré, said: 'Our tests have in no way been harmful to the environment...The real reasons behind the criticism levelled at France are political...No scientific proof whatsoever exists to back up the accusations made against the French government.' This remained the official line, but that year Pompidou had ordered the army to find a suitable location for underground testing in the Pacific.

The five tests in 1973 took place against a backdrop of renewed protest. The New Zealand and Australian governments brought a joint case against France in the International Court of Justice at the Hague, requesting an injunction against the tests. There were mass demonstrations in Europe and peace flotillas invaded the test zone. In November 1973 drilling began on the atolls of Moruroa and Fangataufa in preparation for the underground tests, and the end of atmospheric testing was confirmed by the French to the UN General Assembly on 24 September 1974.

In May 1975, Haroun Tazieff, the French vulcanologist, was sent to Fangataufa at the request of the CEA and the Centre d'Expérimentation du Pacifique (CEP). He reported that 'certain rocks are so porous and...lacking in shock resistance that there is a risk of radioactive leakage...As the tests will be made in spite of all this, the best solution is to request an impartial scientific organisation to supervise them.' (Greenpeace Report, 1981)

On 5 June the Australian government announced that its seismological observatories had registered an underground explosion. The French government confirmed that they had exploded an eight-kt bomb in the core of Fangataufa, at a depth of 623 m.

When the tests moved underground, there was a decline in the scale of the international protest against them. In a public relations exercise called Operation Open Atoll, the French invited 60 journalists to Fangataufa. A military doctor, head of the radiological security service on the atoll, claimed that he had scientifically calculated that it would take 1,500 years for the encapsulated radioactivity to seep into the ocean, by which time it would be harmless. A further 5 to 10-kt test on Fangataufa on 26 November was the last staged until 1988.

On Moruroa, several underground tests were conducted in 1976; on 3 April, 11 July, 23 July and 8 December. The second test in the series resulted in radioactive gas which 'did not escape along the predicted path, and the technicians are still trying to figure out what happened to it,' according to *Le Journal de Tahiti* (27.12.76).

As the testing continued, now shrouded in more secrecy than ever, there were mounting rumours about people's health suffering as a result of the tests. In April 1978 an Auckland newspaper (*8 O'Clock*) reported that French Polynesian people were filtering into New Zealand for cancer and leukaemia treatment. In October that year, Oscar Temaru, the leader of a Polynesian independence party which opposes nuclear testing in the region, claimed 15 Tahitians were being secretly kept in isolation on Moruroa, suffering from radiation sickness.

On 6 July 1979, disaster struck. There was an explosion and fire in one of the concrete bunkers on the surface of the atoll, which are used for detonation experiments designed to study shockwaves. (The Natural Resources Defense Council surmise that these studies investigate the implosion of the chemical high explosive that surrounds the fissile material. They result each time in the release of plutonium — the bunkers are usually abandoned and sealed after each test.) For reasons of economy it had been decided, as an experiment, to try and decontaminate the bunker that had been most recently used. The inside walls were plastered with paper soaked in acetone and these fumes filled the chamber.

When a spark from a boring machine ignited the gas, one man in the decontamination team was killed instantly by the intense heat; another had his chest crushed by a flying door; and four others were badly burnt and were sent to Paris for treatment. This non-nuclear explosion nevertheless scattered plutonium from the blast over the whole atoll and a team of 40 were employed in an attempt to decontaminate it.

Less than three weeks later, on 25 July, a nuclear device was exploded half-way down an 800 m shaft after becoming stuck there. The massive 120 kt explosion that resulted was the largest and most widely recorded explosion between July 1976 and December 1981. It registered 6.3 on the Richter scale. Three hours later an enormous chunk of the atoll's outer wall — equalling one million cu m of coral and rock — fell into the ocean, producing a two to three m tidal wave which spread through the islands, injuring six people on Moruroa's southern rim. The CEA denied any connection between the test and the tidal wave, claiming it was of natural origin.

Pressure from the Tahitian Assembly led France to agree to a more thorough investigation of these events. A hand-picked group of scientific experts — all French — were flown to Moruroa with five Tahitian Assembly members and two government councillors. They spent less than 24 hours there, and were unable to make any detailed observations, take any measurements or visit the test sites, the only glimpse of them being on a helicopter tour of the atoll.

18 December 1970

US Nuclear Tests, Part 4

The Nevada Test Site (NTS) sits 65 miles north-west of Las Vegas, a vast 1,350-sq-mile site with 700 miles of unmarked roads, guarded by armed mobile patrols. Nine thousand people work there — mostly construction workers preparing fresh shafts for upcoming nuclear tests called 'shots' or 'bogies' — and it costs $2 million a day to run.

Between January 1951 — when testing began — and the end of 1988 there have been over 800 known US and British tests here. Test site officials have only recently admitted that in the 21 years and 475 explosions since testing went underground in Nevada, there have been 62 accidents involving a release of radiation. Of these, 53 are classified by the DOE as 'leaks' and 'seeps' — a gradual escape of radiation that usually, but not always, stays on site — and nine as 'venting', which the DOE defines as a 'massive release of radiation'.

PLANNING FOR A 'SHOT'

At the Nevada Test Site, planning for a 'shot' takes about 14 months. A spot is chosen and a 3- to 12-ft-wide hole is sunk from 600 to 5,000 ft deep. The tests are detonated between 1,000 and 2,000 ft down. A six-storey 'bogey tower' is moved into place and the bomb is lowered to the bottom of the shaft. A 200-tonne diagnostic canister, like a miniature physics lab, is lowered on top of the bomb; from this, some 140 leads run to the 'red shack' (an advance post in a trailer close to ground zero), and on to the 'war room' which is the control centre. The shaft is then sealed with sand and rock.

On the day of the test, a small number of scientists accompanied by security guards drive out to the 'red shack' to arm electronically the device or warhead. Two of the scientists carry a special briefcase that contains a bag of tiny cubes with numbers painted on the sides. They pick cubes out alternately and punch the numbers into an 'arm enable' device in the briefcase, generating a random code that is sent to the warhead on a special electrical cable. The scientists return to the control point and send the same code to the warhead to arm it. The explosion creates a cavity underground, which collapses to generate a slump crater on the surface. The millions of tons of surrounding earth are supposed to contain the force of the explosion and absorb the radiation.

Venting occurs in an underground test when the bomb produces more explosive power than expected and part of the covering earth is blown away. This creates a kind of chimney out of which radioactive debris is spewed. The bomb's expanding mass seeks out the lower pressure of the atmosphere and carries radiation with it. Ventings have happened up to two days after detonation, when a shift in the earth allows pent-up pressure to escape.

The worst of these was the test codenamed Baneberry. The detonation of this 10-kt bomb, on 18 December 1970, blew a radioactive plume of three million curies more than 8,000 ft into the atmosphere over a 24-hour period. It created a large hole in the desert floor and a cloud of radioactive debris that was tracked by the USAF as far as North Dakota. The AEC claimed the cloud dissipated quickly and only 'minor levels' of radioactivity were detected off-site.

In 1974, the widows of two test site workers who had been exposed to radiation from Baneberry and subsequently died of leukaemia, filed a legal suit under the Federal Tort Claims Act. The trial was held in early 1979, the judgements following in 1982. Judge Roger Foley stated the government was negligent in its evacuation and decontamination procedures but that the radiation exposures were not sufficient to cause leukaemia.

As retired Air Force Colonel Raymond E. Brim (from 1966 to 1975, ranking Pentagon officer at the Air Force Technical Applications Center, which monitors fallout from the site and from foreign testing of nuclear weapons) commented (*Washington Monthly*, 7.1.81): 'The monitoring of vented radiation tells us much about the government's attitude towards nuclear risks. Whenever an underground test takes place, the Air Force, Public Health Service and some privately contracted planes are standing by to travel alongside the escaped fall-out cloud and track its movements. The existence of the planes is proof enough that the government expects accidents.'

A selection of some of the most significant venting incidents appeared in the *Arizona Daily Star,* (17.11.81):

Eagle (12.12.63) Gaseous radioactive iodine was tracked about 140 miles to central Southern California.

Red Hot (5.3.66) One million curies seeped into the atmosphere over several hours. The radiation was tracked 2,000 miles to eastern Iowa.

Pin Stripe (25.4.66) A 350,000-curie release was tracked to eastern Kansas and Nebraska, where iodine was found in milk and human thyroids.

Umber (29.6.67) A radioactive release dropped fresh iodine in Death Valley and Shoshone, California.

Boxcar (April 1968) The largest 'announced' underground test. A 1.2-Mt yield, equal to 1.2 million tons of TNT.

Schooner (8.12.68) The sixth test in the Ploughshare programme, which was designed to demonstrate peacetime uses for nuclear warheads. Most Ploughshare blasts were called 'cratering events'. The nuclear device was buried only 200 ft down so that, when it exploded, it would dig a wide crater in the earth. Thus its fallout was automatically vented to the sky. Radiation from the Schooner 35-kt blast covered most of the east coast of the USA and registered on radiation monitors in Toronto and Montreal.

MIDAS MYTH AND MIGHTY OAK

Weapons-effects tests — designed to assess what is officially called 'the survivability of US military systems in a nuclear environment' and run by the DNA — are conducted in tunnels bored horizontally into the 11-sq-mile Rainier Mesa, a flat — topped, 200-m-high mountain of hardened volcanic ash on the test site. There have been more than 50 of these tests since 1957.

The bomb is placed at one end of the horizontal tunnel in the zero room which is connected to a 1,000-ft-long steel pipe, containing several test chambers into which components and materials are placed. The pipe is tapered; the zero-room end may be only a few inches in diameter, while the other end may measure as much as 27 ft. The pipe can be evacuated to simulate conditions in space. Sensors to measure radiation and seismic effects are installed both at the tunnel entrance and in a small shaft drilled from the surface of the mesa down to the bomb.

The underground explosion creates a hot ball of gas that vaporises the rock as it expands to create a cavern. Careful calculations must be made to ensure there is enough overlying rock above the cavern, otherwise vertical fracturing will reach the surface. This is exactly what happened on 15 February 1984 — three hours after a test codenamed Midas Myth. Scientists in four or five trailers on the surface were busily disconnecting cables when the ground caved in, forming a shallow depression 20-40 m wide and up to 10 m deep. Men and equipment tumbled into the crater. 12 workers were injured; one of them, Charles Miesch Jr (59), died from his injuries a month later.

Another accident occurred on 10 April 1986 when the $70 million Mighty Oak test destroyed military monitoring equipment worth $15 million and led to an increase in radiation levels in the surrounding area. The steel pipe contains three doors designed to close immediately after detonation, which allow a controlled release of radiation to reach the monitoring and military equipment. In this case, two of the three doors failed to close following detonation, and some radiation was released. Traces of xenon were found 80 km from the test site.

More than a month after the test, monitors in the tunnel were still registering about 25 rad per hour. Two of the 60 workmen employed in clearing the tunnel accidentally absorbed 200 millirem and 70 millirem respectively of radioactive iodine-131 in their thyroids.

1973-1979

Windscale, Sellafield, UK, Part 2

During the 1970s, the Windscale reprocessing plant at Sellafield continued to be plagued by a series of incidents and accidents, starting in 1973 with a major setback to plans to increase the facility's capacity to handle spent fuel from overseas.

The B205 reprocessing plant for Magnox fuel had gone into operation in the mid-1960s, and was soon found to have excess capacity. The plant's operators foresaw a prosperous future in the reprocessing of spent uranium oxide fuel from the new generation of AGRs and light-water reactors, but B205 was not designed to cope with this material in the form in which it arrived. A disused neighbouring building, B204, was therefore proposed as a treatment plant in which the spent uranium oxide fuel would be dissolved in Butex, an organic solvent, in preparation for reprocessing in B205. After extensive decontamination and rebuilding, this 'Head End Plant' went into operation in August 1969 and ran until mid-1972, when building B205 was shut down for one year for repairs.

When the Head End Plant started up again, on 26 September 1973, an unpleasant surprise was in store. Unknown to the operators, previous batches of dissolved fuel had been depositing granules of insoluble radioactive fission products in the process vessel. The heat produced by these granules had boiled away the last of the liquid and had heated up the floor of the vessel.

When the new batch of solvent was poured in, it reacted violently on the hot surface and produced a steam explosion that sent a burst of radioactive gas out into the air of the plant. Alarms sounded, but a report on the incident by the Nuclear Installations Inspectorate (NII) later revealed that these alarms frequently went off for no reason and staff initially ignored them. (The NII was set up in 1960 as a result of the 1957 Windscale fire.)

When senior staff finally realised that there was radioactivity in the 10-storey building, the general order to evacuate was given by personnel running from floor to floor shouting at everyone to get out. During this time, the air was potentially lethal, but since no record was kept of who was working where, it took half an hour to find and evacuate the last of the workers. In all, 35 members of staff were found to have skin and lung contamination, mostly from ruthenium-106. Inside the building, levels of contamination were found up to 100 times the maximum permitted level.

For two years British Nuclear Fuels Ltd (BNFL) continued to claim that all was well and that the plant would soon be back in action, but after four years they had to concede that it could not be used again. The plant that had been designed to handle 300 tonnes of fuel a year only ever processed a total of 100 tonnes.

During exacavation work in 1975, high levels of radiation were detected in the soil — so high that the driver of the mechanical digger had to work in a lead-lined cab. It was finally realised, on 10 October 1976, that radioactive water was leaking from the nearby silo B38, in which Magnox fuel cladding was stored. An NII inquiry, published in 1980, found that the leak may have started as far back as 1972 and had resulted in 50,000 curies of radioactivity escaping.

One of those involved in the excavation work was 21-year-old David Berry, an engineering student on industrial training. 15 months later he died from a type of lymphatic cancer. At the inquest, no one mentioned the excavation work, and an open verdict was returned. BNFL did not even inform the government of the silo leak until a month after his death.

The B38 leak had come at a sensitive time for BNFL. Public outcry at proposals for the expansion of reprocessing facilities and the building of a new Thermal Oxide Reprocessing Plant (THORP) had led to pressure for a public inquiry into the plans, and now Windscale was on the front pages again. In March 1977, the UK Environment Secretary announced the setting up of the public Windscale Inquiry which, under the chairmanship of Mr Justice Parker, sat from 14 June for 100 days.

In the course of the inquiry, during which evidence was given under oath, 194 significant events were revealed to have occurred at the Windscale plant up until 1977. It is known that this list is not exhaustive, but it is more realistic than the 27 incidents that had previously been made public.

The inquiry, which was later accused of ignoring all arguments put forward by objectors, found in favour of the proposed reprocessing expansion, although the government bowed to pressure for a parliamentary debate before a decision was taken. The debate ended with a vote in favour of THORP, but the vote against was the largest ever recorded in opposition to a nuclear decision in Britain.

In 1979, in the course of sinking bore holes to determine how far the radioactivity from B38 had spread, even higher levels of radiation were detected, so high that they could not have come from that silo. Workers digging the bore hole were undoubtedly contaminated. The radiation was initially ascribed to a spillage some 20 years earlier, but further tests revealed that the leak was recent and continuing. It took almost a year to discover the new source.

Despite a 'comprehensive' survey of safety at the Windscale plant in 1976, a tank in the disused building B701 had filled up with high-level radioactive waste accidentally diverted to it in the complex system of pipes. Unnoticed over the years, the liquid had overflowed the tank and filled the sump below. The needle of the sump gauge — which appeared to give a normal reading — was in fact on its second circuit round the dial.

By February 1979, the highly radioactive liquid had been leaking from the sump, through the side walls of the building and into the ground, for at least three years. This major leak was the subject of a damning report by the Health and Safety Executive, who considered that poor management and maintainance were to blame. An NII report in August 1980, which put the leak at 'rather more than 100,000 curies', led to several recommendations, as well as the observation that BNFL had violated several conditions set out in the Sellafield site licence. However, no action was taken against them.

13 November 1974

Karen Silkwood

In the early evening of 13 November 1974, Karen Silkwood was driving in her Honda Civic down a lonely stretch of Highway 74, on her way to the Northwest Holiday Inn near Oklahoma City. Here she was due to meet Steve Wodka, an occupational health expert from the Oil, Chemical and Atomic Workers International Union (OCAW), David Burnham of the *New York Times,* and her boyfriend, Drew Stephens. She was to hand Wodka and Burnham important documents that would prove that defective fuel rods were being made at the Kerr-McGee Company's plutonium production plant at Cimarron, 30 miles from the city. Karen Silkwood never reached her destination. Her car was found in a concrete culvert and she was dead behind the wheel.

The truck driver who first discovered the wreck noticed some papers in the car. Later that night AEC inspectors and Kerr-McGee employees checked her car for radiation and were searching for Silkwood's papers, according to the owner of the garage where the car was taken. But when Stephens went to retrieve the car from the garage the next day, the papers had disappeared. Rick Fagen, the Oklahoma highway patrolman in charge of the case, told Kerr-McGee official Ray King that 'the documents have been taken care of.'

This sinister incident prompted OCAW to hire an expert accident investigator. He discovered a dent in the rear bumper of the car and further evidence that Silkwood was conscious at the time of the crash — the implication being that another vehicle had forced her car off the road, causing her death.

Karen, born in Longview, Texas, in 1946, had come to Oklahoma City in 1974 to make a fresh start after her seven-year marriage had ended in divorce. A major at junior college, she got a job as a laboratory analyst at the Cimarron plant owned by Kerr-McGee. Here plutonium was fashioned into pellets which were then placed in fuel rods, to be used for fast breeder reactors.

The Kerr-McGee Corporation's fortunes were founded first in oil, then in coal and finally in uranium, which the company bought into in 1952, after receiving inside information that there was a bright future for civil nuclear energy. Robert Kerr, the founder, was a poor boy who became an oil baron. He was a former Governor of Oklahoma, who lost his chance to run for president at the Chicago Democratic Convention in 1952, but became leader of the Senate in 1960, a position of enormous power. After Kerr's death in 1963, control of the company passed to his partner, Dean McGee, one of the top geologists in the oil business, who was to build the company into an energy conglomerate. By 1970 it owned one quarter of all known uranium reserves in the US and was the biggest producer in the country. This was the empire that Karen Silkwood joined.

Karen initially became involved with OCAW when her local branch went on strike for higher wages, better training and improved health and safety programmes. The strike failed but Karen became increasingly concerned about procedures at the plant, and was elected to the local committee of OCAW. Her assignment was health and safety. On 26 September 1974 she went to OCAW's Washington headquarters to report that the plant was 'sloppy, dishonest and unsafe'. It was here, for the first time, that she discovered that plutonium was carcinogenic. Wodka persuaded her to gather more documented proof, particularly about her claims that Kerr-McGee made false reports of plant inspection and had doctored X-rays of nuclear fuel rods so that faulty welding did not show up. For the next six weeks she played the double role of technician and clandestine investigator. Then events took a disturbing turn.

On 5 November 1974 Karen went to work as normal but, at the end of her shift, she was found to be contaminated. Contamination is measured in disintegrations per minute (d/m). The AEC safe limit was set at 500 d/m; Silkwood, fully clothed, was registering 20,000 d/m. She was given blood, saliva, urine, faeces and nasal tests, and she then showered to remove the skin contamination. Kerr-McGee never discovered the cause of this contamination.

The next day she discovered she was contaminated again on her right forearm, neck and face. She had only been at the plant for one hour. The following day, when she reported for work, she was checked again and this time she was even more seriously contaminated; levels on and around her nose measured 40,000 d/m.

Kerr-McGee personnel went to her home and found that it was heavily contaminated, especially foodstuffs in the refrigerator. The company's lawyers were later to claim that she had contaminated herself in order to dramatise her campaign against the company.

On 11 November, Silkwood was flown to Los Alamos Scientific Laboratory for detailed medical tests. These tests incorrectly concluded that she had only eight nanocuries of plutonium in her lungs; at the time, this was half the AEC maximum permissible lung burden for workers. Karen was reassured. Later tests would show that she had, in fact, twice the permissible AEC limit leading one doctor to testify that Karen was 'wedded to cancer'.

She went back to work on the Monday and agreed to get the evidence for the *New York Times* ready for Wednesday. Leaving the plant that night, she had a dark brown folder and a reddish-brown spiral notebook. They were never seen again. By the end of that day, Karen Silkwood was dead.

By the spring of 1975 it seemed everyone had given up trying to solve the Silkwood mystery. The Oklahoma Highway Patrol had listed her death as accidental and the FBI had closed its investigation of the case.

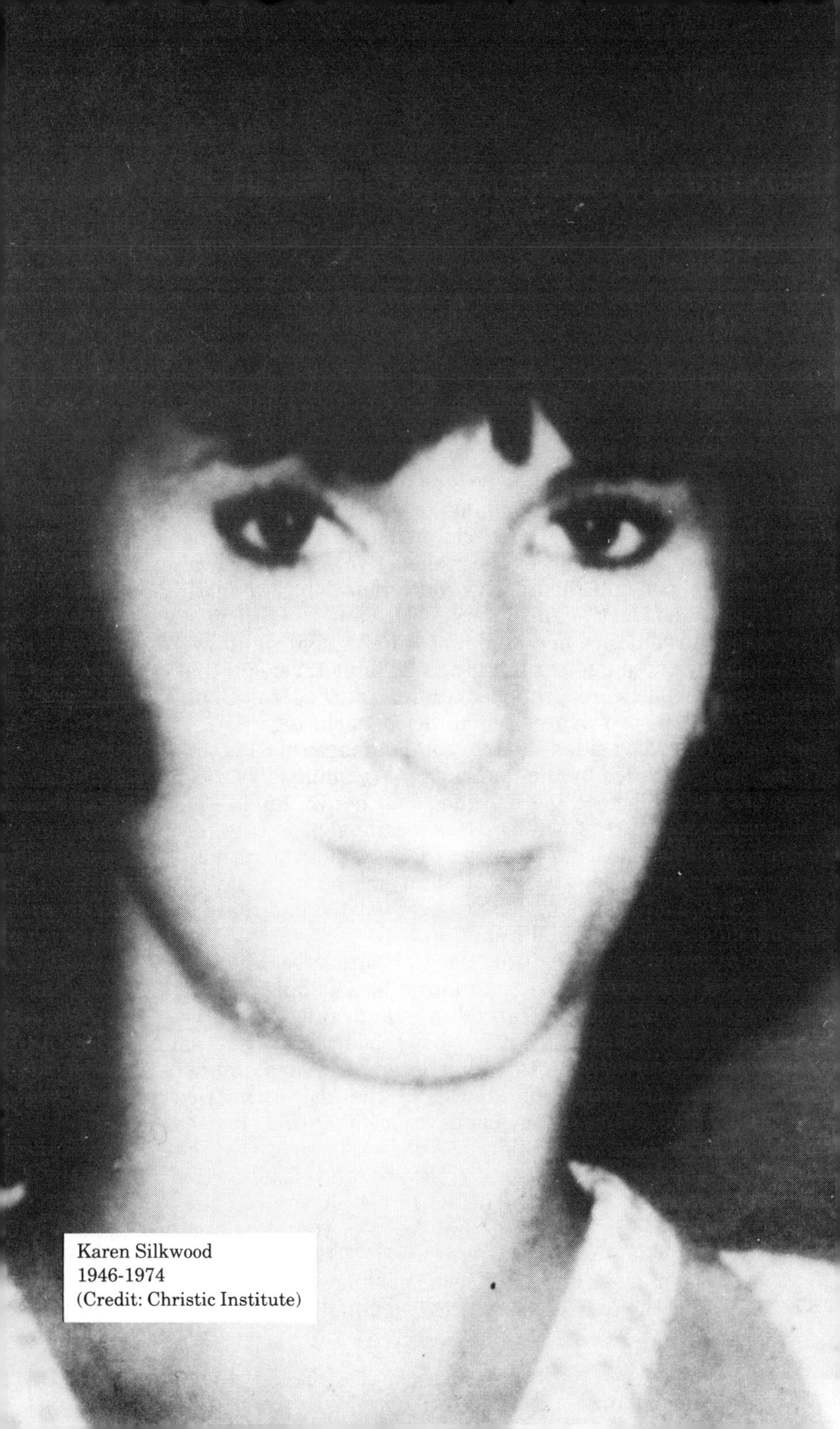

Karen Silkwood
1946-1974
(Credit: Christic Institute)

Then Kitty Tucker, the Washington legislative co-ordinator of the National Organisation for Women (NOW), became interested in the case after reading press reports, and became determined to begin the fight for justice on Silkwood's behalf. She and Sara Nelson, the chair of NOW's Labour Task Force, founded a pressure group, Supporters of Silkwood (SOS), and began lobbying the Justice Department and the US Congress to try and gain support for a proper investigation into the case. It was they who first introduced civil rights attorney Danny Sheehan to Karen's parents, and, in November 1976, Sheehan filed a civil suit, just five hours before the Statute of Limitations on the case was due to run out.

The legal manoeuvrings went on for two and a half years; the case finally coming to court in March 1979. The suit against Kerr-McGee claimed that the company, being the legal custodian of the radioactive materials, was ultimately responsible for Karen's contamination. Two further counts of conspiracy against Kerr-McGee and the FBI involving illegal surveillance and denial of Karen's civil rights were set aside by the judge, on the grounds, later reversed, that only black people were protected under the Federal Civil Rights Act against private conspiracies violating civil rights. The Silkwood side asked for total damages of $11.5 million.

Headed by the flamboyant Wyoming lawyer Gerry Spence, the Silkwood team argued that her contamination and death were part of a Kerr-McGee campaign to frighten her off the investigation, even if murder was not the ultimate intention. Kerr-McGee's case was that she had contaminated herself, that her missing folder never existed and that her death was an act of suicidal frenzy.

Spence introduced several witnesses who described foot-deep leakages of plutonium-laced liquid at Cimarron, improper storage procedures and the emphasis placed by company management on productivity over safety. Kerr-McGee's lawyer talked in detail about 'maximum permissible doses' of radiation allowed by the AEC, then tried to portray Karen as a promiscuous drug-taker who had deserted her husband and children. Judge Theis ruled, however, that 'human failings' were irrelevant, and excluded this evidence.

Later, Allen Valentine, former head of health physics at Cimarron, admitted that he was the only certified health physics officer overseeing Kerr-McGee's 1,400 employees in the nuclear division. He also acknowledged that he had written the company's safety manual but had intentionally edited out all specific references to the 'cancer-causing' properties of even tiny amounts of plutonium.

Pre-trial testimony disclosed that the company did not report more than 100 contamination incidents at the plant and could not account for up to 40 lb of missing plutonium.

It soon became clear that Spence's relaxed, informal style was making a better impression on the jury than Kerr-McGee's reliance on official technical facts and figures. Then, with uncanny timing, several events occurred that brought home the risks of nuclear power: the NRC closed 15 east-coast nuclear plants for safety checks; the film *The China Syndrome* opened in Oklahoma City and was attended by both counsel; and, in Pennsylvania, the Three Mile Island nuclear station at Harrisburg came to the brink of a major nuclear disaster. In addition, on the very day the Cimarron plant' designer was testifying to the plant's ability to withstand any natural disaster, Texas and Oklahoma were swept by the most severe tornado in 50 years.

The trial ended on 18 May 1979. The federal jury found against Kerr-McGee and awarded damages of $10.5 million, almost the full claim, to the Silkwood estate. Kerr-McGee immediately announced that it would appeal. A federal appeals court overturned the ruling on the grounds that the award of damages to Silkwood under state laws constituted 'regulation' of the nuclear industry; a power reserved to the federal government. This decision was reversed by the Supreme Court in 1984, which upheld the trial court's legal authority to impose punitive damages — despite the nuclear plant's compliance with the AEC's inadequate safety standards. On 22 August 1986, Karen Silkwood's family agreed to a $1.38 million settlement of their lawsuit.

Silkwood v. Kerr-McGee was the first time a jury awarded damages to a victim of radiation contamination from a nuclear facility. The case established several precedents:

- Courts can now award heavy punitive damages under state law even though the nuclear industry is regulated by the NRC, a federal agency.
- The nuclear industry cannot evade damages by arguing, as Kerr-McGee did, that radiation leaks were within acceptable limits defined by the federal government.
- The production and handling of radioactive materials is an inherently dangerous activity subjecting the industry to 'strict liability'. If radiation causes damage or injury, the company is liable, whether or not it has been negligent.

1974-1975
Project Jennifer, North Pacific

The existence of a complex and secret US mission, called Project Jennifer, was not publicly discovered until February 1975, when it finally became front-page news. By then it had become the most expensive espionage mission ever, absorbing more than $550 million of the US Navy's funds, employing more than 4,000 people and bringing Howard Hughes and the CIA into partnership.

This mission, the most secretive since the Manhattan Project, actually began on 11 April 1968, one month before the loss of the USS *Scorpion*, and five years, almost to the day, after the sinking of the USS *Thresher*. US surveillance satellites had monitored the departure of a USSR submarine from Vladivostok, and the vessel was then tracked by the Sea Spider acoustic detection system — a 650-mile-radius circle of hydrophones on the Pacific bed which relay their signals back to naval intelligence at Pearl Harbour. These identified it as a diesel-driven Golf-class submarine, which was known to have three Serb-class nuclear-armed missiles on board.

Somewhere in the Pacific, the submarine's crew prepared the craft to surface, in order to recharge its batteries and to expel the stale air and accumulated gases in the ventilation system. As the ballast tanks were being vented, there was an explosion, probably caused by a spark igniting hydrogen from the batteries, and the craft rapidly sank to the bottom of the ocean before any of the 86 crew members could escape. The sonar arrays of the Sea Spider system, which recorded the implosion of the submarine's hull as it sank, pinpointed the wreckage to within a 10-mile area, about 750 miles north-west of Hawaii, in water three miles deep.

Intelligence agencies watched as a Soviet task force spent months searching for traces of the missing vessel but with no success. When the CIA realised only they knew where the submarine was to be found, they reported the matter to the Forty Committee, a secret panel chaired by Henry Kissinger that reviewed all intelligence operations, and it was agreed that the CIA should try to raise the Soviet vessel.

A telling argument in favour of the project was that, if they succeeded, they would be able to tell whether the Soviets were capable of upgrading the nuclear missiles on these older-generation vessels. This information would be critical to the forthcoming Strategic Arms Limitation Treaty talks between the two superpowers. The reverse side of this coin was that Kissinger's parallel policy of *détente* could be irrevocably damaged if the Soviets discovered the plan.

The navy made the first move, sending their secret research ship *Mizar*, packed with sonar, scanners and cameras, to locate the sunken submarine precisely. Once this had been achieved, the CIA turned to Howard Hughes with their plan for the salvage operation.

Hughes, who had long been a major defence contractor, had an interest in deep-sea mining (the plausible disguise for the Jennifer Project), and a passion for secrey. He was also head of a number of companies renowned for their technical excellence, and this was a challenge that would push technology to its limits.

Hughes became fascinated with the idea, as he had previously been with the building of the *Spruce Goose* (the largest aeroplane ever flown). He decided that two vessels were needed for the project.

The main ship was the *Glomar Explorer*. She was 618 ft long, and 115 ft wide; weighed 36,000 tonnes; and was the most advanced deep-sea mining ship ever built. Amidships was a 209-ft derrick capable of raising 800 tonnes. Inside the ship was the Moon Pool, a vast docking well that could be opened to the sea below by means of huge sliding panels in the ship's flat-bottomed hull. When these were opened, the pool half-filled with sea-water.

The other vessel, the HMB-1, was a 160-ft-wide submersible barge that looked like a floating aircraft hangar. This became the secret factory for the construction of the retrieval vehicle — a 180-ft mechanical grab. The barge would be able to submerge and transfer the grab to the *Explorer* without revealing the tell-tale equipment to prying eyes. Once inside the *Explorer*, the grab could be lowered to the submarine on a string of pipes carrying hydraulic fluid to operate it. Once the submarine was raised, it would be dismantled and examined, hidden inside the ship.

The construction of the two vessels and the complex lifting equipment took almost five years. With the grab on board, the *Glomar Explorer* finally set sail on 20 June 1974 and by mid-July had reached the salvage site. There she was able to remain stationary exactly above the sunken submarine by means of a 'dynamic positioning system', an onboard computer linked to huge thrusters located at the bow and the stern, facing to port and starboard. The floor of the ship's Moon Pool was opened, the huge mechanical grab was attached to the steel piping and then slowly lowered to the submarine 17,000 ft below.

Guided by remote control and powered by its own propellers, the grab was positioned precisely and the claws closed around the 320-ft submarine. The 4,000-tonne deadweight was slowly raised from the sea-bed at the rate of six ft per minute, but, half-way to the surface, disaster struck. Three of the grab's five arms broke away and the submarine cracked in two. Video cameras attached to the grab showed the most valuable section, the one containing the missiles and the code room, falling back to the sea-bed. The grab retained only the 38-ft nose section, and this was successfully salvaged and deposited inside the *Glomar Explorer.*

Aware that the bodies of the Soviet crewmen would be recovered along with the wreck, the *Explorer* was equipped with special cooling facilities that could accommodate up to 100 corpses. Only six bodies were recovered, and these were given a funeral service in Russian and English — the ceremony being filmed by the CIA — and then reburied at sea. Amongst the surprisingly well-preserved bodies and the deep-sea crabs that fed on them, the salvage crew found a soggy handwritten journal. This was rushed to the ship's lab for preservation and was soon picked up by helicopter and taken to Washington for analysis.

The salvage crew wore radiation suits, because two torpedoes with nuclear warheads formed part of the crumpled bow, and the metal was giving off a high level of radioactivity. Within a few days, however, they were advised they no longer needed their bulky suits; later contamination checks suggested that many may have received a dangerous dose of radiation as a result.

Throughout the mission, secrecy was of paramount importance, and the CIA spent a great deal of money and effort on hiding the true purpose of the *Glomar Explorer.* Two serious threats to this clandestine mission had to be dealt with.

In October 1973, a trade union took Global Marine, who supervised the *Explorer's* construction, to court for dismissing men after a labour dispute. Senior officers of the company were prevented from testifying in order that they would not be compelled under oath to reveal the ship's true mission.

Then, after six years of the successful cover-up and just two weeks before the *Explorer* sailed, fate took a hand. Four armed men broke into Hughes's headquarters in Hollywood and escaped with some rare vases, $68,000 in cash and two cabinets of secret documents, including a confidential, handwritten memo from Hughes himself concerning Project Jennifer. Within weeks, blackmailers demanded $1 million for the return of the papers and Hughes's company had to inform the CIA. They in turn enlisted the help of the FBI, who met with the Los Angeles police chief.

In an absurd attempt to keep the contents of the documents secret, the police were briefed on what to look for but were instructed not to read the documents if they found them. By now, so many people knew — if only in general terms — about the secret project that a leak was inevitable. In the first week of February 1975, the first story broke in the *Los Angeles Times,* telling of a joint CIA/Hughes plot to raise a Soviet sub — but in the *Atlantic* — and linking this to the burglary.

Then investigative journalist Seymour Hersh, who had been piecing together details of the story since 1973, published his own findings in the *New York Times* shortly afterwards. CIA director William Colby was successful in stifling much of the press coverage but by March, Hersh refused to be stifled and the full story came out. Any plans the CIA may have had to try to retrieve the rest of the submarine were now hopelessly compromised. (Ironically, the burglars at the Hughes HQ had dropped two documents at the time of the raid, and these had been picked up by the security guard. One of them was the Hughes handwritten memo, and its contents were so obviously sensitive that the guard, in a panic, had flushed it down the toilet. The burglars' blackmail demands had been no threat to Project Jennifer and the project's cover need never have been blown.)

By now the Soviets had stationed a tugboat over the wreck site, and the CIA had lost their chance. They had learnt some new information about Soviet torpedoes and about the primitive quality of Soviet welding techniques. But the secret journal proved impossible to crack without the correct naval code. Project Jennifer had proved an expensive failure...or at least that is what the CIA would have us think.

In their book *The Ties That Bind* (1985), Professor Jeffrey Richelson and Desmond Ball claim that the failure story was the product of 'considerable misinformation'. According to them: 'Large sections of the submarine were recovered, including the crushed and battered centre segment containing the three SS-N-5 missiles.' Also saved were two nuclear-armed torpedoes, radio equipment, the submarine's navigation system, the code machine and associated code books.

22 March 1975

Browns Ferry Fire, Alabama

Beneath the control room of Units 1 and 2 at the Tennessee Valley Authority's (TVA) Browns Ferry nuclear power plant near Decatur, Alabama, three two-man teams were working in the cable-spreading room. All the electrical control and supply cables for the Unit 1 reactor are laid out in this room and pass through the wall into the reactor building. Both 1,000-MW reactors were running at full power.

Air pressure in the reactor building is kept slightly below atmospheric pressure so that any leaks in the building's walls will suck air inwards and prevent the escape of any radiation. The electricians were testing for leaks in the wall at the points where the cables, lying in trays, pass through into the reactor building, and they were using the movement in the flame of a lighted candle to locate points where air was being sucked in. One of the men discovered an air current in a corner of the room, and stuffed the hole with polyurethane sheeting. He then moved the candle close to the hole to see whether air was still being drawn in. It was, and the flame ignited the polyurethane.

The current of air quickly drew the fire into the wall and ignited the polyurethane foam surrounding the cable trays. Attempts to put the fire out using carbon dioxide and dry-chemical extinguishers failed, and the fire spread through into the reactor building. Attempts to activate the permanent fire-extinguishing system also failed, because the power had been turned off during the leak tests.

It was now 12:34, and the fire alarm was finally activated: but, unable to use water for fear of further jeopardising the electrical circuitry controlling the reactor, and hampered by the failure of both the lighting and ventilation systems, firefighters' efforts were ineffective until 19:00, when the plant superintendent reluctantly sanctioned the use of water. The fire was out within 45 minutes.

It had caused chaos in the Unit 1 reactor, starting up pumps and other equipment automatically as the electrical circuitry burnt through, and finally closing down every means of providing high-pressure cooling water to the core. Although the reactor scrammed using the control rods, the decay heat still had to be removed. The water covering the fuel began to boil away — but the only pumps still available could supply only low-pressure water, and the pressure in the reactor vessel was too high for them to operate. When the pressure relief valves were opened manually, the water level fell still further, to just 1.2 m above the top of the fuel (the normal level is 5.08 m).

However, with the pressure reduced, the low-pressure pumps were just sufficient to provide adequate cooling to the core until 18:00, when the reactor operators lost control of the manual pressure relief valves: the low-pressure pumps became useless again, and for nearly four hours the fuel rods were kept barely covered by a back-up system supplying an agonising 400 l of water per minute. Not until 21:50 could the operators begin to establish normal shutdown cooling.

The fire was found to have damaged 1,611 cables, along with the trays in which they lay. The accident highlighted a major design flaw, namely that all the cables for the main and the back-up systems were in the same area. Of the damaged cables, 628, more than one third of the total damaged, were safety-related. Repairs cost in excess of $10 million and TVA had to pay out a further $10 million a month to buy in replacement power for their customers while the reactor was out of action.

Since the Browns Ferry fire, there have been several more reported accidents at the plant. One of the most serious occurred in June 1980 when, during a routine maintenance shutdown at the Unit 3 reactor, 75 of the 185 boron control rods that halt the nuclear reaction failed to descend into the reactor core. After three manual attempts to shut down the reactor had failed, the automatic scram finally moved the rods and a serious accident was narrowly averted. Referring to the success of the automatic scram, the NRC reported: 'This, however, was apparently a coincidence in that a manual scram should have produced the same result.'

In 1989 the US pressure group, Public Citizen, issued a report, entitled 'Nuclear Power Safety 1979-1989', which showed that Browns Ferry had filed 548 licensee Event Reports with the NRC, making it one of the top 10 most accident-prone US power plants.

21 November 1975

USS Belknap/USS J F Kennedy

While involved in night-flight exercises in rough seas 70 miles east of Sicily in the Mediterranean, the US aircraft-carrier *John F. Kennedy* and the cruiser USS *Belknap* were involved in a dramatic collision. Following a signalled change of course by the *Kennedy,* the *Belknap* failed to alter course and passed under the carrier's angled flight deck, which sheared off the *Belknap's* superstructure. The collision severed jet fuel lines aboard the *Kennedy* and electrical wiring aboard the *Belknap,* resulting in an immediate blaze aboard both ships and a series of explosions aboard the *Belknap,* which killed six of the *Belknap's* crew and one man aboard the *Kennedy*. A further 25 men were seriously injured and 50 sailors, who were pitched into the sea by the collision, had to be rescued by escort ships and helicopters.

As the fires raged, the Commander of Striking Force for the US Sixth Fleet, Rear Admiral Eugene Carroll, sent out a Broken Arrow message, declaring a possible nuclear weapons accident. His message alerted higher naval authorities to the 'high possibility that nuclear weapons on USS *Belknap* were involved in fire and explosions subsequent to collision.' The cruiser was carrying Terrier surface-to-air missiles armed with W45 one-kt nuclear

warheads. The missiles were not in fact involved in the fire, although the blaze came within 40 ft of the missile magazine. The explosions were caused by the detonation of conventional ammunition which rained shrapnel on the nearby ships. Despite assistance from other naval vessels, it was not until 08:00 the following morning that the last of the fires, which repeatedly flared up throughout the night, were finally extinguished.

The *Belknap*'s superstructure, including the bridge and engineering rooms, was completely destroyed. She was taken to Sicily to have her weapons transferred to another ship, and was then towed across the Atlantic to the Philadelphia shipyard for repairs that took four years to complete.

Although details of the collision and fires were released to the public by the US Navy at the time, neither the DOD nor the navy has ever made any mention of the possible involvement of nuclear weapons, nor of the Broken Arrow message, and the *Belknap* collision is absent from the DOD's 'comprehensive' list of nuclear weapons accidents.

The USS *Belknap* pictured before (left) and after (below) the 22 November 1975 collision with the USS *Kennedy*. The black circle on the right-hand picture indicates the location of the nuclear weapons on board. (Credit: US DOD).

30 August 1976

Hanford, Part 2, 'Atomic Man'

The worst worker-contamination accident in US nuclear history occurred in the early hours of 30 August 1976 when 65-year-old Harold McCluskey was working the night-shift at the americium-241 unit at the Hanford plant. Americium, a by-product in the production of plutonium, is an alpha-emitting isotope and one of the most toxic man-made elements. McCluskey's job was to stand by a glove-box and monitor the isotope's chemical extraction.

A few hours into his shift on the night in question, he heard a hissing sound and saw dense fumes inside the sealed glove-box. In the next instant, there was a massive explosion and McCluskey's rubber respirator was ripped from his face by the force of the blast. Hundreds of pieces of radioactive metal, leaded glass, and rubber were embedded in his skin. Temporarily blinded by nitric acid, he struggled to escape but in the next few minutes he inhaled the largest dose of americium ever recorded.

The maximum permissible 'body burden' of americium for an entire lifetime is .05 microcuries. In that short period, McCluskey's face alone was estimated to have received more than 300 microcuries.

Nine other workers who were contaminated were back at work within days; McCluskey, however, was so 'hot' that he set Geiger counters clicking 50 ft away and he was rushed to the Emergency Decontamination Unit in nearby Richland for what was to be a lengthy period of medical treatment.

He was first placed in a shower to wash off the radiation. McCluskey recalls, 'There were eight doctors hovering round me. Four of them gave me a 50:50 chance to live. The rest wouldn't say anything.' The water from the shower contained 3,000 microcuries of radiation.

To avoid contamination, McCluskey spent the next five months in a steel and concrete isolation tank, tended by nurses wearing respirators and protective clothing. Unable to hear or see for much of this time, he was injected with an ex-

The scarred face of Harold McCluskey, a year and a half after an explosion at Hanford's Z plant exposed him to a massive dose of americium - the worst worker-contamination accident in US nuclear history. (Credit: *Tri-City Herald*).

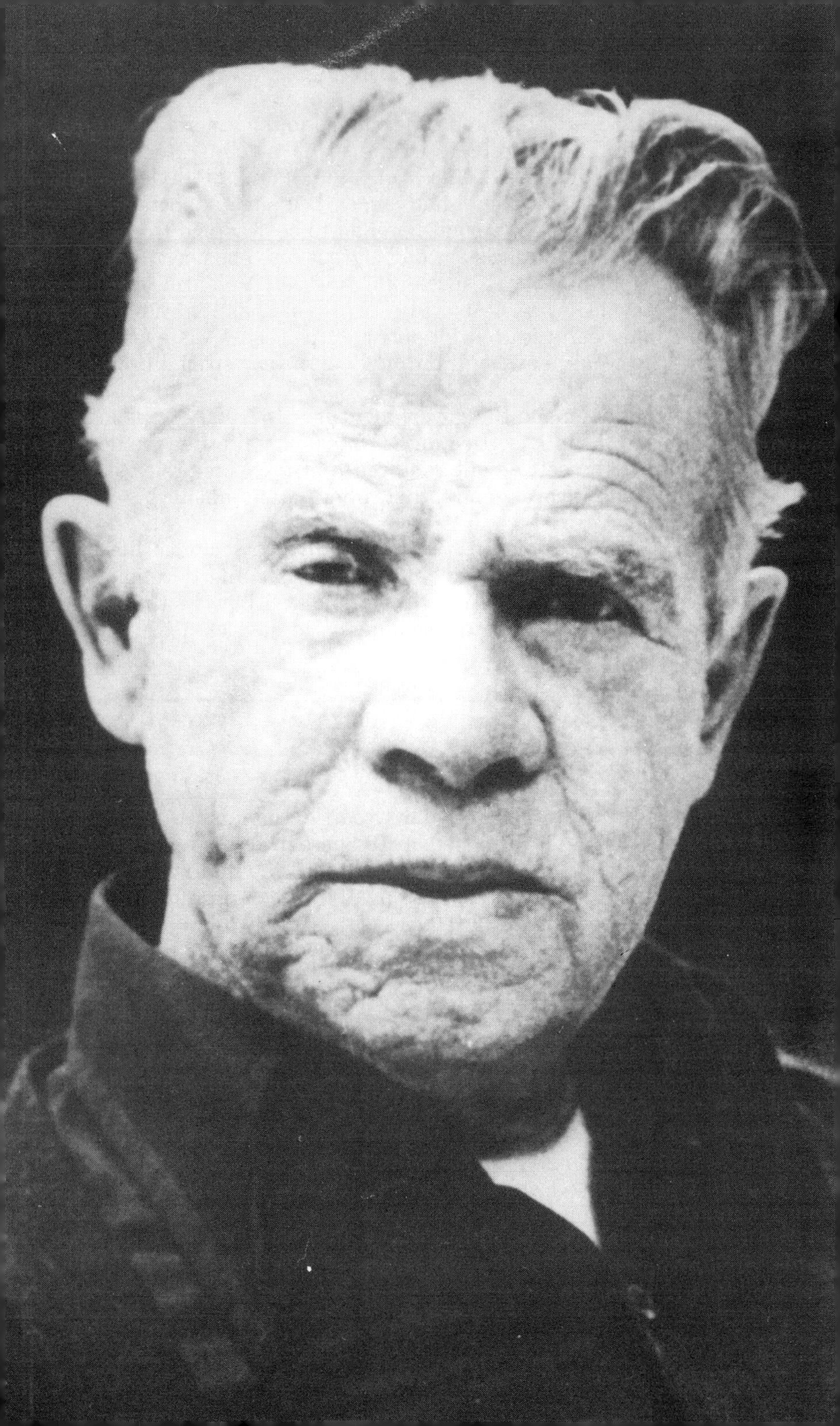

perimental zinc compound that bonded to the americium and enabled him to excrete it in his urine at the rate of .02 microcuries per day. This compound (zinc diethylenetri-amine-penta-acetic acid) had never been tested before on humans, and special permission had to be obtained from the Food and Drug Administration before it could be used.

The treatment was successful and, on Valentine's Day 1977, McCluskey was able to return home, as his radiation count had been reduced by about 80 per cent. His eyesight remained weak, however, and he had to wear special protective glasses.

He set out to sue the government for $975,000 compensation but eventually settled for $275,000 plus lifetime medical expenses. The settlement required him to permit release of his medical records and other information about his experience.

For the rest of his life, McCluskey lived a secluded life in the town of Prosser and spent his days listening to religious tapes, taking walks with his wife and visiting the three or four doctors who treated his illness. He died on 17 August 1987.

24 January 1978

Cosmos 954, Canada

Cosmos 954, a Soviet Radar Ocean Reconnaissance Satellite (RORSAT) launched on 18 September 1977, was 46 ft long, weighed 8,000 lb and travelled at 1,000 mph. On board was a nuclear reactor, fuelled by 100 lb of uranium-235. It was a companion satellite to Cosmos 952, which had been launched two days earlier. The pair would provide comparative signals on US ship movements.

At the end of its short life the satellite would typically have been boosted into a higher orbit, where it would remain for centuries while its radioactivity decayed. However, on 29 October it was clear that Cosmos 954's orbit was decreasing; by 7 November it was further confirmed that its nuclear reactor had not been jettisoned — the normal procedure in such a case. On 28 November, the US State Department was informed that Cosmos 954 was out of control. The impact date was estimated to be April 1978 but the impact site was a matter of pure conjecture. Then, on 6 January, the satellite began to tumble and a new re-entry time of 23 January was predicted.

President Carter's National Security Adviser, Zbigniew Brzezinski, held his first meeting with the Soviet Ambassador, Anatoly Dobrynin, a former aeronautical engineer, on 12 January; the aim was to confirm whether Cosmos 954 did have a nuclear reactor on board and, if so, what potential threat it posed. The Soviets confirmed it was nuclear-powered and was out of control but insisted that it could not reach critical mass and explode. Further meetings between the two sides followed and the US secretly informed its NATO allies and the relevant authorities in Japan, Australia and New Zealand of the situation.

Up until the day before the satellite finally crashed, the impact point remained uncertain; though by now NORAD was predicting that Canada was the most likely site. US and Canadian emergency teams were on standby ready for an instant response.

Just before 07:00 on 24 January, Cosmos made its 2,089th orbit before finally burning up over the Queen Charlotte Islands and dispersing radioactive debris over Canada's Northwest Territories. To observers on the ground it looked like a comet or a meteorite.

Operation Morning Light, as it was codenamed, swung into action. As Canada had no immediate capability for searching for, or locating, nuclear debris, US assistance was formally requested from the DOE. A USAF C-135 and a U-2 high altitude reconnaissance aircraft made aerial sampling surveys but detected no radiation signature.

In the next few days, four C-130 Hercules cargo planes carried more than 100 US scientists, technicians and soldiers to the area. In addition a 22-man Canadian nuclear-accident response team was sent to Yellowknife and a 44-man USAF task force of specialists arrived in Edmonton to set up a crisis control centre. US intelligence teams also joined the search. Having defined an initial search area, calculated from data obtained from the satellite's final orbit and from on-the-ground visual observations, a non-stop series of aerial and ground sorties began, scanning a huge area of 50,400 sq km of harsh wilderness in temperatures way below freezing. Even with advanced technology and extensive manpower, finding the satellite was to prove a difficult task.

The first genuine trace was detected on 25 January by a Canadian team. It proved later to be a piece of charred metal, nicknamed the 'hotplate' because it was emitting 200 roentgens per hour. The Canadian Minister of National Defense, Barney Danson, told the press, 'It's either a piece of debris or the greatest uranium mine in the world.' (Morrison, 1982). However they were subsequently unable to relocate it and the press were told it was a false alarm. (Further embarrassment followed when it proved later to be a genuine find.)

The first major piece of the satellite, nicknamed the 'antlers' because of its distinctive shape, was found on 28 January. As chance would have it, the find was made not by an official search party but by a team of scientists (five Americans and one Canadian) who had set out the previous spring to retrace the steps of a 1926-27 pioneer expedition, led by the English adventurer, John Hornby. Having been overtaken by the first snows, they were working from an old log cabin in the region, from which they made forays into the surrounding wilderness.

At 15:00, in the afternoon, two of them — John Mordhorst and Mike Mobley — were travelling by dogsled along the Thelon River when they saw a crater in the snow from which were projecting prongs and struts. When they returned to the cabin to report their discovery, their friends were discussing the satellite's re-entry. Convinced they had found the wreckage, they radioed the authorities.

SOVIET SATELLITES

The Soviet RORSATs monitor maritime traffic with a long-antenna radar. Most satellites derive their power from solar cells but these are not powerful enough for RORSATs (a solar panel produces only one kW). The other disadvantage is that the panels increase the drag of the satellite's already low orbit — which is necessary to obtain better image resolution — and make it more vulnerable to the gravitational pull of the earth. Two main kinds of nuclear-power sources are used by the Soviets:

- Radio-isotope Thermoelectric Generators (RTGs). These use the given heat produced by the natural radioactivity of a radioisotope to generate electricity with thermocouples (devices in which a temperature difference between two parts of a circuit produces a current). RTGs are extremely reliable because they have few moving parts but they cannot produce as much electricity as reactors.
- Small, fully automatic nuclear reactors, which transform the heat energy from their core into electricity either by activating a liquid metal coolant to operate a turbo-generator, or by activating an array of thermocouples.

The first Soviet RTG-powered satellite may have been launched as early as 1965 and more than 30 have been launched since. The majority have been nuclear-reactor-powered RORSATS.

Soon after, 13 Canadians arrived by Chinook helicopter, and the entire exploration team and their dogs were quickly airlifted out for medical check-ups. Mordhorst and Mobley, it turned out, had been exposed to the equivalent of two chest X-rays. The others had a completely clean bill of health. Meanwhile, with some US back-up, the Canadians set out for the crater, which proved to be only mildly radioactive, indicating that the main reactor had fallen elsewhere.

Two days later, more debris was found on the snow-covered ice of the Great Slave Lake. Some of these chunks were so radioactive that they had to be handled with long tongs by men standing behind a 1,600 lb lead shield — all this in a 35 mph wind and sub-zero temperatures. A survey of the site found abnormal concentrations of rubidium, zirconium and niobium; balloons also detected traces of enriched uranium in the atmosphere.

By 4 February the large-scale search was over. Five radioactive pieces of the satellite had been safely recovered and preparations were made for the next stage of the operation — a fine-grid search using airborne detection equipment in order to make sure the area was 'clean'. This was followed by an even more extensive ground and helicopter search to try and recover literally thousands of minute pieces of the reactor which were scattered over an area of more than 160,000 sq km.

The Canadian Forces alone logged some 5,000 hours of flying time during this period. Two hundred 'hits' had been recorded, of which 88 had led to pieces being recovered. By 21 April, Operation Morning Light was officially over and control of the situation was taken over by the Canadian Atomic Energy Control Board, who conducted a summer search and final clean-up between July and September 1978.

Under the 1972 UN Convention on International Liability for Damage Caused by Space Objects, Canada claimed just over $Cdn 6 million in damages from the Soviets. Negotiations between the two parties ran from January 1979 to December 1980, when a payment of $3 million was agreed.

As a result of the Cosmos 954 incident, President Carter pledged in 1978 that the US would pursue a ban on nuclear power in space. This position was later abandoned. The same year, US Energy Secretary James Schlesinger stated that 'there are serious risks and I regard it as inappropriate to have nuclear reactors orbiting the Earth'. (Aftergood, 1988.)

The other result of the accident was that, for a two-year period, no new Soviet nuclear-powered satellites were launched and the RORSAT was redesigned. These satellites are now intended to separate into three parts when their operating life is over. The instrument platform first separates from the reactor. Then the booster fires to raise the reactor to a higher orbit, en route to which the core is ejected from the reactor housing.

Subsequently the Soviets have had to deal with two further accidents involving these craft. On 28 December 1982 they hit problems with Cosmos 1402; the instrument platform did not separate from the reactor, so it was not able to reach a higher orbit and it began spiralling back into the earth's atmosphere, re-entering above the Indian Ocean on 7 February 1983.

Opinions differ as to whether anything like a complete burn-up was achieved. Much of it may have gone straight into the sea in solid form.

Then, on 10 April 1988, Soviet controllers lost contact with another of their satellites, Cosmos 1900. After months of speculation that it would crash and shower the earth with radioactive debris from the 50 kg of uranium-235 on board, the Soviets announced that sensors on board the satellite had detected its decaying orbit, shut down the reactor, and activated its boosters, thereby successfully pushing it into a higher orbit.

This successful first-time test of the new safety systems built into the craft was hailed by analyst Nikolay Zheleznov of the Soviet news agency *Tass*. He remarked that the success made it 'possible to substantially reduce restriction on the scope of application of nuclear power plants in outer space.'

Coincidentally, in May 1988, on the same day that the Soviets announced that the Cosmos 1900 satellite's orbit was decaying, the Federation of American Scientists and the Committee of Soviet Scientists Against the Nuclear Threat issued a joint proposal.

This demanded a halt both to the USSR's nuclear-powered surveillance programme, and to the development of orbiting reactors planned under the US Strategic Defense Initiative (SDI) programme.

Roald Z. Sagdeev, chairman of the Committee of Soviet Scientists and former director of the Soviet Institute of Space Research, wrote to US Senator J. Bennett Johnson that a worst-case accident could 'be comparable to the long-term consequences of Chernobyl'.

US SPACE NUCLEAR POWER PROGRAMME

The US began to develop small nuclear-power sources for use in space in 1955 as part of the SNAP (Space Nuclear Auxiliary Power) programme. The first space nuclear-power source was launched in 1961, and 22 spacecraft powered by RTGs, and one reactor-powered satellite — for both military and civil missions — were launched between then and 1973, when the US space reactor programme was terminated.

There have been a number of accidents involving these satellites:

1964, April 21: US Navy 5BN-3 *Transit* navigation satellite, carrying a SNAP 9-A RTG powered by 2.2 lb of plutonium-238, failed to achieve orbit and the RTG disintegrated in the atmosphere at an altitude of 50 km. It sprinkled radioactive particles of plutonium over a wide area of the southern hemisphere, a fact confirmed by soil samples taken from 65 sites around the world between October 1970 and January 1971. The 17,000 curies of plutonium-238 released by this one accident increased the total world environmental burden of plutonium, principally from atmospheric weapons tests, by about 4 per cent.

1968, May 18: following a launch failure, a US *Nimbus B-1* weather satellite was aborted and its two SNAP-19A RTGs fell into the Santa Barbara Channel, off southern California. They were recovered intact after a five-month search.

1970, April 17: as part of the dramatic rescue of *Apollo 13*, the lunar module, carrying a SNAP-27 plutonium power supply, was jettisoned at 18,000 km over the South Pacific. It splashed down near the island of Tonga. It was never recovered but US officials claim that radiation levels in the area indicate that no plutonium has been released.

After a ten-year hiatus in space nuclear-power development, the US is currently developing a new SP-100 reactor as part of the SDI programme. Such reactors are considered an essential component of directed energy weapons.

NASA's Project Galileo spacecraft, aimed at Jupiter and scheduled for launch on board the space shuttle in October 1989, is to be powered by two RTGs, each containing 11 kg of plutonium-238. They will produce 4.2 kW of heat which will be converted into 280 watts of electricity. Plutonium-238 is 270 times more radioactive than plutonium-239; therefore Galileo will have 274,000 curies of plutonium on board. An accident on the launch pad during lift-off or during an accidental re-entry of the spacecraft could release large amounts of radiation into the atmosphere.

In January 1989, a team of six leading Soviet scientists disclosed at a conference in the US that they had staged two successful flight tests of an advanced new class of Topaz reactor. These are fuelled by about 50 kg of enriched uranium and are capable of producing 10 kW of electrical power. The scientists claimed that these tests were the first step towards the development of nuclear-powered engines for a manned flight to Mars and that this technology would be offered for sale to the West.

In February 1989 Congressman George E. Brown Jr of California, the ranking Democrat on the House Science, Space and Technology Committee, introduced a bill in Congress aimed at banning nuclear-powered satellites from space. He commented to *Time* magazine (2.2.89): 'If we don't stop the use of nuclear-power sources travelling over our heads, we're likely to wake up one day with a nuclear reactor landing *on* our heads.' His argument runs counter to the view that what the US needs is an Anti-Satellite Systems (ASAT) programme to disable these military 'eyes in the sky'. This, Brown claims, would be self-defeating as the US, even more than the USSR, is enormously dependent on military satellites for communications, early-warning, navigation, and intelligence gathering. Such actions would endanger this vital network.

Gamma-ray pollution from orbiting reactors is emerging as a very serious issue in the astronomical community. The US *Solar Maximum Mission* detected gamma rays from RORSATs in 1980, but this fact was classified information until late 1988. Orbiting reactors produce gamma rays directly (at close quarters) and also by positron emission. Positrons from a reactor move along the earth's magnetic field lines for thousands of miles, generating gamma rays if they strike a satellite. The positive side of this is that, because of gamma-ray emissions, a ban on such satellites would be verifiable.

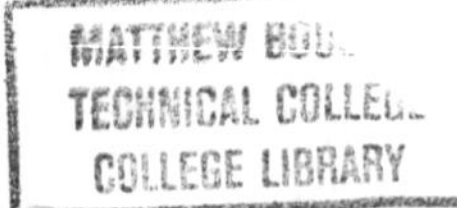

30-31 December 1978

Beloyarsk Reactor, USSR

The accident at the Beloyarsk nuclear complex, USSR, has been described as 'the most serious in the world involving nuclear reactors before Chernobyl'. Yet the Soviet government has never officially admitted to the accident, and it was never reported to the International Atomic Energy Agency (IAEA). This silence was finally broken by the Soviet magazine *Socialist Industry,* which described how a fire had come near to destroying the complex's three reactors. 'One is nagged by a recurring thought,' concluded *Socialist Industry.* 'Why didn't they tell people publicly about those who'd saved the Beloyarsk station and about the lessons of the accident? If they had done that, maybe the Chernobyl events would not have taken place...'

The Beloyarsk nuclear complex lies some 50 km from the city of Sverdlovsk, and houses two Chernobyl-type RBMK reactors and a newly-installed BM600 fast-breeder reactor. According to *Socialist Industry,* the emergency began with a fire, caused by an electrical short circuit, in the complex's machine hall. What happened over the next day and night, recounted by eyewitnesses, reads like the scenario for a Hollywood disaster movie.

The complex's operators were first alerted to the fire by a loud explosion in the machine hall, recalled Deputy Chief Electrician S. Morkhov: 'I dashed into the machine hall. I saw the sky and the stars. For a moment I didn't realise that the roof had gone. Unable to withstand the temperature, the steel girders and reinforced concrete of the roof had collapsed into the centre of the hall, leaving a huge hole above the No. 2 generator.' The fire had already shorted out cables, causing secondary fires and neutralising the fire-control system. As workers fought to control the fire, the lights went out.

Morkhov and his colleagues faced a major problem: emergency procedures called for *both* RBMK reactors to be shut down. But they were in the middle of the Russian winter, and the temperature outside the reactor buildings was nearing -50 °C.

If they were to shut down both RBMKs, the reactor cooling systems would freeze and the cores would overheat — which could lead to a double meltdown. A second real concern was the consequence of depriving the Beloyarsk district of power and heat in the sub-zero conditions.

Since Reactor No. 2's generator was covered in debris from the collapsed machine hall roof, it was decided to shut that RBMK down and keep Reactor No. 1 and its turbine running. But, despite the fire raging around it, the temperature within the machine hall was -40 °C and the Number 1 turbine was frozen: 'Turbine-generator No. 1 was covered in frost and snow,' added Morkhov. 'But it had to be restored to normal working conditions, to generate electrical energy to remove heat from reactors 1 and 2.'

The fires had spread extensively throughout the station's cable ducts and the operators could no longer remain at their controls. Fortuitously, the first of 1200 firemen arrived: 'Accompanied by firemen wearing breathing apparatus and holding torches,' continued Morkhov, 'engineers made their way to the controls and took readings of equipment still working, then rushed to the exits to breathe...Others rushed in to adjust the controls by hand. Several lost consciousness and had to be dragged out. Minutes later they were back at work. There was no one else capable of doing it.'

In the reactor building, temperatures had risen to 80 °C. Sparks were flying from shorting cables all around the firefighters, who were severely hampered by the lack of breathing equipment. By now, the inferno had reached the room where the station computers were located, and it looked as though they would lose control over the reactors: 'With 17 hours of 1978 left,' said *Socialist Industry,* 'people thought that the end was upon them. No one had much strength left. The specialists started talking about the active zone itself being in danger because of the intense heat. Malashev, the director, formed an emergency staff, and preparations were begun to evacuate the adjoining settlement.'

At the last moment, the firefighters regained control of the crippled complex, the prospect of a double meltdown was averted and the Sverdlovsk region was spared the world's worst nuclear accident. Miraculously, the reactor containment systems were intact. No radioactivity escaped from Beloyarsk. As a result of the fire, safety regulations at the plant were tightened.

According to *Socialist Industry,* 84 firemen and plant operators were awarded medals for bravery. These awards were never announced by Soviet authorities. Neither were any casualty figures.

28 March 1979

Three Mile Island, Pennsylvania

Three Mile Island lies in the middle of the Susquehanna River, some 10 miles downstream from Harrisburg, state capital of Pennsylvania, a town of 60,000 people. This area of the Susquehanna Valley, administered by Dauphin County, has a population of a quarter of a million people. Barely 200 miles from New York City, Dauphin County is an area of townships and farming communities. Much of the land has been farmed by the same families for 200 years or more, conservative people who trace their ancestry back to the German settlers now known, via a mistranslation, as the Pennsylvania Dutch.

Three Mile Island is as long as its name but only a few hundred yards wide. There are several townships within five miles of the island: Middletown, Royalton, Falmouth, Newberry, York Haven and Goldsboro. It was the people of these townships and farms, and the other residents of Dauphin County, who were to suffer most from the accident at Unit 2.

Unit 2 was one of a pair of 1,000-MW Pressurised Water Reactors (PWRs) built by the engineering company Babcock and Wilcox. Both units were operated by the Metropolitan Edison Company (MetEd), a subsidiary of the plant's owner, General Public Utilities Company (GPU), later renamed GPU Nuclear Corporation. In a PWR, two circuits containing water under high pressure take heat from the reactor's nuclear core and use it to drive turbines which produce electricity. These circuits are designed to keep the reactor's temperature within safe limits.

The primary cooling circuit removes heat directly from the core to a steam generator. Since it flows around the reactor core, water in the primary circuit is intensely radioactive and scaldingly hot. To keep it from flashing into steam, which can lead to serious problems, water in the primary circuit is kept under tremendous pressure.

Inside the steam generator, water in the primary circuit runs through thousands of metal tubes immersed in uncontaminated water under lower pressure. Water in this secondary cooling circuit flashes into steam, which drives the PWR's turbines. Steam is thus both the PWR operator's friend, since it drives the plant's turbines; and his enemy — if pressure inside the primary circuit drops and water boils into steam, the reactor is in danger of overheating. In the worst-case scenario, the reactor core can become exposed and may eventually melt.

At the time of the accident, Unit 2 had been in operation for less than three months. In its short life, the new reactor had already experienced a number of breakdowns. But MetEd had a great incentive to hurry: putting the reactor into operation at the end of 1978, rather than the new year of 1979, gave the plant's owner, GPU, important tax advantages. Thus, when Unit 2 came on-stream on December 30, GPU saved $40 million in tax benefits and qualified for a rate increase worth nearly $50 million.

The sequence of events that led to America's worst commercial nuclear-reactor accident began early on the morning of Wednesday, 28 March 1979. Just after 04:00, with the Unit 2 plant running at high power, a water pump in the secondary cooling system failed and the reactor's turbine automatically shut down. Although Unit 2 was no longer generating electricity, the reactor was still running at full power. As the temperature within the primary cooling circuit rose, two more automatic devices tripped. First, a PORV (Pilot Operated Relief Valve) on the primary system opened to reduce pressure — just like the valve on a pressure cooker. Second, the reactor 'scrammed' and shut itself down.

Neither of these events, which happened in seconds, were unusual. Unit 2's operators had experienced turbine and reactor 'trips' before. They knew that as the surge in pressure abated, the PORV would close and the system would return to normal. However, they were unaware that the PORV had jammed open, continually reducing pressure inside the primary cooling system, rapidly filling a drain tank. After several minutes this tank was full and the containment vessel was flooded with radioactive water. This was the first in a sequence of equipment malfunctions and operator errors which would lead to disaster.

Soon after the initial pump failure, alarms in Unit 2's control room sounded again. The operators knew that the plant was now in trouble, but had no indication of the cause — a console light indicated that the PORV had closed normally.

The open valve was not their only problem. Back-up systems, which should have pumped water into the secondary cooling system, had also failed to operate as planned. They had been closed for maintenance — and the tags which showed their status were blocking console warning lights. Again, Unit 2's operators were unaware that a vital system was malfunctioning.

With the emergency back-up systems closed, steam generators ran dry. The primary system, losing pressure rapidly through the PORV, was now the only way to take heat away from Unit 2's reactor. (Even with the reactor shut down, radioactive decay heat was still producing six per cent of normal reactor power). At this point, an emergency system was activated, pumping thousands of gallons of high-pressure water into the primary system.

This should have brought the water temperature down and increased cooling system pressure. But again, human error overrode a supposedly foolproof system: operators incorrectly interpreted the situation and turned off one of the emergency pumps. Within a matter of minutes, Unit 2's primary coolant level dropped so low that water flashed to steam. Temperatures inside the reactor core, which was now partially exposed, soared, and fuel rods ruptured. The operators were unaware of it at the time, but Unit 2 was beginning to melt down. By now, radioactive gases were pouring into the skies above Dauphin County.

The immediate aftermath was one of confusion. For days, GPU officials minimised the extent of the accident and gave out conflicting information. Said the then chairman of the US Nuclear Regulatory Commission Joseph Hendrie, in the NRC's emergency mission office in Washington two days after the accident: 'We are operating almost totally in the dark. His [Pennsylvania Governor Richard Thornburgh's] information is non-existent, mine is ambiguous. It's like a couple of blind men staggering around making decisions.'

It took two days for Thornburgh to decide to evacuate 3,500 children and pregnant women living within five miles of the plant. Another 200,000 people in the surrounding communities, mistrusting the official line that Unit 2 was under control and that little radioactivity had been released, voluntarily fled the area. 'Everyone was in a state of panic, and no-one knew what to do,' said Patricia Longenecker, a resident of Elizabethtown, some seven miles from the plant. 'It was fuelled by the sirens, and by broadcasts through the town. people just began leaving quickly, some not even knowing where they would go. By Sunday, Elizabethtown had become a ghost town.' By then, the worst of the radiation had already passed overhead.

While the terrified citizens of Dauphin County fled a real-life reactor accident, audiences across the United States packed cinemas to watch a fictional account of such a disaster. *The China Syndrome,* starring Jane Fonda, Michael

Douglas and Jack Lemmon, opened just 12 days before TMI and, when news of the accident broke, drew sell-out crowds: 'The boom started the day of the accident,' said a Columbia Pictures president. 'Everyone heard they'd get an inside look at what went wrong at Harrisburg.'

The film described a near-disaster at a California nuclear plant, beginning with pump shutdowns and faulty gauges. (The title refers to a tongue-in-cheek nuclear industry term for a serious core meltdown, in which tons of molten nuclear fuel burrow through the earth in the general direction of Asia.) 'The scene that always grabs them,' said a cinema manager at the time, 'is where Jack Lemmon [playing a troubled nuclear engineer] says that a California meltdown could wipe out an area the size of Pennsylvania. The audience laughs — then starts to clap.'

In the fictional event — made nine months before the TMI meltdown — the reactor is saved and disaster averted. There are no radiation leaks, no core breakdown and no evacuation. But a central theme of the film — faulty reactor design and construction, incompetent plant management and attempts by plant officials to cover-up the extent of the damage — predicted with uncanny accuracy the events which predated, and may have caused, the disaster at Three Mile Island.

The film was loosely based on a number of past mishaps at American nuclear reactors, including those at the Rancho Seco plant in California, a PWR also built by Babcock and Wilcox. According to US government documents released by the Union of Concerned Scientists (UCS) shortly after the accident, TMI operators had been warned of serious shutdown problems at Rancho Seco following a turbine trip in August 1978. The Californian plant had also experienced sticking valves, leading to dangerous drops in coolant pressure, instances where emergency cooling pumps had failed to operate, and faulty gauges on the primary system. Former NRC engineer Robert Pollard detailed repeated problems with Babcock and Wilcox cooling systems, and alleged that the US government had allowed Unit 2 to go on-stream 'despite the red flags'. In fact, reported the UCS, an NRC inspector had formally recommended, two months before the accident, that Babcock-and-Wilcox-built PWRs be examined for a wide range of problems. His report was rejected.

During an NRC inquest in the months after the accident (which resulted in MetEd being fined $155,000, the maximum permitted under law) former Unit 2 control room operator, Harold Hartman, told investigators that the crippled reactor had suffered problems prior to the accident, and that they had been concealed. He alleged that operators, with the knowledge of some MetEd managers, had systematically falsified primary coolant system leak figures in order to keep the plant running. Documented records also showed a history of failures in the primary coolant system. These problems, it was alleged, had desensitised operators to warning signals and may have damaged the PORV that stayed open during the accident. In 1983, the US Department of Justice indicted MetEd for criminal falsification and destruction of safety data. At the hearing, the US Attorney stated that 'There were a number of self-serving observations made in the defendant's [MetEd's] response. One is a fairly grand promise to do in the future what in my judgement the company should have been doing the whole time it was operational in 1978 and 1979; that is, to follow the technical specifications, follow the regulations and operate the plant the way they promise to operate it in the future. And for all the weeks and months that followed, they didn't. That is an undeniable and undisputed fact...' The company pleaded guilty to some of the counts and no contest to the others.

And, in response to charges that, like their ficitional counterparts in *The China Syndrome,* TMI officials had attempted to cover up the true extent of the accident, an official US House of Representatives committee report, published two years after the accident, declared that '...TMI managers did not communicate information in their possession that they understood to be related to the severity of the situation...[which] prevented State and Federal officials from accurately assessing the condition of the plant. In addition, the record indicates that TMI managers presented State and Federal officials misleading statements that conveyed the impression the accident was substantially less severe and the accident more under control than what the managers themselves believed and what was in fact the case.'

'The bottom line is that nobody really knows how much radiation escaped.' Dr Ernest Sternglass, Professor Emeritus of Radiological Physics, University of Pittsburgh School of Medicine.

The first estimates of radiation released by Unit 2 during the accident came from the plant itself. Operators indicated that heavy amounts of fallout, up to 40 rads an hour, could be coming down on Goldsboro, just two miles from Unit 2. But, as the seriousness of the accident was downplayed, so were the radiation levels. MetEd advertising after the accident compared the collective dose to that of an X-ray given to everyone in the locality.

MetEd and the NRC argue that external monitors around the plant and helicopters and aircraft flying over Three Mile Island detected only minor emissions, and that surveys of milk and produce from nearby farms showed no high fallout levels. These findings have become the basis for the official line on Three Mile Island; which is that no significant amounts of radiation were released from Unit 2 during or after the accident. However, the amount of radiation vented from the crippled plant, and the amount of fall-out which descended on Dauphin and other nearby counties, has been under dispute for 10 years.

During the accident, monitors on the vent stacks, through which much of the radiation escaped, became saturated with moisture, rendering them useless. Records of radiation releases for the crucial first two days of the accident, claimed MetEd, were lost. Moreover, because of complications with the massive clean-up operation, a 'core inventory' — the technique of comparing known amounts of radioactivity inside the core before and after the accident — was never undertaken.

An independent study of radiation monitoring during the accident, published in 1984, conducted by Dr Jan Beyea, senior energy scientist with the National Audubon Society, found that 'the monitoring network in place, both inside and outside the plant, did not perform adequately,' that 'environmental sampling...was insufficiently coordinated,' and that 'a great deal of crucial data does not exist, or is unreliable.'

'There wasn't enough radiation equipment,' said Beyea, 'half of it wasn't working; and a large amount went off scale, because they [nuclear industry officials] never anticipated such large releases.' Other studies independent of the nuclear industry found that there were elevated radiation levels in small mammals taken from the area; that radioactivity could have escaped through plant openings other than the vent stacks; that a quarter of a million gal of highly radioactive water were dumped into the Susquehanna River without warning; that one external monitor did record a high radiation reading in precisely the direction critics charge the worst plume was blowing in the early stages of the accident; and that other plumes could have passed undetected, through gaps in the monitoring system.

'People were exposed to higher doses than were acknowledged,' says the Institute for Energy and Environmental Research's (IEER) executive director, Dr Bernd Franke, who has worked to reconstruct dose levels from the accident. 'There were "hot spots" in the area which were not detected, simply because the plume does not always do you the favour of landing on a detection instrument.'

'People around here are dropping like flies. Every day, it gets worse and worse. It's no longer a question of whether that plant has had an impact. It's just a matter of who gets hit next, and when.' Local resident, Jane Lee.

Marie Holowka had farmed near Three Mile Island for over 50 years. At the time of the accident, she was milking cows when the barn started to shake. She heard a noise that came from deep beneath the ground 'like water boiling in a pot', and assumed it was an earthquake. At daybreak, as she was leaving her barn, 'my eyes started to pinch me and there was a funny taste in my mouth...I couldn't see for more than 10 feet.' On the way back to her farmhouse, Marie Holowka felt so weak she fell three times. Though she tuned in to several local radio stations, Marie heard nothing until 09:00 when a station in Philadelphia, a hundred miles away, announced a major accident at Three Mile Island and advised people to stay indoors.

Two hours later, a visiting county official told Marie, her sister Eva and brother Paul to leave their dairy herd enough fodder for three days and prepare for evacuation. That was the only word they heard. They heard no news on local radio, and no evacuation took place.

Many of the younger people left of their own accord, but the older ones, including the Holowkas, stayed because they had nowhere to go.

Marie's weakness and sore eyes were followed by a rash that wouldn't heal. Six months after the accident, she came down with a thyroid infection, then cancer. She died in 1989, at the age of 73, having undergone six operations and 39 radiation treatments. Many of her neighbours have died from cancer since 1979: 14 in 1987 alone.

From the time TMI started up, the Holowkas lost more than 200 cows from their dairy herd. Vets could find no evidence of poison or disease in the carcases: 'It used to be when a calf was born, you knew it was going to grow up,' she said. 'But not anymore. They just lay down and die.'

Marie Holowka's story is similar to those of hundreds, if not thousands, of local residents. In the days following the accident, many people experienced symptoms associated with radiation exposure: a strong metallic taste, hot sensations on the skin, burning and tearing of the eyes, nausea, vomiting and diarrhoea. Some broke out in rashes that wouldn't heal; others had high white-blood-cell counts.

Humans were not the only ones affected: eggs would not hatch; farm animals aborted or gave birth to dead, deformed or retarded offspring; the leaves and buds of plants grew oversized, or deformed. Botanist Dr James E. Gunckel, an expert on the effects of radiation, found 'a number of anomalies entirely comparable to those induced by ionizing radiation' in plant specimens taken from the area.

In the 10 years following Three Mile Island, there have been only four major physical health studies of people living near the plant. The official line is that only small amounts of radiation were released by the accident and that therefore there is little point in monitoring the health of hundreds of thousands of people.

Two of the four studies were conducted by Dr George Tokuhata, Director of Epidemiology for Pennsylvania State Department of Health. The first, in 1981, concluded that no significant physical health effects were expected from the accident; the second, four years later, found no increase in cancer rates around Three Mile Island. Both studies were criticised as flawed by independent scientists including Dr Robert A. Huttguist, a Pennsylvania State University Professor of Statistics, and Dr George Hutchinson, a Harvard University Professor of Epidemiology.

But a study in 1980 by Pennsylvania State Department of Health Secretary, Dr Gordon MacLeod, found an abnormal number of babies born with serious thyroid problems — often a tell-tale sign of radiation exposure — and dramatic increases in infant mortality. MacLeod, an outspoken critic of the Department's handling of health aspects of the accident, was fired that year.

Marie Holowka and her brother Paul used to win prizes for showing their livestock. After the Three Mile Island accident, most of their cows — and many of their neighbours — died. Marie was diagnosed as having cancer in August 1980. She died shortly after this picture was taken in 1989. (Credit: Brian Jaudon).

Four years later, Marjorie Aamodt, a former Bell Laboratories psychologist, and her husband, who lived 50 miles from Three Mile Island, initiated a neighbourhood study of health effects in the region. They found that cancer death rates in areas around the plant were several times the national average. Their study, dismissed by the nuclear industry, was backed by several independent scientists.

The most chilling assessment of the health effects from Unit 2 came from Dr Ernest Sternglass, Professor Emeritus of Radiological Physics at the University of Pittsburgh's School of Medicine and an internationally acknowledged authority on radiation and health. Sternglass, who arrived at Harrisburg the day after the disaster with a Geiger counter which showed elevated levels of radiation, later studied official US health statistics and discovered that an infant mortality peak during 1979-80 had disappeared from later publications. Similar figures, showing a drop in infant mortality while Unit 2 was shut down, also vanished at a later date. Sternglass concluded that the statistics were deliberately 'massaged' to disguise the true health implications of Three Mile Island: 'There is no other way to explain the changes,' he said.

'Why wasn't there core on the floor?' Department of Energy engineer during Unit 2 clean-up.

In the year after the accident, 1,700 clean-up workers began the long task of decontaminating Unit 2. This process, MetEd estimated, would take a couple of years at most, at a cost of perhaps $140 million. A decade later, and with well over $1 billion spent, Unit 2 is still hot and the decontamination process is not completed. With the possible exception of Chernobyl, the clean-up operation at Unit 2 has proved to be the biggest, the most problematic, and the most expensive in civil nuclear history.

In 1982, when radiation inside the containment had reached levels low enough for workers to enter, nuclear engineers took the first step towards assessing the true extent of damage to the reactor's core. A remote-controlled TV camera placed inside the reactor vessel revealed a five-ft deep cavity where the tops of the fuel rods should have been. It was only then that the watching engineers realised that temperatures inside the reactor had risen high enough to burst open the fuel assemblies. According to one eyewitness, 'the mood was one of shock'.

Further shocks were to follow: three years later, engineers jacked up the reactor's 55-ton lid and lowered a second camera to the bottom of the vessel. There they saw a hardened lava bed of rubble, including some 20 tons of once-molten uranium dioxide fuel. The engineers now knew that the temperature inside the reactor had risen past 5,100 ^{0}F, the melting point of uranium dioxide.

The extent of damage to Unit 2 was far more serious than had ever been imagined in 1979. About 53 per cent of the reactor's core had melted, coming perilously near to breaching the eight-in-thick carbon steel reactor vessel. As one engineer put it: 'Why wasn't there core on the floor?'

As a result of probing deep into the heart of the crippled reactor, nuclear engineers have now evolved a new scenario for the final operating hours of Unit 2. During the first two hours of the accident, cooling water fell to its lowest level and the exposed core's temperature rose high enough to rupture cladding around the fuel pellets. Two and a half hours into the event, burning cladding ran down the fuel assemblies, followed by partially molten fuel. Three hours into the event, as more water was pumped into the system, the water level rose and overheated fuel assemblies shattered, like a hot glass held under ice-cold water. Just before four hours had elapsed, the molten core streamed towards the bottom of the reactor, reaching the bottom of the vessel in under a minute. Only the small amount of coolant water in the reactor prevented a breach, say the nuclear engineers. Luckily long-term cooling was established at about this time. Unit 2 barely escaped a total meltdown and the release of radiation on a scale possibly approaching that of Chernobyl.

Removing the tons of radioactive slag which was once the reactor core became the main task of the Unit 2 clean-up. 'Defuelling', as the process is known, is hazardous, time-consuming and expensive. Working on a steel and lead platform above the crippled reactor, defuellers use everyday tools — pliers, chisels and shovels — mounted on long metal poles with which they pry small chunks of rubble away from the main mass. Each piece is brought up through an 18-in-wide slot in the work platform and chopped up with pneumatic drills until the debris is small enough to fit inside a eight-in-wide can, a process described by one journalist as 'smashing up radioactive rocks with big sticks'.

The defuelling operation runs seven days a week, five shifts a day. To keep exposure levels down, workers are only allowed on the platform for four hours at a time, working one week in every six. Even though they wear plastic boots, two radiation suits, a plastic smock, several pairs of gloves (changed every 15 minutes) and a battery-powered respirator, defuellers are still exposed to high levels of radiation. On average, say defuellers, there are two contamination incidents every week. On one occasion, a worker fell up to his waist into the water surrounding the ruined core.

Ten years after the Unit 2 meltdown, defuellers have removed 150 tons of radioactive rubble from the reactor vessel. Sometimes bacteria thrive in the hot water which now covers the crippled core, turning it into a radioactive soup. When this happens, or when sediment swirls reduce visibility, the clean-up operators work by feel.

The canned rubble is shipped by rail more than 1,800 miles to the world's largest known 'hot shop': an 890-sq-mile nuclear dump at Idaho Falls, Idaho, which contains the remains of 52 experimental reactors and other garbage of the atomic age (including the highly-radioactive hulks of experimental nuclear jet engines). Here, the cans are stacked together in a storage pool, a temporary haven until the US constructs a new national nuclear-waste cemetery.

Defuelling is not the only problem at Unit 2. During the accident, contaminated water flooded the basement of the reactor building, and it is still too radioactive to enter. Instead, two robots roll around the area, blasting the walls with high-pressure water and vacuuming up hot sludge.

Nobody knows when the clean-up will finally be finished, nor how much it will eventually cost. GPU Nuclear Corporation have consistently underestimated the time it would take to decontaminate Unit 2. In 1987, for example, the company confidently predicted decontamination would be completed by the end of the year. In 1989, the company predicted completion by 1990.

Sometime around 2020, when TMI-1 comes to the end of its life, both reactors will be dismantled together. So far, the clean-up has cost over $1 billion. Additional costs include nearly $200 million for the rubble stored in Idaho, for which the US government pays, using federal tax dollars. The remaining clean-up will cost around $200 million; entombment $500 million (the hulk will be monitored by 50 engineers); and decommissioning a further $200 million. The total, a conservative $2.1 billion, will be three times as much as the plant cost to build in the first place.

The human costs of the accident at Unit 2 are harder to define. In 1989 there were over 2,000 lawsuits pending in the courts, representing, as one reporter described it, 'an enormous catalogue of human suffering'. Plaintiffs allege miscarriages, birth defects, kidney and thyroid problems and cancers of all types. These cases have been waiting for up to five years for a hearing. The hold-ups are technical: can people file a lawsuit 10 years after the accident? Can an heir to an estate file on behalf of the deceased? And which court should the cases be held in — county, state or federal? Some lawsuits have already been settled out of court. GPU Nuclear, who say the settlements are solely to avoid the costs of going to court and do not admit guilt, have made them conditional on the plaintiffs remaining silent.

The psychological problems facing the citizens of Dauphin County are more easily proven. Several studies have shown that local people suffer from abnormally high levels of anxiety, depression and hostility as a result of the accident, the years of cleaning up Unit 2 — during which more radiation was released into the Susquehanna River and into the air — and the eventual restart of Unit 1. Many people, some with roots stretching back generations, have left Dauphin County for ever. Those who remain, says local activist Kay Pickering, are unconvinced by assurances that Three Mile Island is now safe, and feel powerless in the face of the proposed evaporation of millions of gal of radioactive water and the continued discharge of radiation from Unit 2. They also have to face the fact that, like many of their neighbours, they and their children may develop accident-spawned cancers in the future. 'The local mood,' she says, 'is "I just have to learn to live with it." People aren't fighting that strongly.'

When Unit 2 was first commissioned in 1979, there were 94 nuclear plants under construction in the US. A decade later, there were just two. These had been ordered before 1974 and were still under construction; the continued delays in completion were due to stricter NRC licensing regulations and engineering incompetence.

The economic and political fallout from Three Mile Island hit the US nuclear industry like a hard rain. Jimmy Carter's Presidential commission into the accident put it starkly: 'To prevent nuclear accidents...fundamental changes will be necessary in the organization, procedures and practices [of the] nuclear industry.'

New regulations covering everything from control room layouts to operator training and engineering quality control were written and enforced: a further financial burden on an industry already suffering from massive cost overruns. Rocketing costs, new safety regulations, opposition to nuclear power, and a slow-up in the demand for electricity resulted in the cancellation of billions of dollars' worth of nuclear plants, many already under construction.

The survivors of this purge suffered massive cost overruns. In 1989, the average US nuclear plant cost $4 billion. Some rose to as high as $6 billion — a construction liability which pushed the cost of nuclear electricity to twice that of conventionally generated electricity. The American dream of nuclear electricity 'too cheap to meter' had turned into the reality of Three Mile Island.

POSTSCRIPT

In the mid-80s, a Toronto newspaper reported that the late Admiral Hyman Rickover, father of America's nuclear navy and the PWR, had made an amazing confession. Rickover's daughter-in-law, Jane Rickover, swore an affidavit to say that he told her he had used his personal influence with President Jimmy Carter to suppress the most alarming aspects of the Presidential Commission report into Three Mile Island and, instead, release it in a watered-down form. The report, if published in its entirety, continued Rickover, would have destroyed America's civilian nuclear power industry, 'because the accident at Three Mile Island was infinitely more dangerous than was ever made public.' As a career naval officer, specialising in nuclear engineering, Carter had served under Rickover. Some of GPU's officers had even served on Rickover's staff. (President Carter sped to TMI with an entourage of experts the day after the accident, to quell growing fears, using his personal expertise to the full. Carter later became a considerable nuclear sceptic.)

Two workers wear special protective 'radiation suits' for decontamination operations at Three Mile Island. The suits have taped seams to prevent the entry of radioactive dust. (Credit: Alexander Tsiaras/Science Photo Library).

REED

Additional Stories

1970 January 10 During training off Jacksonville, Florida, the aircraft carrier USS *Shangri-La* suffered a fire when an A-4 Skyhawk aircraft ignited on the flight deck, killing one person. (**Neptune**)

1970 January 10 The Italian liner *Angelina Lauro* reported hitting an unidentified object in the Bay of Naples. A few days later, a Soviet Foxtrot-class submarine was seen with 20 ft of her bows missing. If nuclear torpedoes were on board at the time, it can be assumed that some of them sank inside the bow section. (**Neptune**; **Bradford**)

1970 January 10 The refit of the nuclear-powered attack submarine HMS *Dreadnought* at Rosyth in Scotland was delayed by several months as a result of serious difficulties during the refuelling of her nuclear reactor. This was the first time that such a refuelling had been carried out at a British shipyard. (**Neptune**)

1970 January 29 The submarine USS *Nathanael Greene* ran aground in thick fog in Charleston harbour, South Carolina. It took seven hours to refloat the vessel. (**Neptune**)

1970 January or February A very large explosion occurred at the Gorki submarine yards in the Soviet Union. The Volga River and the Black Sea estuary were contaminated by the incident. (**Bradford**)

1970 April 12 A Soviet November-class nuclear-powered submarine sank in the Atlantic, 300 miles north-west of Spain.

The submarine had been sighted stationary in the water the previous day with her crew attempting to rig towlines to two Soviet ships, but on the morning of 12 April, US naval patrol planes found just two oil slicks. Soviet survey vessels guarded the area for several months afterwards. It is believed that the accident was related to the nuclear propulsion system. (**Neptune**)

1970 May 28 The Polaris submarine USS *Daniel Boone* collided with the Philippine merchant ship *President Quezon* during sea trials off Cape Henry, Virginia. Damage to the submarine was minor, but the *President Quezon* was severely damaged. (**Neptune**)

1970 June 13 The cruiser USS *Little Rock,* flagship of the sixth fleet and armed with Talos missiles, collided with the Greek destroyer *Lonzhi* off the southern coast of Greece during the NATO exercise 'Dawn Patrol 70'. (**Neptune**)

1970 June 16 The destroyer USS *Eugene A. Greene,* armed with ASROC nuclear anti-submarine missiles, collided with the US fuel ship USS *Waccamaw* in the Mediterranean, causing minor damage. (**Neptune**)

1970 November 4 A boiler explosion on board the destroyer USS *Goldsborough,* armed with ASROCs and Tartar missiles, killed two crew members and injured four others. The ship was north-west of Taiwan. (**Neptune**)

1970 November 29 In Holy Loch, Scotland, a fire broke out in the stern of the submarine tender USS *Canopus,* which was carrying Polaris missiles. Two US Polaris submarines, the *Francis Scott Key* and the *James K. Polk,* were moored alongside. Three men were killed in the fire, which took four hours to bring under control. (**Neptune**)

1971 February 2 The French nuclear-powered ballistic missile submarine *Redoutable* collided with a trawler off Brest in France. The trawler was holed and her crew had to be rescued by a French naval vessel. (**Neptune**)

1971 December 29 The nuclear-powered submarine USS *Dace* accidentally discharged 500 gal of water used as reactor coolant into the Thames River at New London, Connecticut. (**Neptune**)

1971 December 31 On two occasions in 1971, distress buoys were accidentally released from US nuclear-powered ballistic missile submarines. The buoys shot to the surface, signalling that the submarines had been sunk by enemy action. Critics claimed that this set off a major US alert and raised the threat of accidental war, but a spokesman for the Pentagon, while admitting that there had been two involuntary releases in 1971 — in the Mediterranean and the North Atlantic — said that home bases had been immediately informed of the releases and there had been no kind of alert. (**Neptune**)

1972 It was revealed for the first time that since 1946 the AEC had dumped enormous quantities of radioactive waste in some 50 ocean dumps off the US coast.

One of the largest of these dumps was near the Farallon Islands, 37 km from San Francisco's Golden Gate, which was the final resting place for 47,500 55-gal drums. An Environmental Protection Agency oceanographer revealed in 1980 that an estimated 25 per cent of the barrels had broken open and that low level radioactive waste had leaked out into an area where fish such as deep sea sole and the commercially important sable fish had been observed. Radioactivity in bottom sediments was 2,000 times greater than background.

The AEC had also licensed the dumping of 28,800 barrels in the Atlantic Ocean. Only a few of these barrels were found to have broken open, but radiation levels at the site were found to be 260,000 times greater than the background level.

News of the leaking canisters prompted a former US Navy Lieutenant Commander, George Earle IV, to reveal for the first time that, in October 1947, he flew three secret missions to drop half a dozen large metal canisters, each containing 2-3 tons of radioactive waste, into the sea 160 km from Atlantic City, New Jersey. (Rosalie Bertell, *No Immediate Danger*, The Women's Press, 1985; K.A.Gourlay, *Poisoners of the Seas*, Zed Books, 1988; Ron Claiborne, 'Bares '47 air drop of atom waste off Jersey', *New York News* 2.1.81)

1972 February 24 A Soviet Hotel II-class submarine was spotted on the surface by a US Navy P-3 Orion patrol plane. The nuclear-powered vessel had apparently lost all power as the result of an accident in which several members of the crew are believed to have died. The disabled submarine was towed home across the Atlantic by Soviet ships in stormy conditions, reaching the waters of the White Sea five weeks later. (**Neptune**; **Bradford**)

1972 March Following tests at the Oconee Unit 1 reactor in Seneca, South Carolina, inspection revealed extensive damage to the reactor coolant system. This was found to have been caused by loose metal parts inside the system; mainly in-core instrument nozzles. Of the 52 nozzles, 21 had broken loose, damaging tube ends and welds, and denting and scratching parts through the reactor vessel and the steam generator. (**Bertini**)

1972 April 11 The Polaris submarine USS *Benjamin Franklin* collided with a tugboat off Groton, Connecticut, sinking the tug. The submarine was undamaged. (**Neptune**)

1972 April 12 During a test run of the 670-MW Wurgassen boiling-water reactor near Kassel in West Germany, a relief valve in the main steam line opened accidentally, allowing steam to flow into the suppression pool at the rate of about six and a half tonnes per minute. The valve could not be closed.

Scramming the reactor immediately would have created severe stresses in the reactor pressure vessel, so the reactor power was gradually reduced. Meanwhile, the temperature in the suppression pool steadily rose and vibrations, caused by powerful pressure pulses in the condensation chamber, damaged the pool, allowing water to leak out into the containment vessel.

The rate of power reduction was increased and the relief valve finally closed. However, because of the loss of water from the suppression pool, a relief valve soon had to be opened in order to reduce the pressure in the reactor.

This final, deliberate opening of the relief valve, and the accompanying rapid temperature reduction, led to severe stresses in the reactor pressure vessel, after all. Had the pressure vessel failed, no safety system could have prevented a catastrophic accident. This was the most severe accident to date in a German BWR, and a core meltdown was a real possibility. In the five days following the accident, more than 1,000 cu m of radioactively contaminated water were discharged into the River Weser.

In 1981, this reactor, and three others of the boiling water type, were shut down by order of the federal Minister of the Interior. The order came after the Bonn Reactor Safety Commission found corrosion and cracking of the primary circuits — the huge steam pipes linking the reactors to the turbines. These had to be replaced completely at a cost of $160 million per power plant. (Dr Helmut Hirsch, Gruppe Ökologie Hannover, original contribution, Dec 1988; WISE, 29.2.81)

1972 June 9 While workmen were in the basement of the Quad Cities reactor complex in Cordova, Illinois, modifying valves in the steam cooling system, a 10-ft butterfly valve slammed shut.

The shock ruptured a seal in the circulation system that carries water from the Mississippi River to cool the steam, and river water flowed into the basement of the turbine building, flooding it to a depth of 15 ft. (**Bertini**)

1972 July 27 Two men at the Surrey Unit 1 reactor in Gravel Neck, Virginia, were attempting to locate a problem in steam release valves that had failed to function. They were severely scalded when a valve released steam into the building, and were rushed to hospital, but both men died four days later. (**Bertini**)

1972 September 1 At the Millstone Unit 1 reactor in Connecticut, water from the Atlantic is used to condense steam from the boiling water reactor. After excessive chlorides were detected in the primary coolant and the reactor had been shut down, it was found that the salt water had corroded the tubes in the condenser and entered the primary system. All 120 of the instruments that measure the power in the reactor were damaged and had to be replaced. (**Bertini**)

1972 October 6 The nuclear-powered attack submarine USS *Tullibee* collided with the West German merchant ship

Hagen while cruising just below the surface in stormy weather some 150 miles off Cape Hatteras, Massachusetts. Neither ship was seriously damaged. (**Neptune**)

1972 October 29 Three people died in a fire on board the aircraft carrier USS *Saratoga* while in port in Singapore. (**Neptune**)

1972 December According to raw CIA intelligence reports, a Soviet nuclear-powered submarine, situated off the east coast of the USA, suffered a nuclear accident when material leaked from a nuclear-armed torpedo. Reportedly, 'Doors were immediately secured in accordance with regulations and some crew members were trapped within the space where the nuclear radiation leakage occurred.' Another raw CIA intelligence report several weeks later reported that a crippled Soviet submarine, having been towed across the Atlantic for six weeks, reached the USSR. 'The crew members trapped in the forward space initially consumed dry rations that were permanently stored in the compartment and later they received food through a small opening from the weather deck. Upon arrival in Severomorsk, crew members were permitted to debark the submarine. Several men died shortly after the accident, others later...The majority of the submarine crew members suffered from some form of radiation sickness.' (**Neptune**)

1972 December 13 A fire in the main machinery room of the aircraft carrier USS *Ranger*, while stationed off the coast of Vietnam, took two hours to bring under control. (**Neptune**; **Bradford**)

1972 or 1973 This record was obtained from CIA files under the Freedom of Information Act by David Kaplan of the Center for Investigative Reporting Inc., San Francisco. It concerns hearsay information on a nuclear incident at the Soviet nuclear testing grounds at Semipalatinsk: 'Between May 1975 and November 1976, source heard from several extended service NCOs who served in the early 1970s at the [illegible] airfield...that a nuclear accident at the Semipalatinsk nuclear testing ground killed an entire company of soldiers who were responsible for maintaining the testing facilities.

Allegedly, the accident took place in 1972 or 1973. The explosion at the testing grounds caused panic at a garrison located about 25 km from the testing ground...The troops, fearing radiation, abandoned the post and ran away into the desert. Eventually, the troops were rounded up by security police and returned to the garrison.'

1973 March At the Japanese Mihama 1 reactor, in the prefecture of Fukui, a disturbance in the flow of coolant broke the top 70 cm off two fuel rods, scattering pellets of uranium oxide through the cooling system in the reactor core. The uranium fuel may have partially melted. The owners, Kansai Electrical Power Company, did not consider the incident to be worth reporting officially and kept it secret for three years.

The Mihama 1 reactor had previously been plagued by leaks in the steam generator tubes and had been repeatedly shut down for repairs. By 1973, one fifth of the tubes were found to be defective and had to be repaired. Further repairs were carried out in early 1974 but, after its restart, the reactor ran for only 42 days at 40 per cent power before radiation leaks were again detected. Inspection of the Mihama 2 and Mihama 3 plants revealed similar problems.

Damage to the tubes was given as the sole reason for the final shutdown of the Mihama 1 plant in July 1974, but two years later a group of workers from the plant disclosed that damage to the core, caused by the 1973 accident, was the true reason. The owners and the government denied this, but a team from the Ministry of Trade and Industry examined the plant and publicly admitted that there was indeed a problem in the core.

In December 1976, the local anti-nuclear group discovered that the broken rods were to be transported by road to the government research station at Tokai-Mura before radiation had fallen to the required level. Trade unionists and political leaders joined a sit-down protest to block the gates of the plant, and 250 riot police were called in to break up the demonstration. (J.Takagi, original contribution; **Cook**)

1973 April 21 The nuclear-powered submarine USS *Guardfish* suffered a leak of primary coolant while running submerged off the west coast of the USA. The vessel surfaced for decontamination and repairs, and four crew members were taken to the Puget Sound Naval Hospital for monitoring. (**Neptune**)

1973 September 5 The US DOD reported that a Soviet Echo II class submarine was seen in the Caribbean south of Cuba with an eight-ft gash in its port bow deck, presumably as the result of a collision with a Soviet cruiser seen in the area with a scrape on its hull. (**Neptune**)

1973 October 7 While shadowing the British aircraft carrier *Hermes* during NATO exercises in the North Sea, a Soviet Kanin-class, nuclear-capable destroyer suffered a fire in a torpedo tube and accidentally released a torpedo. Other torpedoes had to be jettisoned from tubes near to the fire. (**Neptune**)

1973 October 24 DEFCON 3 (see box, p. 173) came at the end of the Middle East War. That afternoon, with the cease-fire breaking down and the Egyptian Third Army under threat of destruction, Sadat called on the Soviets and the Americans to send troops to secure the cease-fire. The US government was determined to resist the introduction of Soviet troops into the Middle East, by force if necessary. Global strategic nuclear forces were placed on alert. (Scott D. Sagan, 'Nuclear alerts and crisis management', *International Security*, Spring 1985)

1973 December 10 An operator at the Surrey reactor in Gravel Neck, Virginia, entered an airlock which keeps pressure in the reactor building below atmospheric pressure, without closing the door behind behind him or depressurising the passageway. The pressure within the reactor building was therefore considerably lower than the pressure in the area in which he was working. As he examined a safety hatch on an upper level for leaks, it suddenly opened. He was sucked through the hatch, flew 20 ft through the air and crashed into a crane. When pressure in the reactor building rose as a result, the reactor was shut down and engineers began to search for the cause, but they were unable to open the door out into the airlock until pressure in the reactor building had been raised to equalise the pressure on both sides of the door. This took about half an hour. The operator was finally discovered and taken to hospital with serious injuries. (**Bertini**)

1973 December 11 Six crew members of the USS *Kitty Hawk* were killed in a fire in the main engine room. The aircraft carrier was positioned 700 miles east of the Philippines at the time. (**Neptune**)

1974 The first Soviet plutonium breeder reactor was built near the town of Shevchenko on the eastern shore of the Caspian Sea. It not only produced electricity but its steam was used to desalinate water from the Caspian. The cooling system, which carries the heat from the reactor to the steam generators, contains liquid sodium metal, an element that reacts violently with water. In 1974, one year after the reactor went into operation, a weld on one of the tubes in a steam generator failed, and 125 gal of water mixed with one ton of sodium. The hydrogen produced by the resulting reaction exploded into a flash fire.

The Soviets claim the accident was not serious enough to shut down the plant, despite US claims that their spy satellites had witnessed the incident. Extraordinarily, Mikhail Troyanov, deputy director of the Institute of Physics and Power Engineering, told US science writers who were in the USSR on a trip arranged by co-operation between the Soviets and the Atomic Industrial Forum (a Washington-based trade organisation made up of major nuclear suppliers): 'This failure was so insignificant that not only was it impossible for sputniks to see it, but the local populace didn't even notice it.'

However, the ruptured coolant tube was not repaired until 1979, when a new steam generator and coolant loop were installed. In the meantime, the plant was forced to operate at 65 per cent of its rated power. (Thomas O'Toole, 'Russia pushes on with breeder reactor', *International Herald Tribune* 6.10.78; Reuter, *Sunday Times* 23.11.88)

1974 January 8 The nuclear attack submarine USS *Finback* collided with the USS *Kittiwake* at the Norfolk Naval Base in Virginia, causing minor damage to the hull of the *Kittiwake*. (**Neptune**)

1974 June An operator at the Canadian Pickering-A nuclear power station, on the northern shore of Lake Ontario, accidentally triggered the emergency core cooling system of the Unit 4 reactor while it was running at full power. The resulting 'water hammer' blew out a gasket and flooded a sump — which should have provided emergency cooling water in the event of a loss-of-coolant accident — to a depth of nine feet, putting the reactor out of action for several months. Essential indicators in the sump which should have warned that the level was rising were not working because all of them had become clogged up. Checks on the other Pickering-A units revealed that the level indicators on these sumps were also clogged.

The Pickering-A station comprises four CANDU (CANadian Deuterium Uranium) reactors, which are cooled and moderated by heavy water. This is the only type of reactor in use in Canada. A loss-of-coolant accident in a CANDU reactor can lead to a runaway chain reaction and a meltdown unless the emergency shut-down system is activated within seconds.

When the station went into operation in 1971, the plant's owners, Ontario Hydro, declared the CANDU reactor to be a model of safety and reliability, placing their faith in the three in-built safety systems — emergency shut-down to stop the chain reaction; emergency core-cooling to remove the decay heat; and containment to restrict any escape of radiation to the reactor building. They were so confident of the reactor's safety that they told the Ontario Royal Commission on Electrical Planning that 'the likelihood of having a CANDU meltdown is...one in ten million or less'.

In the 10 years to 1983, the Pickering plant alone submitted 1,400 'significant event reports', and despite official predictions that the chance of an uncontrolled chain reaction taking place was once in 100 years, such an event occurred six times in the first four years of operation. (Paul McKay, *Electric Empire*, Between the Lines, 1983)

1974 June 26 A CH-47 helicopter ferrying nuclear weapons from Long Island to New Jersey made a forced landing on Jones Beach, Long Island. (**Bradford**)

1974 July At the H. B. Robinson Nuclear Power Plant in Hartsville, North Carolina, a contract employee who had been plugging the steam generator tubes decided to clean up using a vacuum cleaner. After putting on protective clothing but before putting on his respirator, he attempted to start the vacuum cleaner. It didn't work so he opened it to try and fix it. He later found he was contaminated with radioactive dust. A skin cleansing agent did nothing to reduce his level of contamination and later medical examination showed he was suffering from significant levels of internal contamination. After this incident, all vacuum cleaners in the plant were locked away. (Pollard, *Nugget File*, UCS 1979)

1974 August 20 While the Beznau 1 pressurised water reactor at Beznau in Switzerland was running at full power, a disturbance in the external electrical supply caused one of the two turbines to shut down. A valve which should have opened to allow the steam from the reactor to bypass the turbine remained closed and caused the pressure in the steam system to increase. This in turn caused the temperature and pressure in the primary system to rise. Two pressure relief valves opened, but, when the pressure dropped, only one of them closed again, allowing steam and water to flow out into the drain tank and causing the pressure in the primary system to fall too far. The reactor scrammed and the second turbine shut down, but water continued to flow out of the primary system until the operator realised that the relief valve was stuck open and manually closed another valve which sealed it off.

This incident was included in the **Bertini** report because it was 'considered to be a precursor to a more serious accident, such as that which occurred at Three Mile Island in March 1979'. (**Bertini**)

1974 September 27 The *New York Times* reported that a nuclear-capable Soviet Kashin-class guided missile destroyer exploded and sank in the Black Sea whilst on sea trials. At least 275 crew members are believed to have died. (**Neptune**)

1975 January 10 At the Tsuruga-I reactor complex in Japan, 13 tons of radioactive waste liquid leaked from a crack in a storage tank and onto the floor of the building. Decontamination teams were exposed to high levels of radiation. (J.Takagi, original contribution)

1975 May The collapse of seals in a primary coolant pump at the H. B. Robinson Unit 2 reactor led to the escape of 132,000 gal of primary-coolant water, flooding the floor of the containment building to a depth of one ft. (**Bertini**)

1975 November 5 A valve in the off-gas system at the Cooper boiling water reactor in Brownville, Nebraska, was accidently left closed, diverting hydrogen, oxygen and steam through a sump. Two workmen were investigating the cause of a pressure rise in the sump when a spark from an electrical air sampler being turned on ignited the mixture of gases, injuring both men and releasing some radioactivity at the site of the explosion.

Two months later, while the reactor was operating at 88 per cent of full power, an alarm indicated that the airflow from the chimney carrying the off-gas into the atmosphere was reduced. Switching on fans in the system made no difference, so the shift supervisor and an operator went to investigate. On entering the off-gas building, they detected an unusual smell and noted that radioactivity levels were off the scale of their meters. 'They evacuated the building forthwith, and shortly thereafter the building exploded. It was completely demolished. The reactor was shut down immediately.'

It seems that a plug of ice had formed in the chimney and prevented the gases from escaping. The explosion was probably caused by a spark from machinery. Some radioactivity was released in the area around the explosion, but levels at the site boundary were 'below maximum permissible concentrations throughout the accident'. In eleven days the building had been rebuilt with a modified stack and the reactor was back in action. (**Bertini**)

1976 early May A Norwegian fishing vessel off Murmansk snagged the fin of a Soviet nuclear-powered attack submarine that was travelling 450 ft below the surface. The submarine surfaced and the Soviet crew cut the cables away using hammers and chisels. Soviet rescue ships towed the submarine towards Murmansk. Two months later another Norwegian fishing vessel was towed backwards for about a mile in the Barents Sea, close to the Murmansk Base, when its nets were snagged by the bow of a Soviet November-class nuclear-powered attack submarine. (**Neptune**)

1976 August 28 In the Mediterranean, a Soviet Echo II class nuclear-powered cruise-missile submarine struck the USN frigate *Voge* on its port quarter, splitting the *Voge's* bulkheads, buckling plating and damaging the frigate's propeller. The *Voge* had to be towed 150 miles north-east to its base at Souda Bay, Crete. The Soviet submarine suffered damage to its conning tower. (**Neptune**)

1976 September 14 An F-14 Tomcat aircraft armed with a Phoenix missile rolled off the deck of the aircraft carrier USS *John F. Kennedy* and sank in 1,890 ft of water, 75 miles north-west of Scapa Flow, Scotland. (The aircraft was recovered on 11 November.) On the same night, the *John F. Kennedy* collided with the US destroyer *Bordelon* during an underway refuelling operation. (**Neptune**; **Bradford**)

1976 October According to CIA intelligence reports, a Soviet nuclear submarine suffered a fire in its launch compartment, killing three officers. (**Neptune**)

1976 October 8 The nets of a Japanese fishing vessel snagged the conning tower of a Soviet Charlie-class cruise missile submarine 160 miles off the coast of the Kamchatka Peninsula, dragging the trawler backwards. The submarine surfaced and the nets were cut free. (**Neptune**)

1976 October 25 Swedish seismographs registered an earthquake in the sea passage between Finland and Estonia. Suspicions that the cause of the tremor was an accident at the nearby Soviet base of Paldiski — where nuclear-armed submarines are housed in pens cut into chalk cliffs — were raised when copies of the newspaper *Sovietskaya Estonia* reached Sweden. The four issues of 26-29 October recorded 20 deaths, a number of which were described as occurring 'suddenly'. All of the casualties had Soviet names. They included: Alexander Napokjuks, a major of the army medical corps reserve; Maria Zjurba, a woman doctor of the state police; Bartholomey Blintsov, a member of the police school in Tallinn, and Yuri Semyonov, a member of the marine officers' college.

The last time an earthquake was recorded in the area was 1953. Finnish authorities found no traces of radioactivity.

The Soviets said nothing, and the incident is still shrouded in secrecy. (Lars Persson, 'Soviet quake disaster rumoured', *Sunday Times* 21.11.76; Colin Narbrough, 'Nuclear "Earthquake"', *Observer* 21.11.76)

1976 December 19 An F-14 Tomcat aircraft, landing on the deck of the USS *Enterprise,* struck two other planes, veered out of control and crashed into the South China Sea. (**Neptune**)

1976 or 1977 Some time during this period, a fire aboard the nuclear-powered ballistic missile submarine HMS *Repulse* caused £200,000-worth of damage. (**Neptune**)

1977 CIA intelligence reports say that a Soviet nuclear-powered submarine was forced to surface in the Indian Ocean to fight a fire on board. The fire took several days to extinguish, and the submarine was then towed to Vladivostok by a Soviet trawler. (**Neptune**)

1977 March 20 An operator at the Rancho Seco reactor in California was replacing a bulb in a control console when a loose bulb fell into the electrical circuitry below. This caused a short circuit, cutting off the electricity to about two thirds of the non-nuclear instruments, whose readings inform the integrated control system (ICS) which controls valves and other equipment.

Responding to incorrect signals, the ICS shut off the supply of feedwater to the steam generators so that they no longer performed their function of cooling the primary cooling circuit. The temperature and pressure in the primary circuit rose and a relief valve vented the excess pressure, but attempts to switch in a supply of feedwater to the steam generators were bedevilled by the ICS which was still responding to erratic and erroneous signals caused by the short circuit. Normal functioning was eventually restored after 70 minutes.

A similar incident, when a system associated with the steam generator feedwater supply partially shut down for reasons that are still unknown, occurred at the Davis-Besse reactor at Oak Harbor, Ohio, six months later. (**Bertini**)

1977 July 27 In Japan, a routine inspection revealed 1-in-deep cracks on the inside of the Fukushima I-I reactor pressure vessel, raising questions of safety at all Japanese boiling-water reactors. Workers carrying out repairs inside the vessel suffered radiation exposure. (J. Takagi, original contribution)

1977 September 20 The nuclear attack submarine USS *Ray* struck the sea-bed in the Mediterranean south of Sardinia, causing three crew injuries. The *Ray* then cruised to La Maddalena naval base on Sardinia with an escort. (**Neptune**)

1977 October 11-13 Sea-water accidentally entered the cooling system of the Hunterston-B AGR on the River Clyde after technicians had rigged up a temporary pipe and then left it in place. Six thousand gal of salt water poured into the core of the reactor, leaving salt deposits on the stainless steel walls of the pressure vessel. Fortunately the reactor was undergoing maintenance at the time and had been shut down. Repairs took two and a half years and cost £15 million. The South of Scotland Electricity Board put the cost of buying in electricity, while the plant was out of action, at about £50 million. (The Safety of Nuclear Power Plants, Uranium Institute)

1978 February 3 A B-52 bomber at Robins AFB, Georgia, was vandalised by base personnel. The plane was on alert at the time and was reportedly armed with short-range attack missiles and nuclear warheads. One missile had been tampered with and 'looked like somebody beat on it with a hammer'. (**Bradford;** *Washington Star* 6.2.78)

1978 June 18 At the 800-MW Brunsbuttel BWR near Hamburg in West Germany, a leak in a steam line allowed two tonnes of radioactive steam to escape into the atmosphere, and emissions of iodine-131 exceeded permissible limits. In spite of the leak, and in contradiction to the rules laid down in the operators' handbook, operation was continued for 2 hours 41 minutes after the leakage began. An automatic system would have shut down the reactor after five minutes, but operators had switched this off for reasons of economy. It was later learned that this was a common practice. (Dr Helmut Hirsch, Gruppe Ökologie Hannover, original contribution, 1988)

1978 August 19 A Soviet Echo II class cruise missile submarine was sighted dead in the water near Rockall Bank, 140 miles north-west of Scotland. Power loss was attributed to problems with the nuclear propulsion system. The following day it was observed under tow south of the Faeroe Islands. (**Neptune**)

1979 In 1979 there were 2,310 mishaps at the 68 US nuclear power plants. One quarter of these were due to human error. (**Public Citizen** 1979-87)

1979 Several valves at the USA Palisades plant were found to have been locked open for a year and a half. Had an accident occurred, these would have allowed the release of radioactive materials. The NRC fined the plant's owners,

Consumers Power Company, $450,000, but later halved the fine. (**Public Citizen** 1979-87)

1979 February 5 The United States guided missile destroyer *Farragut,* while on a training cruise in the Caribbean, fired eight training rounds. These landed about four miles from a Soviet research vessel which had strayed into the gunnery range area while towing a Soviet Foxtrot class submarine. (Byers)

1979 March 7 Entangled in the nets of a Scottish trawler in the Sound of Jura off the west coast of Scotland, the nuclear-powered submarine USS *Alexander Hamilton* towed the fishing vessel backwards for about 45 minutes, until the nets were cut by the trawler's crew. (**Neptune**)

1979 May 2 A faulty electrical system scrammed the Oyster Creek BWR in the USA, shutting down the coolant recirculation and feedwater pumps and preventing coolant from entering the system to remove the decay heat. The core had to be cooled using the isolation condensers, and the reactor was brought to a cold shutdown. (The Safety of Nuclear Power Plants, Uranium Institute)

1979 May 11 Primary coolant water leaked from one of the two nuclear reactors aboard the USS *Nimitz.* (**Neptune**)

1979 June 20 The submarine USS *Hawkbill* suffered a leak in the reactor primary cooling system. The US Navy stated, 'The leakage was caused by normal wear of inside parts of valves. Such leaks happen occasionally.' (**Neptune**)

1979 December 3 A loosened plug in the primary coolant circuit at the Japanese Takahama 2 reactor caused 80 tons of coolant to leak out over a period of nine hours. (J. Takagi, original contribution)

УЧАСТНИКУ
ЧЕРНОБЫЛ

80S

1980 onward

French Nuclear Tests, Part 3

On 23 March 1980, the same day that France exploded a 50-kt nuclear device on Moruroa, French Defence Minister Yvon Bourges gave a press conference in Tahiti. He said of the underground tests: 'At the moment of explosion, there is a process of vitrification which happens and the products of the explosion become fused in the chamber, that is to say, in the rock...we are taking precautionary measures to a ridiculous degree.'

In May 1980, a report in the *Washington Post* (October 1979) alleged that underground testing had left the Moruroa atoll like a 'Swiss cheese' — a remark described by the French Defence Ministry as 'grotesque'.

During the 1970s, Moruroa had been hit by a number of tropical cyclones. The 1981 test series was already under way when, on 11-12 March, a particularly severe storm ripped off layers of asphalt covering a radioactive waste dump and debris on Moruroa's northern beach and swept it into the ocean. News of this was not disclosed until November when three CEA engineers, members of the Socialist Confédération Française Démocratique du Travail (CFDT), the trade union that represents technicians employed at Moruroa, broke the French Defence Secrets Act and revealed their story to the French press. Their testimony, that radioactivity levels on the atoll had doubled in the past four months, substantiated the demands of 2,500 civilian technicians on Moruroa who, the previous month, had threatened to strike unless the radioactive waste released by the storms was cleared up. (There were three other cyclones in 1981 alone which may have dispersed radioactive material across the atoll. From 1983 onwards, the test schedule was adjusted so that all tests occurred outside the December to April cyclone season.)

François Mitterand had been elected President in May 1981. On 4 August that year, France's new Socialist Minister of Defence, Charles Hernu, released the following statement: 'The press must be informed about the security problems. If there is an accident, it is better to let the truth be known than to let all sorts of rumours spread. Nothing must be hidden that affects the health of the population. When New Zealand and Australia ask for information about these problems, we shall supply it.' (Greenpeace 1981)

This failed to stop Polynesians; French trade unionists and politicians; and New Zealand and other Pacific countries campaigning for an end to the testing.

In July 1981, the French magazine *Actuel* reported that Polynesians suffering from cancer were being flown out in secret by military planes to military hospitals in France. There were 50 such cases recorded in 1976, 70 in 1980 and 72 in the first six months of 1981. This prompted the French government to send an eight-man scientific team, headed by French vulcanologist Haroun Tazieff, to Moruroa in late June 1982. After just three days on Moruroa, during which they witnessed what was claimed to be the smallest ever nuclear test — less than 1 kt — they confidently told the media they could find no evidence of radiation leakage. (However, the scientists were forbidden to visit Moruroa's north shore, where the nuclear waste dump was located.) This satisfied no one and protests continued to mount.

On 23 February 1983, Moruroa was hit by Cyclone *Orama*. On 23 March, Hernu announced that the test series, already late that year, was to be postponed for 'technical and meteorological reasons'. Cylones *Reva* and *Veena* swept the area in March and April. A week later, an article in the *New Zealand Herald* (30.3.83) reported that CEA officials had admitted that 'Moruroa is beginning to show signs of wear and tear from the tests of the past decade.'

On 19 April, a 50-kt device was exploded at Moruroa whilst Hurricane *William* was passing through the region. By June, a controversial map of the atoll was circulating in Tahiti. The map, alleged to be the responsibility of French trade unionists, showed four major cracks around the 60-km coral rim of the atoll; a severely contaminated area, apparently where nuclear waste has been stored; and 'forbidden entry' zones.

The continued concern expressed by Pacific countries, that testing at Moruroa was less safe than officially claimed, led to a further mission to the atoll by five scientists from Australia, New Zealand and Papua New Guinea for four days in October 1983. Conducted under strict French supervision, the Atkinson mission (led by Hugh Atkinson, Director of New Zealand's National Radiation Laboratory) was not permitted to observe any tests; inspect any test sites; visit the dump site; or collect any sediments from the lagoon for testing.

Their report, released on 8 July 1984, was claimed by the French as evidence that their test programme was safe. In

October they announced that testing would continue for at least another 15 years.

The controversy continues unabated. In June 1987 Jacques-Yves Cousteau and the crew of the *Calypso* spent five days at Moruroa during which they were allowed restricted access. On 10 November 1988, the Cousteau Foundation presented its preliminary scientific findings in Paris: principally, that analysis of water, sediment, and plankton samples did not show significant radioactive contamination. The report concluded that the tests posed no health risk to French Polynesians.

However, the underwater film Cousteau and his crew shot off Moruroa confirms the existence of spectacular cracks and fissures, submarine slides and subsidence. The French military officials subsequently announced their plans to stage the largest underground tests at Fangataufa, to prevent further damage at Moruroa.

The last test in the 1988 series of eight blasts was duly conducted at Fangataufa, but this transfer raises even more problems. Fangataufa is an even smaller atoll than Moruroa, only six by 10 km in area. Shafts cannot be drilled on the island because it remains contaminated from previous atmospheric tests there. So a drilling platform has been set up to enable the French to drill shafts into the base of the atoll, below the waters of the lagoon.

In October 1988, the local government in Tahiti petitioned Paris for the formation of an investigative medical commission to look into the health effects of nuclear testing in the region.

In February 1989, the European parliament narrowly rejected a resolution calling for an international, independent commission of scientists to investigate the environmental impact of the French testing programme. (The French socialist government put pressure on Spanish and Greek socialists to block the resolution.)

The National Resources Defense Council (NRDC) report, *French Nuclear Testing 1960-1988*, summarises their organisation's key findings as follows: 'France has conducted 172 nuclear tests since 1960, almost ten per cent of the nearly 1800 nuclear tests conducted worldwide since 1945, yet has produced less than one per cent of the approximately 100,000 warheads produced worldwide in that same time period...For each French warhead type, some 20 nuclear tests have been conducted. Since 1963, France has produced approximately 800 nuclear warheads of eight major types.'

CIGUATERA

Since the 1960s, there has been a sharp increase in the number of Pacific islanders suffering from *ciguatera*, the commonest form of food poisoning. According to a study by Tilman A. Ruff of the Monash Medical School in Victoria, Australia (*The Lancet* Vol 1, 1989), this is largely caused by the effects of nuclear tests and related support operations on coral reef ecology.

Damaged reefs support abnormally high numbers of a species of plankton called *Gambierdiscus toxicus*. The fish who eat them absorb their toxin but remain unharmed; humans who eat the fish suffer from vomiting, diarrhoea, loss of balance and co-ordination. There have even been fatalities.

The disease organism was named after the Gambier Islands in French Polynesia, where it was first recorded in 1968 after the construction of a military base. Between 1960 and 1984, the average number of *ciguatera* attacks per islander was 5.7.

On Hao, *ciguatera* was unknown before 1965 — when the French first developed the island as a support base for the nuclear testing programme — but by mid-1968 it had affected 43 per cent of Hao's 650 inhabitants.

At its peak in the early 1970s, *ciguatera* poisoning affected 12 out of every 1,000 inhabitants of French Polynesia; the annual incidence rose 10-fold between 1960 and 1984. Similar effects are reported from the Marshall Islands.

1980 onward

The Indian Nuclear Programme

TARAPUR

The Tarapur Atomic Power Station (TAPS) was one of the first nuclear power plants in Asia, and was supplied by the US corporation General Electric and its contractor, Bechtel, on a turnkey basis. This meant that the Department of Atomic Energy (DAE) was not involved in any aspect of the design, engineering, construction or testing of the reactors. Consequently, when the plant began to run into problems, soon after it started commercial power generation in 1969, the DAE was unable to cope.

According to a 1986 report by the Centre for Science and Environment (CSE) in New Delhi: 'This jewel of the Indian nuclear crown is also the world's worst performing nuclear power station. It is estimated to be the most polluted atomic power plant in the world...The toll it extracts from its workers, measured in total human-rems, is the highest known anywhere in the world in relation both to its wattage and energy generation.'

The first major problem in Tarapur's boiling-water reactors was fuel failures, as the CSE explains: 'When a fuel bundle fails, the zircaloy cladding develops leaks and the highly radioactive nuclear fuel waste spills into the coolant tubes. This creates high radioactivity regions — "hot" regions — within the reactor and as the radioactive material also spreads into the coolant circuits, their tubes and valves, and finally to the steam generators, condensers and turbines, the entire power station is quickly contaminated.'

But fuel failure was by no means the only problem at the plant. The CSE claims: 'The station bristles with problems: steam generator tube leaks, recirculation pump failures, control valve leaks, failure of control rod drives (needed to regulate and shut down the reactor), numerous cracks in system piping, extensive corrosion, leaking condenser tubes, and low vacuum in condenser, large leaks in primary to secondary water line, in-core high flux, feedwater pump failures and — very important — numerous failures of the electronic monitoring and control systems, and of instrumentation.' In addition, says the CSE, the TAPS reactors are subject to problems generic to boiling-water reactors — 'intergranular stress corrosion cracking' — which affects dozens of steam pipes, elbow joints, valves and welds.

As a result, the Tarapur plant experienced 38 'unusual occurrences' in the first five years of commercial power generation. By 1980, the number of incidents had risen to 344. More than 70 per cent of these were due to equipment or design deficiency.

The worst accident of all occurred in March 1980, when a routine check revealed a small crack in the primary coolant tube. It was decided to remove the two-ft-long cracked tube and weld in a new piece of pipe. This was a far from easy operation, as the Indian magazine, *Sunday* (1980), explains:

'The primary coolant tube carried hot radioactive water from the reactor core and a small crack is enough to drain out sufficient radioactive water, contaminating the area in a flash of a second. Slow but steady leakage would invariably lead to a fall in the water level of the reactor core, causing the temperature of the core to shoot up relentlessly. This would bring up the temperature to an awesome several thousand degrees — the fission energy pumping up the temperature even further. A core meltdown would then become inevitable. The standard procedure to check the outpour of hot radioactive water is to seal off the tube for a short while and weld in the crack. The sealing in Tarapur was done by using an "ice plug" of liquid nitrogen, penetrating deep inside the tube, hopefully withholding the pressure of water for a sufficiently long time.

'The "accident" on March 14 at 8 a.m. occurred when the "ice plug" did not remain inside the coolant tube long enough and was pushed out by the gushing water, after the bad pipe was cut out and before the new pipe was welded in. A temporary plug of a sort was then somehow pushed in and held in position by chains. The gush was controlled but the trickle wasn't. In the initial gush, an awful lot of hot radioactive water came out on the hall floor, causing panic and hysteria. The level of water in the core by that time had already gone down, and this was the moment when all hell broke loose...A routine check, some careless decisions, mistiming, lack of precision, a slightly longer delay than allowed by a reactor handbook, within minutes can transform a simple routine job to a nerve-shattering, horrifying crisis.'

The CSE claim that 'The DAE got hold of unskilled and illiterate labourers from a nearby village, and made them get into the highly radioactive water in order to retrieve the plug. No one knows what radioactivity doses the men received but it is reasonable to conclude that they must have been very high.'

Up until the end of 1982, a total of 10,806 of these unskilled workers had been used at TAPS to work in highly radioactive areas. According to one Indian journal quoted in *Nature* (7.6.79): 'Tarapur is so heavily contaminated...that it is impossible for maintenance jobs to be done without the personnel exceeding the fortnightly dose of 0.4 rems in a matter of minutes. Thus the maintenance worker...holding a spanner in one hand and a pencil dosimeter in the other, turning a nut two or three rotations and rushing out of the work area, is a common phenomenon at Tarapur.'

RAJASTHAN

The Rajasthan Atomic Power Station (RAPS) had the worst performance of any CANDU reactor in the world. The Unit 1 220-MW pressurised heavy-water reactor had to be shut down 251 times in its 10-year life — roughly once a fortnight.

In 1982, abnormal radiation conditions were detected around the reactor vessel and an inspection revealed serious cracking in the vessel's end shield. Unable to enter the area because of the high radiation levels, engineers spent four years using remote-control equipment to try and weld the cracks, but failed.

The problem is believed to have been caused by unexpectedly rapid radiation embrittlement of the carbon steel plates that make up the end shield.

Construction cost overruns for Unit 1 were 76 per cent and for Unit 2, over 90 per cent. Both reactors have also suffered from heavy water leakage — up to 80 kg a day. In addition, there have been turbine failures, valve leaks, pipe breakdowns and malfunctions of the fuelling machines. Large numbers of workers at RAPS have been exposed to ionising radiation and tritium. RAPS-1 was shut down in March 1982 and has not been started up since. Two further 235-MW units are currently under construction.

NARORA

Controversy surrounds the CANDU reactors being built at Narora in western Uttar Pradesh. The two units planned are only about 60 miles from Delhi, in a geologically unstable area just 56 km from the epicentre of a major earthquake which took place in 1956. Survey holes for the foundations of the plant were to be drilled to 300 m, three times the normal depth, but even at that depth, no solid rock bottom could be found. The twin reactors are also near the banks of the River Ganges and in an area of dense population.

Dr Dhirendra Sharma, who has led the protest against the plant, has interviewed people in Narora who say sub-standard building materials have been used in its construction. He comments 'If Bhopal had occurred at an atomic power station at Narora, millions living a thousand kilometres down-river in the fertile plains of Uttar Pradesh, Bihar and Bengal, would have suffered irreversible genetic damage. Not for one generation, but for thousands of years, the area would become inhospitable.' (1986)

The plant at Narora became operational at the end of 1988. Its opponents argue that it is not only unsafe but unnecessary, since the 470 MW of electricity that it is designed to produce could be obtained at a lower cost through hydroelectric schemes. (India has two 235-MW CANDU reactors in operation in Madras and four other units being developed — at Kakrapar and Kaiga.)

HEAVY WATER

The production of sufficient heavy water, used as the moderator in CANDU thermal reactors, has proved to be the biggest problem for the Indian nuclear programme. Water contains 0.015 per cent deuterium, the heavy form of hydrogen, and it takes 10 tons of water to extract 1 lb of heavy water. Water is treated with compounds like hydrogen sulphide or ammonia and then undergoes elaborate processes of electrolysis or distillation.

The Indians use the ammonia-hydrogen exchange process and their heavy-water production plants at Baroda, Tuticorin and Talcher work in tandem with fertiliser plants. The ammonia is produced from coal which is also the source of ammonia fertiliser.

The $60 million Talcher plant, situated 350 miles south-west of Calcutta, was built in 1972 to a West German design, but only began producing heavy water in December 1985. Its construction was delayed by two years when two towers, en route from West Germany, fell into the sea off Portugal.

On 30 April 1986, a major fire broke out after a gas pipeline broke. Hundreds of workers and families in the area fled from the blaze, which was brought under control by seven fire companies after three or four hours. The plant's control room and pumping room were destroyed in the fire.

At the Baroda plant, design faults led to an explosion during trial runs in 1977, and it was put out of action until 1981. The plant at Kota was shut down on 23 October 1984 after a leak of hydrogen sulphide resulted in the death of an engineer and injuries to two others.

The Tuticorin plant has had many technical problems and has so far only achieved a third of its capacity.

Science journalist Nagesh Hegde has commented (1987): 'The tragedy of the CANDU-type reactors is that they are dependent on heavy water which in turn depends on smooth running of fertilizer plants which in turn have to work with erratic supply of electricity if they rely on nuclear power plants. It's a perfect vicious circle.'

INDIA'S BOMB

The architect of India's nuclear programme was Homi Bhabha, who had received an education in nuclear physics in Europe and then used his family ties with the Tatas, owners of the largest industrial empire in India, to establish an Atomic Energy Commission in 1948. Bhabha was chairman of the Commission; a post he held until his death, in an aircraft near Geneva, in 1966.

With Nehru as its patron, the Indian AEC was insulated from democratic checks and continued to work in utmost secrecy until the 1970s, when its cover was breached for the first time. Bhabha's long-term strategy was based on the fact that India has the world's largest deposits of thorium. His plan was to build natural uranium reactors to produce plutonium. This would be burned in breeder reactors, using a thorium blanket, to produce uranium-233 which, in turn, would fuel thorium breeders.

Bhabha made an agreement with the Canadians in 1956 which gave India its first plutonium-producing heavy-water reactor, and he later used US technology to build a reprocessing plant to extract plutonium from spent fuel rods. When China exploded its first bomb, Bhabha began preparations for India's first nuclear test. Some 20,000 people were relocated from an area of the Pokhran desert in Rajasthan and, on 18 May 1974, a 15 kt device was exploded there 100 m below the surface. The coded message sent to New Delhi to signal the success of the blast read: 'The Buddha is smiling.'

3 June 1980

NORAD Computer Glitches Colorado

During the early 1960s, a series of caves were blasted out of the rock in Colorado's Cheyenne Mountain, located at the southern end of the front range of the Rocky Mountains near the city of Colorado Springs, to form the North American Aerospace Defense Command Center (NORAD).

The steel buildings constructed inside this installation are connected to a worldwide net of sensors composed of early-warning satellites in outer space and ground-based radars. The data from these, and from other intelligence sources, is processed by 87 computers and evaluated by dozens of technical experts. The aim: to determine whether the Soviet

Union has launched a missile or bomber attack against North America.

At 02:26:00 Eastern Daylight Time on 3 June 1980, inside the underground post at SAC Headquarters at Offut AFB, Nebraska, the alarm went off on one of the two video screens the Warning System Controller (the 'WISC') uses to track the path of incoming missiles. Two Soviet SLBMs were inbound from the North Atlantic.

In addition to calling the Pentagon for more information, the senior controller also alerted SAC's 76 B-52 bomber crews, eight two-man crews of the FB-III nuclear bombers, and the 240 missile crewmen. Then he consulted the NORAD alert crew in their Cheyenne Mountain bunker. The incoming missile counter was clicking up missile tracks so fast that, within seconds, hundreds appeared to have been launched, and NORAD duty officers searched their early-warning system for some confirmation of the attack. When none was found, the SAC controller ordered his bomber crews to shut off their engines but stay on alert.

The missiles continued to appear, always in twos, at SAC, NORAD, and also on the computers at the National Military Command Centre (NMCC) in the Pentagon. The three controllers held a threat assessment and agreed the problem was a fault in NORAD's computer relay system. The alert ended at 02:29:12. It had lasted three minutes and 12 seconds.

Following the 3 June incident, NORAD began an investigation into its cause. In order to do this, they tried to duplicate the error by running the system in the same configuration, hoping to reproduce the same erroneous data and thus isolate its source.

On 6 June at 15:38 the mistake was reproduced on screens at SAC and NMCC, which registered indications of an ICBM attack — but only on the process data coming from NORAD. A thorough search of the system led NORAD to the conclusion that the failure was caused by a faulty integrated circuit — a tiny, 46-cent silicon chip manufactured in Taiwan that had short-circuited in a communications multiplexer.

The multiplexer takes the analysis from the NORAD computers and puts it into message form for transmission to the other command posts. It forms the message which NORAD transmits continuously to all command posts in order to have a continuous check on the conditions of the circuits. According to Hart and Goldwater (1980): 'Normally, in that part of the message which indicates how many SLBMs or ICBMs

THE MISSILE ATTACK WARNING SYSTEM

The Missile Attack Warning System consists of three major parts: 1. satellite and radar sensors to detect missile launches; 2. computer centres and communication links to analyse and distribute the data from the warning sensors; 3. command posts where the implications of the warning are assessed and appropriate actions are taken.

Because the time between launch and impact of Submarine Launched Ballistic Missiles (SLBMs) is so short — a matter of minutes — data from the satellites and from the radar system dedicated to SLBM detection (Pave Paws, as it is known) is fed directly to all four major command posts. These are located at NORAD, SAC headquarters, NMCC at the Pentagon, and the Alternate National Military Command Center (ANMCC) at Fort Ritchie, Maryland. Any other data is fed only to NORAD where it is analysed and the results transmitted to the three other command posts.

When there is any indication of a real threat, the four command posts begin a formal conferencing procedure to evaluate and assess the situation. There are three types: missile display conferences, threat assessment conferences and missile attack conferences.

From 1977 to 1982 inclusive, there were 15,206 missile display conferences, of which 14,502 were routine and 744 were to evaluate possible threat. The commander-in-chief of NORAD makes a judgement on each, and, if there is a possible threat, a threat assessment conference is convened.

At a threat assessment conference, more senior people than the duty officers are drawn into the evaluation, the nature of the threat to North America is established, and decisions are taken on the preliminary steps to 'enhance force survivability'. In 1979 and 1980 there were four threat assessment conferences, including the 3 June incident. They were called on the following occasions for the following reasons: 3 October 1979: an SLBM radar at Mount Hebo picked up a rocket body that was in a low orbit, close to decay, and generated a false launch and impact report; 9 November 1979 : a 'war game' simulation tape was accidentally loaded into the NORAD computer system and generated false indications of a mass raid; 15 March 1980: four SS-N-6 SLBMs were launched from the Kuril Islands in the Pacific as part of a Soviet troop training manoeuvre. One of the launchers generated an unusual 'threat fan'.

The last step in the assessment process is the missile attack conference. According to Hart and Goldwater (1980): 'If the threat persists, the final action taken is to convene a missile attack conference which brings in all senior personnel including the President. No such conference has ever been convened.'

have been launched, the display will indicate zeros. The effect of the failure of this particular integrated circuit was to fill some of those zeros with the number 2 and to do it on a random basis both as to which command post received the data and to which fields of data had the 2's in them.'

Later, NORAD was to claim that the chip was fine, and that the problem lay in the circuit board, designed by Ford Aerospace. Some computer experts remained sceptical of these explanations and believe that there was a fundamental mistake in the design of the communications device, namely that it didn't have an adequate error-checking capability. NORAD's response to outside criticism was to prevent information about the number of false alarms the system was generating from leaking out.

Weaknesses in NORAD still exist. According to Babst et al. (1985): 'A major factor which contributed to the failure of the modernization program was the 1970 decision by the Joint Chiefs of Staff to force NORAD to use (for internal computer communications) the computer equipment of the WWMCCS, Worldwide Military Command and Control System [see box, p. 258], which was inadequate for NORAD's requirements...This failure has resulted in the expenditure of more than $1 billion in a command and control system that has been judged by the General Accounting Office (GAO) as incapable of meeting a crisis condition.'

The GAO has stated that moving ahead with the $281 million computer-equipment purchase presents 'a very high risk to overall communications for Cheyenne Mountain.' The agency claims the proposed system has 'unstable software' which has prevented Air Force officials from issuing commands through the system and which engineers have been unable to fix. 'The problem could recur at any time, causing loss of communications,' say the GAO. 'A failure of this kind during actual operations could seriously affect the Air Force's ability to satisfy mission requirements during a crisis.'

In addition, the system's wiring is incompatible with the other electronics at Cheyenne Mountain, a fact the Air Force knew when issuing specifications for the system to replace the ageing computer communications inside the nerve centre.

WIMEX

The Worldwide Military Command and Control System (WWMCCS), pronounced Wimex, was conceived in 1963 when President John F. Kennedy expressed interest in a computerised warfare system after he had found himself unable to keep track of events during the Cuban missile crisis.

The $15,000 million computer-based system, which connects all US defence centres, surveillance satellites and early warning radar stations, has already cost each American taxpaying family an average of $160. Of this, about $1,000 million has been spent on the 35 computers which are 'the brains of the system's giant global web of sensors, telecommunications links, intelligence tie-ins and 26 major command centers worldwide.' (*San Francisco Chronicle* 4.11.79)

The Wimex project is in 'disaster mode'. The entire network is built around the Honeywell 6000 series of computers, first manufactured in May 1964 by General Electric. They are two full generations behind the latest computer technology. They are so slow and imprecise they would take 10 to 15 minutes to plot the trajectories of just 10 incoming enemy missiles. Wimex's best shot at pinpointing the missiles' targets would leave a margin of error of 4,000 miles. In 1977 an exercise called *Prime Target* attempted to test inter-computer communications between Wimex and other key defence computer systems. Overall, the computers worked only 38 per cent of the time.

18/19 September 1980

Titan II Missile Fire, Arkansas

The 103-ft-long Titan II missile first went into operation in 1963. At the peak of its deployment, in 1978, there were 58 of them in underground silos around three US AFBs: Davis-Monthan in Arizona, Little Rock in Arkansas, and McConnell in Kansas. Its nine-Mt nuclear warhead, with a yield-to-weight ratio some 400 times greater than that of 'Fat Man', was the largest in the USAF arsenal. The US Office of Technology Assessment estimated that a Titan II airburst over Leningrad would result in 2.4 million fatalities and 1.2 million injuries.

The accident in silo complex 374-7, near Damascus, Arkansas, began at 18:45 on 18 September when a maintenance man, dressed in a clumsy protective suit, dropped a detachable seven-kg socket head from the end of his spanner.

It fell 20 m down the silo, ricocheted off a concrete mounting, and slammed into the half-cm-thick aluminium skin of the Titan II missile; punching a 3.5-in hole in the first-stage pressurised fuel tank and releasing a vapour stream of highly toxic and inflammable aerozine-50 fuel into the silo.

The maintenance crew quickly evacuated the silo and the missile launch crew, housed in a three-storey underground shelter, closed the 750-ton reinforced concrete and steel silo doors and the exhaust ducts. Temperatures inside the silo began to rise, creating pressure on the remaining fuel tanks. The emergency spray system cut in, releasing 100,000 gal of water, but this failed to halt the leak or absorb the fuel.

At 22:00, SAC headquarters ordered the launch crew to evacuate the site, after first cutting power supplies in order to prevent an unauthorised launch, and ordered the evacuation of homes within three km of the silo.

Titan II missile silos were equipped with giant exhaust fans designed to start automatically when deadly fuel or oxidiser vapours reach a certain level. These fans, however, had been disconnected several months before, on orders from SAC headquarters following a previous mishap. (In April 1980, an oxidiser leak in a silo at Potwin, Kansas, had set off the fans, sending the toxic red vapours into the night air before farmers could be warned and evacuated. SAC subsequently directed that these fans should only be used when the surrounding area had been cleared of civilians.) As the vapour built up in the silo, it was suggested that the fans be reconnected, but the proposal was rejected.

The silo was also equipped with remote-controlled valves to allow venting, but since the power supply had been cut off, these could only be operated manually from within the silo. Experts from the USAF and from the company that designed and built the missile advised against sending men in to do the job; this advice was disregarded by SAC vice-commander Lieutenant-General Lloyd R. Leavitt Jr.

At about 02:00, twin two-man entry teams and a support crew of 10 other airmen moved up to a perimeter fence only 150 ft from the smoking missile silo. The first team, wearing protective suits and breathing apparatus, and carrying portable vapour detectors, went in to assess the situation. They made their way down four flights of stairs and found limited traces of vapour, but had to return to the surface when their oxygen supply ran low.

The second two-man team, David Livingston and Jeff Kennedy, retraced the first team's steps, but when they opened a blast-proof door 40 ft down inside the silo, their vapour detectors gave a maximum reading. An explosion was extremely likely and they were ordered to leave the silo. At 15:00, just as they reached the surface, the gases in the silo exploded, and blasted the two men 40 ft into the air, fatally injuring Livingston and severely injuring Kennedy. Of the other USAF personnel, 21 were injured.

The explosion blew off the silo door and catapulted the missile's 2,700-kg re-entry vehicle, containing its nine-Mt nuclear warhead, some 600 ft into the air. A Defense Department official later claimed that the warhead — which landed 450 m north of the silo, creating a crater 80 m across — was slightly dented, but had not leaked radiation.

A reporter from the *Arkansas Democrat*, who was an eyewitness to these events, was convinced that the first explosion would be followed by a nuclear detonation. He wrote: 'I knew I was going to die. I knew we were all going to die. I only wanted to somehow make sure I'd die first. I didn't want to have to witness the horror of so many others dying...I looked up to see yellow-white chunks of debris firing from the silo into the darkness overhead. The chunks gently floated several hundred feet up, then rained back down into the silo and on to the surrounding ground. As the giant sparks fell, a mass of pinkish-orange flames rose out of the silo, over the treetops and high enough to be seen for miles.' (Quoted in *The Times* 20.9.80.)

This was neither the first nor the worst accident involving the Titan II. That occurred on 9 August 1965, at Searcy, Arkansas, 55 miles north-east of Little Rock AFB. Silo No. 373-4 had been out of operation for six weeks, undergoing repairs to its air conditioning, plumbing and exhaust systems. The missile was still in place inside its 174-ft-deep concrete and steel-plated 'gun barrel' but its nuclear warhead was in storage. As the 56 workmen involved in the renovation re-entered the silo after lunch, there was a blast and a flash of flame. A fire had started in a diesel-engine room three levels down and had triggered off an explosion in a water-chilling room used for the air conditioning system.

The resulting power failure stalled the elevator and made it impossible to move the 700-ton steel and concrete lid that sealed the silo airtight. As the silo filled with smoke, the trapped workmen attempted to escape using a ladder which was now the only way out. In their panic, two workers became wedged together on a narrow opening between levels, preventing others from escaping. Only three men, working near the surface, survived. Gary Lay (18) had to walk through the flames, searing his arms and legs in the process. Hubert Sanders reported: 'I looked down and saw smoke coming up. I heard a man crying, "Help me, God help me!" All of the 53 men who remained in the silo died.

An immediate investigation into the disaster was ordered by President Johnson. An unnamed Air Force officer, quoted in *Newsweek* (28.8.65), said, 'These things are not supposed to happen. We have many, many safety factors.'

Another Titan II accident occurred on 24 August 1978, at a silo near the town of Rock in Kansas. A 'propellant transfer' team was loading nitrogen textroxide into the missile when a teflon seal that should have been removed, jammed a valve open. 13,000 gal of the toxic oxidiser — which converts to nitric acid when it escapes into the atmosphere — was released, killing two maintenance workers, seriously injuring another and causing minor injuries to a further 20 men. The population of Rock, two miles north of the missile complex, had to be evacuated.

The Titan IIs were steadily withdrawn from service from October 1982 onwards. They had originally been due to be scrapped in 1971, but were kept in use when the USSR refused to retire some of its own large missiles.

6 January 1981

Cap la Hague, France

At the end of the 1960s, France and Britain began to develop plants for the reprocessing of spent natural uranium fuel. The French reprocessing plant was built at Cap la Hague in Normandy. Facilities to reprocess uranium oxide fuel went into operation in 1978 and were soon handling 250 tonnes of the oxide fuel every year, but plans were soon put forward to expand the plant.

In Britain, similar plans met with opposition from environmental groups and led to a full public inquiry. In France it was the trade unions that opposed the plans, and the debate was intensified by a spate of accidents at the Cap la Hague plant.

In January 1980, routine analysis of algae and sediment close to the pipeline carrying radioactive waste out to sea led to the discovery of two splits in the pipe which were allowing the waste to escape less than 25 m from the shore. The outflow had to be halted for a total of 20 days for repairs.

In April 1980, the waste treatment plant had to be shut down and personnel evacuated when a fire in an electrical transformer cut the power supply to a building in which fuel is recycled, stopping the ventilation system. The back-up electrical system failed because it uses the same transformer room. All the plant's control systems were out of action for about 12 hours and mobile generators had to be rushed in to provide power to cool and agitate vessels containing a solution of highly radioactive fission products. These would otherwise have boiled within hours and caused a major radiation leak. Had the plant been engaged in normal production at the time, a major release of radioactivity would have been inevitable. As it was, low levels of contamination were detected in two buildings.

On 22 September 1980, the failure of a pump caused a tank containing radioactive water to overflow into surrounding fields. The authorities denied that the nearby Ste Hélène river had been contaminated. Three months later, samples of sand were taken from the riverbank by a local environmental group for analysis, and they revealed up to 52 times the normal level of caesium, which has a half-life of 30 years. The river water is drunk by cattle in this dairy-farming area.

On 4 December 1980, a breakdown in the part of the plant in which uranium and plutonium are separated from fission products caused a leak of several litres of liquid, containing one g of plutonium per litre.

On 6 January 1981, the most serious known accident at the plant gradually made its presence felt. At 04:00, raised levels of atmospheric radioactivity were discovered in three buildings at the plant. Three hours later, despite increased ventilation, radiation was at the maximum permissible level as prescribed for workers, and protective masks were put on. By 09:00 analysis had revealed that the air contained caesium-137, and by midday, levels at parts of the site had reached 38 times the maximum permissible level. By 14:00 it was clear that the source of radiation was a spent-fuel dry-waste storage silo which had caught fire. Uranium and magnesium had ignited cotton waste, soaked in solvent, that had been used in a decontamination operation there several weeks earlier. Fire hoses pumped in water, but this turned to steam, increasing the leakage of radiation. The fire was finally extinguished using liquid nitrogen.

A steady drizzle during the afternoon deposited radioactivity on the ground inside the plant, and during the following night large areas had to be hosed down to decontaminate them. Workers who had been at the plant during the day were allowed to go home in their own cars without being checked for contamination. In fact, radiation detectors at the main gate had been turned off, because radiation levels were so high that the alarms would have sounded continuously.

The accident caused immediate controversy; workers claimed that they had not been informed about the seriousness of the situation and demanded full medical examinations. Several were found to be contaminated and one had received in excess of a year's maximum permitted radiation dose (11 times the annual permitted dose for a civilian).

Despite the fact that a 50 km per hour wind had been blowing while the radiation was escaping, the management claimed that there had been no contamination outside the site. However, a sample of milk from a farm two km from the site was found to contain 25,000 picocuries of radioactivity per litre. The normal level is 20-50 picocuries per litre.

Workers demanded that in future they should be informed immediately of any accident. On 15 January, they discovered that another accident had occurred on 11 January. Three cu m of water, containing acid, 26 g of uranium and three kg of plutonium, leaked into the extraction section of the plant when a pump broke down.

On the night of 20 October 1982, a radiographer who had been X-raying welds in part of the plant received a large dose of radiation while transferring the radioactive source from the X-ray machine to a protective container. He received up to 50 rem whole-body dose and up to 250 rem to his hands. The maximum permissible annual dose is five rem whole-body, and 60 rem to the hands.

The level of irradiation set a new record for the plant, but this record was broken on 20 May 1986 when five workers were irradiated by an accidental surge of radioactive liquid in a pipe during a decontamination operation. The men carried on working because their personal alarms failed to sound. One worker received a dose of 272 rem to his hands.

2 November 1981

LX-09 Explosive

At Holy Loch, near Glasow in Scotland, a US Poseidon C3 SLBM, containing 10 50-kt W68 nuclear warheads, was being winched between the US submarine tender USS *Holland* and a ballistic-missile submarine when the winch ran free and the missile fell 5 m. The automatic brakes on the winch brought it to rest just short of the submarine's hull and the missile smashed into the tender's side.

Although there was no risk of a thermonuclear explosion, there was a serious risk of detonating sensitive chemical explosives in the missile's warhead trigger-system. This could have ignited the missile's propellant fuel and engulfed the ship, the submarine, and the nearby town in a cloud of radioactive debris.

The accident was classified as a 'Bent Spear', the second most dangerous class of nuclear accident. (It may be worth noting that, shortly before this incident, some crew members of USS *Holland* had been sentenced for trading in LSD, cocaine and amphetamines.)

The Poseidon C3 involved in the incident was one of several hundred still equipped with the unstable high-explosive LX-09, developed in the 1960s in response to demands for a light, compact, chemical high explosive for the new W68 multiple warheads. Machined into specially shaped blocks, LX-09 was later found to behave erratically.

On 31 July 1974, the Lawrence Livermore weapons design laboratory in California released a summary of tests carried out on LX-09. The explosive had been found to display some 'very undesirable properties', including 'low threshold velocity for reaction and rapid build-up to violent reaction. Any accidental mechanical ignition has a large probability of building to a violent deflagration or detonation.' In half the experiments in which it was dropped from a height of less than one ft, it exploded. (Pacific News Service 8.10.81)

Despite this warning, production was not halted until 1977, following an accident on 30 March that year in which three workers were killed in an explosion at the Pantex nuclear weapons assembly plant near Amarillo, Texas, while working in a bay in which LX-09 was being machined on a lathe. The accident, which the DOE blamed on a machining or handling error, caused $2.5 million in damages and hurled debris more than 320 ft.

Faced with the inherent dangers of LX-09, the Pentagon had two choices: either to 'recall' all warheads immediately, thus disrupting the deployment of all 31 Poseidon submarines; or to play down the problem and replace the warheads over a longer period. They took the second option. A programme to re-equip Poseidon warheads with a safer substitute was set in motion about a year after the Pantex accident, and was scheduled to take six years. There has been no further use of LX-09 in any other nuclear weapon.

25 January 1982

Ginna Reactor Leak New York State

The worst US nuclear-power-plant accident of the year occurred at the 450-MW Robert E. Ginna (pronounced Gin-nay) plant in New York State. At 09:25, with the plant running flat out, a delicate tube in the steam generator ruptured, forcing 760 gal per minute of radioactive core coolant water into the 'clean' steam circuit that drives the turbine.

The operators saw the pressure in the reactor drop suddenly and began an emergency shutdown of the plant, but the leak of high pressure coolant water raised the pressure in the steam circuit, which caused a relief valve to open several times, venting radioactive steam into the atmosphere in five-second bursts totalling several minutes.The valve then failed to close properly and leaked continuously for about 50 minutes. Rochester Gas and Electric, the operating company, estimated that 485.3 curies of radioactivity were released in the form of noble gases krypton and xenon, as well as 1.15 millicuries of radioactivity as iodine-131. Some 45,000 people live within 10 miles of the Ginna reactor, but the NRC said that the radiation exposure of people near the plant was no more than background levels during normal operations and was the equivalent of a few chest X-rays.

A pressure relief valve in the primary circuit, deliberately opened to reduce the pressure and slow the leak, also stuck open and poured primary coolant into a holding tank which then overflowed, spilling about 5,000 gal of contaminated water into a sump. More importantly, the loss of coolant lowered the pressure in the reactor and caused a steam bubble to form in the reactor pressure vessel. Had the steam bubble migrated downwards, it could have exposed the fuel rods and caused the core to overheat. Fortunately the operators recognised the problem and pumped in more primary coolant in time. By the following evening, the plant was fully shut down, and it remained that way for the next four months while repair work was carried out.

The sequence of mishaps and malfunctions bore a striking similarity to events at Three Mile Island, and demonstrated that the lessons of TMI had still not been learned. Most significantly, the accident drew attention to an endemic problem in PWRs — severe corrosion of the steam generator tubes.

TUBE LEAKS

'It's a miserable fact that these damn tubes keep corroding.' Professor H.W. Lewis, US National Science Foundation Study Group.

In a PWR like Ginna, water takes heat from the reactor core to the steam generator via a 'primary cooling circuit'. This water is extremely hot, under great pressure and intensely radioactive. Inside the steam generator, primary cooling water flows through thousands of thin metal tubes immersed in uncontaminated water under lower pressure. This 'secondary coolant' instantly flashes into steam, which drives the PWR's turbines.

If a tube ruptures, radioactive water under high pressure escapes into the secondary cooling circuit. The minimum effect of a tube leak is the release of radioactive steam into the atmosphere through safety valves — as happened at Ginna. Large single tube leaks, or mutiple tube leaks, can cause a host of problems which may potentially lead to a meltdown of the reactor core.

The NRC concedes that tube leaks caused by corrosion are a fundamental design problem of older PWRs. An NRC report published in 1982 revealed that 40 PWRs suffered from corroded tubes, causing high operating costs and radiation exposure for staff.

The cost of 'sleeving' — inserting new, smaller tubes inside defective steam tubes — the Ginna reactor ran to over $10 million. The temporary workers, recruited from rundown neighbourhoods, who sleeve tubes are known as 'jumpers' or 'sponges' since they can only spend a minute or so inside the 'hot' steam generator.

'Considering all of the costs involved,' says the US pressure group, **Public Citizen** (1987), 'nuclear plant steam generators are likely to go down in the annals of American business as the biggest product failure ever.'

1983

Radioactive Scrap, Mexico

Three junkyard workers stole an old cancer therapy machine from a warehouse in the Mexican city of Juárez, close to the Texan border. The obsolete three-tonne radiotherapy machine, built in 1963, was previously owned by a Methodist hospital in Lubbock, Texas. The hospital should have paid $2,000 to have it dismantled at a low-level radiation dump. Instead, in 1977, it was sold to a dealer who sold it on to a Mexican clinic which had no licence to handle radioactive materials and, in fact, never used it.

The machine had a rotating head in a lead shield which enclosed 6,010 metal pellets, each a mm in diameter, and each containing about 70 microcuries of cobalt-60. When the workers threw it on their truck, the lead split open and the pellets spilled everywhere. Some stuck to the tyres of passing cars, others to people's shoes. Children played with them, unaware of the dangers.

On 6 December (the precise date is known because the yard's paperwork went radioactive on that day), the machine was taken to the Jonke Fenix scrap yard in Juárez, which employed 60 people, and from there to two Mexican steel foundries where it was smelted down and incorporated in 500 tonnes of building reinforcement rods and 17,000 table supports, most of which were exported.

The accident was only discovered by a strange twist of fate. On 16 January, a truck loaded with the radioactive rods took a wrong turning near the Los Alamos National Laboratory, a major centre for US nuclear weapons research. It happened to pass over a special radiation sensor, designed to prevent radioactive materials leaving the site, and set off an alarm. But for this chance occurrence, a much greater disaster would have resulted.

All 50 US states and several federal agencies have worked together since then to track down the hot steel. By late February, all but six per cent of the table supports were accounted for, but the construction rods proved more of a problem, as many of them had already been installed in buildings.

In Arizona, private houses, a state prison and a medical centre had to have their foundations ripped out and re-done. In Mexico, at least 20 homes were found to be radioactive and some had to be dismantled.

More than 60 radioactive pellets were found by the Mexican authorities, scattered along the roadways. The radioactive pick-up truck was discovered in El Paso, Texas — the city opposite Juárez on the Rio Grande — and it was removed to the Juárez jail, where the back and the driver's cab were filled with concrete in an attempt to cut down the radiation it was emitting. This was so strong that it registered on a Geiger counter 300 yd away.

At least 200 people received significant doses of gamma radiation ranging from 1 to 50 rem. One Mexican received a dose of 10,000 rem to his hands. The three men who had stolen the machine and two of the junkyard workers were found to have been exposed to as much as 500 rem whole-body radiation. At least 3,000 Mexican workers from the steel mills were exposed to between 300 and 450 rem. Most swiftly recovered from the short-term effects but many of those affected may develop cancer later.

The Mexican government closed the junkyard and buried contaminated scrap, some 2,000-6,000 tonnes of contaminated steel they had recovered, and the pick-up truck in a specially prepared site between Juárez and Chihuahua.

RELATED INCIDENTS

In March 1962, a boy named Henry Espindola found seven pellets of radioactive cobalt in the streets of Mexico City and brought them home and kept them in a cookie jar in the Espindola house as a decoration. Henry died on 29 April, and his mother on 19 July. The cause of their deaths was recognised and the cobalt was removed on 22 July. Henry's sister died on 18 August and his grandmother on 15 October. Only his father, who visited on weekends, survived.

In February 1983, a foundry in Auburn, New York, discovered by chance that the molten steel coming out of the furnace was radioactive. Workers noticed that the steel thickness gauge, which used a caesium-137 source and detector, was not working properly, and when a repairman with a Geiger counter arrived, he began to register radiation the moment he walked in the door. The steel was found to contain 25 curies of cobalt-60 which had been included in the scrap for melting. The origins of this were unknown.

14 November 1983
Sellafield, UK, Part 3

Windscale's worst known accident of the 1980s happened to have witnesses present. On 14 November 1983, four Greenpeace divers from the ship *Cedarlea*, who were taking samples of silt near the end of the pipe through which radioactive waste is routinely discharged from the plant, suddenly saw the readings on their Geiger counters leap from the normal four to 10 counts per second to more than 500, and then go over maximum on the dial. They had been contaminated by an oily radioactive slick that had apparently come from the pipe. The divers left the area, but more than 24 hours later their clothes and inflatable dinghy were still giving high readings. The NRPB asked them to have a check-up and found traces of radioactive ruthenium on the dinghy. Their clothing had to be destroyed, but the dinghy was decontaminated and returned to them.

Five days later, heavily contaminated items were found on the beach. In the meantime an inspector from the Department of the Environment (DoE) had visited the plant and been told of an oil slick, but not that it was radioactive, and the beach had been declared free of pollution.

On 24 November, after a review of BNFL's monitoring techniques, the DoE asked for further checks, and these led to the discovery of radioactivity near the surface of the sea and of radioactive debris on the beach. The public were warned against using a 200-m section of the beach.

On 9 December, the DoE extended the closed area to 40 km, having found that radioactive debris was being washed up north and south of the pipeline and that contamination levels were higher than they had been the previous month, between 100 and 1,000 times the normal level. It was also revealed that the Ministry of Agriculture, Fisheries and Food (MAFF) had withheld information it had received concerning the contamination of fish and solid waste in the area.

In Parliament, the Secretary of State for the Environment announced that, as the leak involved a possible breach of the law on discharges, the matter was being referred to the Director of Public Prosecutions. Scientists were finding that the rate at which contaminated material was coming ashore, instead of falling, remained at the same level.

On 14 February 1984, reports by the Radiochemicals Inspectorate of the DoE and the Health and Safety Executive detailed the events that had led to the radioactive discharge. There had in fact been three separate discharges — on 11, 13 and 16 November — during washing-out operations.

Tanks which have held highly radioactive waste are washed out with solvents, and the mixture of water, solvent and radioactive 'crud' is normally returned to the original tank for separation. The less radioactive watery liquid is then discharged to the sea.

However, on this occasion the mixture was passed directly to the sea tanks and the valves were opened for discharge. Alarms alerted BNFL to what was happening, but they assumed that the crud was floating on the surface of the water and discharged some of the water from the tanks to the sea to isolate the crud. The rest of the mixture was then pumped back through the system, but radioactive material stuck to the sides of the pipes, and when BNFL resorted to flushing the pipes out, as they did on the dates above, radioactive crud was discharged into the sea.

The Radiochemicals Inspectorate condemned the Sellafield management outright, maintaining that existing controls were inadequate and accusing BNFL of not considering the consequences of discharging highly active particles of waste into the sea and of failing to notify the appropriate government departments as soon as possible, even when there was a potential danger to the public.

The beaches remained closed for almost nine months after the incident; until the end of July 1984, when the DoE finally gave the all clear, though monitoring would continue and further contaminated material would be removed. In August, the Director of Public Prosecutions announced that BNFL were to be prosecuted under the Radioactive Substances Act 1960 on two counts and under the Nuclear Installations Act 1965 on four counts, the first time that criminal proceedings had been brought against the British nuclear industry.

In June 1985, a jury found BNFL guilty on a number of these charges — failing to keep discharges as low as possible, failing to keep records of discharges, failing to take all reasonable steps to minimise exposure of persons to radiation, and failing to keep adequate operational records (the company had pleaded guilty to this last charge). BNFL were fined £10,000 plus costs.

In the meantime, Greenpeace members had been taken to court for defying a High Court injunction and attempting to stop the discharge of radioactive waste by plugging the BNFL pipeline. Greenpeace was fined £50,000 plus costs.

The decade had started badly for the Windscale plant. In February 1981 the Health and Safety Executive had published a report entitled 'Windscale: the management of safety', the findings of an inquiry instigated by the government in the aftermath of the serious leaks in 1976 and 1978.

The report pointed out that almost half of all incidents at British nuclear installations during the preceding two years had occurred at Windscale, and went on to say, 'By the early 1970s the standard of the plants at Windscale had deteriorated to an unsatisfactory level.' Nonetheless, the report expressed confidence that BNFL would succeed in improving safety at the plant.

The first action that BNFL took was, in May 1981, to change the name of the plant from Windscale back to the original name of the site — Sellafield. Critics claimed that this was only done to rid the installation of the negative associations of the name 'Windscale'. If so, the ploy was largely unsuccessful. Within months the newly-named Sellafield site began to build its own unenviable reputation.

On 4 October 1981, radiation monitors at Sellafield detected higher than normal levels of iodine-131 in the air, and the plant was shut down for 24 hours when it became clear that a leak of more than 300 times the normal daily release of iodine-131 had contaminated milk on two farms within two miles of the plant. The release was caused by the reprocessing of fuel which had not been left to decay for long enough. Unwisely, BNFL waited several days before reporting the matter, by which time rumours were rife.

The contaminated milk was not destroyed but was distributed and sold as usual. BNFL's Director of Information said that members of the public would receive less than 10 per cent of the permitted maximum annual exposure laid down by the ICRP.

On 30 October 1983 Sellafield was again in the spotlight. A Yorkshire Television documentary, entitled 'Windscale — The Nuclear Laundry', alleged that the incidence of leukaemia among children in the nearby village of Seascale was 10 times the national average, and that plutonium dust had been found in houses in Cumbria.

Public concern was so great that the then Social Services Secretary, Norman Fowler, set up an independent inquiry under the chairmanship of the former President of the Royal College of Surgeons, Sir Douglas Black, to examine the incidence of cancers around the Sellafield plant. (It was during this period of public anxiety that Greenpeace divers discovered the radioactive slick for which BNFL were taken to court and fined.) The Black Report was inconclusive. Whilst recognising that a link between a confirmed high incidence of cancer, particularly leukaemia, and the presence of the Sellafield plant could not be ruled out, the report did not feel that the connection had been proved.

Sellafield suffered a further series of incidents in 1986. On 23 January, as the result of a faulty evaporator, almost half a tonne of uranium accumulated in one of the two tanks that discharge diluted waste into the sea. (This accidental accumulation bore a strong similarity to the conditions that led to the 1983 discharge, though BNFL had said that such a thing could not happen again.)

BNFL informed the DoE, and it was agreed that, since the plant is allowed to pump three tonnes of uranium into the Irish Sea every year, even this large quantity was still within permitted discharge safety limits. The tank was then emptied directly into the sea. BNFL admitted that it would have been possible to recover the uranium from the tank, but said that both the NII and MAFF had also agreed that the discharge was acceptable.

On 1 February a small fire broke out at Sellafield's low-level waste dump site at Drigg. BNFL said that early tests on the smoke showed 'no significant increase in radioactivity'. Then, four days later, for the first time since the Head End Plant accident in 1973, Sellafield was put on 'amber alert', indicating an emergency in one building and a threat to the rest of the plant. The building concerned was B205, next door to the abandoned Head End Plant. Workers had been repairing a valve through which plutonium nitrate was passed for testing, but before the valve could be isolated in a plastic container, a mist of the plutonium compound leaked into the air, and 71 workers had to be evacuated when monitors registered the contamination. The plant was shut down and staff took more than two hours to locate the source of the radioactivity.

BNFL were quick to state that the mist had been contained within the building, that there were no implications for the public, and that no one had been contaminated, but the following day it was announced that two workers had been contaminated and that an estimated 50 microcuries may have escaped into the atmosphere. The facts were further revised the following week when BNFL admitted that 11 workers had been contaminated and a further four may have been affected. One worker had received the maximum permitted dosage for a whole year. The changing story did little to enhance public confidence.

On 13 February there was a 40-minute fire at Sellafield's Drigg waste dump when low-level waste was tipped into a disposal trench, and on 18 February three workers were contaminated while repairing a 'sludge removal' pipe which had broken and allowed radioactive water to escape. On 1 March a weld on a glove-box failed and seven workers were contaminated with plutonium.

Later the same month, the House of Commons Environment Select Committee, in its report entitled 'Radioactive Waste', stated: 'The UK discharges more radioactivity into the sea than any other nation. As the Ministry of Agriculture confirmed to us, Sellafield is the largest recorded source of radioactive discharge in the world. The anxiety and controversy which this arouses in the UK is well known. It also creates anxiety in other nations. We found, for example, that the Swedes could identify radioactive traces in fish off their coast being largely attributable to Sellafield, greater even than the contamination from adjacent Swedish nuclear power stations. Similar experiences were reported to us by the Isle of Man government. That the UK, with a comparatively small nuclear industry, should be so dramatically out of step is a cause for concern.'

The succession of accidents in the first months of 1986 prompted the Health and Safety Executive to launch yet another investigation into Sellafield's safety record, and its report 'Safety Audit of BNFL Sellafield 1986' appeared in October of that year.

Building B205, in which the plutonium mist had been released, came under particular scrutiny for never having had a written 'safety case', the basis upon which maintenance, operation and emergency procedures should be founded. In so many words, BNFL were given 12 months to put their house in order, or shut the plant.

The year ended with a statement by the Secretary of State for the Environment, in response to a question in the House of Commons, to the effect that the Dail of the Irish Parliament had, on 3 December, passed a resolution calling for the closure of Sellafield.

In January 1987, 12 men were contaminated by a leak of plutonium during the removal of a pressure valve in an area in which fuel rods are made for the Dounreay fast breeder reactor, and in February the reprocessing plant was closed temporarily after a leak of high-level radioactive waste.

Ironically, Sellafield has now become a major tourist attraction. In 1987, 104,000 people took up the BNFL invitation to look around the site, and the English Tourist Board made an award in recognition of this achievement. In early 1989 Sellafield won the coveted Best Loo in the Country Award and then, in May 1989, BNFL received the Institute of Public Relations' top prize, the Sword of Excellence, for the company's acclaimed public relations programme 'Restoring Confidence in Sellafield'.

In the same week, an article in *Time Out* (24.5.89) revealed the contents of a draft letter written in 1987 by the chairman of the CEGB, Lord Marshall of Goring, to the then UK Energy Secretary, Peter Walker. The letter had suggested that Britain's 14 Advanced Gas Cooled Reactors (AGRs) would be forced to a standstill unless the rate of reprocessing at Sellafield was improved and the problems of spent fuel rod storage were solved.

At present, all spent British-produced Magnox fuel rods and most of the more modern uranium oxide fuel rods from the AGRs go to Sellafield, where they are put in storage awaiting reprocessing and the extraction of plutonium. The long-term plan has been that the plutonium would fuel the projected fast breeder reactors, but the virtual abandoning of the fast breeder programme casts doubt on the need for Sellafield's civil reprocessing capacity.

No spent uranium oxide fuel has been reprocessed since the 1973 accident at the Head End Plant, and Sellafield has a huge backlog of spent fuel in its underwater storage pools. These may soon be filled to capacity, forcing the AGRs to stop production. Moreover, the spent rods are steadily corroding and contaminating the water with radioactivity. The CEGB regards the situation as serious enough to warrant building its own dry storage facility, and doubts the viability of underwater storage for long periods of time.

In his draft letter, Lord Marshall says 'We are under attack from environmentalists and our critics for storing Magnox fuel in water. The attack is difficult to answer because it is basically correct...'

25 August 1984

Mont Louis Sinking, North Sea

On 25 August 1984 in the English Channel, the 4,201-tonne French freighter *Mont Louis* sank in 14 m of water, 18 km off Ostende, Belgium, after colliding with the West German cross-channel ferry *Olau Britannia*, which was carrying 935 passengers. No one was hurt in the collision.

Greenpeace discovered that the freighter's destination was Riga in the Soviet Union and that it was a sister ship to the *Borodine*. For 10 years the *Borodine* had been shipping uranium hexafluoride (known within the industry as hex) — the basic raw material from which nuclear fuel is made — to the USSR, where it was enriched and then shipped back for use in French nuclear reactors (see box p. 276).

On Sunday 26 August, the Greenpeace office in Paris released what information it had about the *Mont Louis* to the media and suggested that this ship's cargo was also nuclear material.

Later that day this suggestion was supported by a French trade union — the Confédération Française Démocratique du Travail (CFDT).

Confirmation of its cargo was not forthcoming for 24 hours, when it was revealed that the ship was indeed carrying hex, 375 tonnes of it in 30 reinforced containers, valued at $20 million. The shipment belonged to the French Compagnie Générale de Matières Nucléaires (Cogema) and the Belgian power company Synatom. Although unenriched hex is only mildly radioactive, it is highly toxic; if the containers were punctured, the hex would react violently with water. The cargo also included 22 empty barrels, and several of these were later washed ashore on the Belgian coast. It was also revealed that this was the first nuclear shipment on the *Mont Louis* and that the crew were not experienced in handling such hazardous cargoes.

The Greenpeace vessel *Sirius* arrived in Ostende on 10 September and a press conference was held on board at which Jim Slater, general secretary of the UK National Union of Seamen, called upon the International Maritime Organisation to ban such shipments until stricter international safety regulations had been introduced.

Two days later the *Sirius* sailed close to the wreck of the *Mont Louis* and observed empty barrels floating in the sea; a minesweeper and a tugboat were spraying detergent on an oil slick that had formed around the boat, and six warships were patrolling the area.

After a difficult recovery operation, hampered by bad weather conditions and technical problems, the last barrel was extracted from the wreck on 4 October. Fortunately, none of the containers was damaged, but the whole incident drew attention to the international trade in uranium and provided a warning of the potential dangers.

URANIUM HEXAFLUORIDE AND NUCLEAR TRANSPORT

In naturally-occurring uranium, the isotope uranium-235 forms less than one per cent of the total. The percentage has to be increased for use as fuel in certain kinds of nuclear reactor; for use in nuclear weapons, the proportion of U-235 has to exceed 90 per cent. The uranium therefore has to be 'enriched', and uranium hexafluoride, or 'hex', forms a key stage in this process.

The uranium is 'hexed' at hexing plants, and then transported to enrichment plants, often over long distances as a result of complex international contracts. In 1982, the UK Secretary of State for the Environment made a statement in the House of Commons (*Hansard*, 30.7.82) saying that the amount of hex being sent to the USSR by the Central Electricity Generating Board (CEGB) was 'in the order of 170 tonnes per annum', part of a 10-year contract with the USSR signed in 1980.

The Euratom Supply Agency's Annual Report for 1987 put the number of transported consignments of 'natural' (unenriched) hex at 1,840, and of enriched hex at 368, for Europe alone. The figures worldwide were 5,898 consignments and 1,180 consignments respectively for the same year.

Three days after the *Mont Louis* sinking, and following reports that Panamanian authorities had refused port facilities to the *Pacific Fisher*, which was carrying spent nuclear fuel rods from Japan to Sellafield, BNFL stated that such shipments had been passing through the Panama canal 'for years'.

The *Mont Louis* represented just one link in the chain of maritime nuclear transports. She carried a much-travelled cargo, consisting of uranium from Canada that had been hexed at Cogema's plant at Pierrelatte in south-west France and was now on its way to the USSR. The picture that emerged from enquiries following her sinking was one of countless cargoes of radioactive material being shipped across the Atlantic and through the busy shipping lanes of the English Channel, the North Sea and the Baltic. A report in *The Times* (13.9.84) alleged that 'perhaps nine million separate movements of nuclear cargoes are notified to the International Atomic Energy Agency annually.'

11 January 1985

HERO Accident, West Germany

At a training ground near the town of Heilbronn in West Germany, US troops were unloading the first stage of a solid-fuel rocket motor from its crate before coupling it onto the rest of a Pershing II intermediate-range nuclear missile when it caught fire. Three soldiers were killed, seven were severely burnt, and a missile transporter and truck were destroyed. The nuclear warhead of the missile was not involved. The incident caused a political uproar. Six German civilians filed suit in West Germany's highest court claiming the Pershing II posed the same danger to civilians as an unsafe nuclear power station.

The US Army Accident Investigating Board, after three months studying the facts, reported that 'on the basis of all the evidence...a discharge of static electricity within the rocket propellant was the cause of the accident.' (Axelrod et al. 1988) The other services have different names for this kind of incident but in the US Navy they are called HERO accidents, standing for Hazards of Electromagnetic Radiation to Ordnance. (See box, p. 278)

The trigger on most conventional, and nuclear, bombs and rockets is an electro-explosive device (EED), an electrical wire surrounded by explosives. When the device is energised it becomes like a hot lamp filament, and when it exceeds its threshold it goes off — regardless of the source of the signal. A nuclear weapon may have hundreds of EEDs. EEDs also respond to unintentional electric currents produced by lightning and other forms of electromagnetic radiation.

The military have attempted to overcome the HERO problem by including filters, containers, shielding and grounding; these measures are far from foolproof and have themselves, on occasion, triggered HERO incidents.

Dr Theodore Taylor has worked on the development of nuclear weapons for 40 years. He was Deputy Director of the Defence Atomic Support Agency (the predecessor to DNA) and was appointed by President Carter to the President's Commission on the accident at Three Mile Island. He stated: 'I am deeply concerned about the possibilities of accidental explosions of military ordnance, including nuclear weapons.' (Axelrod et al. 1988)

HERO ACCIDENTS

Some 260 weapons systems are considered HERO unsafe, and suspected HERO accidents between 1945 and 1980 include:

1959: Six Thor and Polaris missiles accidentally exploded at Cape Canaveral. HERO was suspected.

1967: During the Vietnam War, a Zuni five-in anti-aircraft rocket warhead shot across the deck of the US aircraft carrier *Forrestal* and slammed into a plane, setting its fuel tank on fire. Other bombs on board exploded, causing fires in which 134 people were killed and 63 planes were destroyed.

1981-87: Five UH60 Black Hawk army helicopters (costing $6 million each) crashed during this period, killing or injuring 22 army personnel. Army investigators suspect the accidents were caused by the helicopter's susceptibility to radio waves from microwave towers, radio antennae, and radars which result in interference with the helicopter's own electronics. The helicopter was not tested for HERO until after the fifth one crashed. It failed the test but is still operational.

1984, mid-May: A massive explosion occurred at a Soviet naval ammunition depot at Severomorsk, about 900 miles north of Moscow. This was one of a number of such detonations in the USSR over a seven-month period. They are believed to have been triggered by radar transmissions.

1987, March 26: A $78-million Adrian-Centaur rocket, carrying an $83-million communication satellite, blew up 51 seconds after take-off, when electromagnetic interference from lightning scrambled its electronic brain.

1987, June 10: Three small NASA weather rockets were accidentally launched from Wallops Island, Virginia, after being struck by lightning.

5 January 1986

Gore, Oklahoma

Just over a mile south of Gore, Oklahoma, stands the Sequoyah Fuels Corporation, a uranium processing plant owned by the Kerr-McGee conglomerate. The facility processes uranium hexafluoride from yellowcake and ships it to the US DOE uranium enrichment plants, which manufacture it into fuel for nuclear reactors.

Inside the plant, 26-year-old James Neil Harrison was standing by the side of a cylinder that had been overfilled with 29,500 lb of uranium hexafluoride — 2,000 lb over the limit. It was being heated to reduce the contents, an extremely hazardous practice that was apparently common procedure at the plant.

At about 11:30 the cylinder burst, spewing the highly toxic and caustic 'soup' towards Harrison, engulfing him. Blinded and barely able to breathe, Harrison stumbled about as hydrofluoric acid — created when the uranium hexafluoride reacted with the moisture in the air — began to eat away his lungs.

Kerr-McGee's emergency plans for medical treatment and evacuation were virtually non-existent. Co-workers managed to rescue him but there was no oxygen — essential for his survival — on hand at the plant and they had to drive him eight miles to get a canister from a nursing home. They then drove a further eleven miles to the Sequoyah Memorial Hospital which was unable to cope with his case. By the time they got him to the Regional Medical Centre in Fort Smith, Arkansas, 21 miles away from the plant, Harrison was virtually dead.

Fumes from the explosion also filtered into the lunchroom through the plant's air conditioning system. More than 100 people were treated for breathing difficulties and 34 were detained in hospital. Uranium in the uranium hexafluoride formed particles of uranyl fluoride which were carried in the escaping gas cloud. Kerr-McGee claimed that because uranyl fluoride is a fairly heavy material, it fell only on the plant and its immediate environs. Yet urine samples of residents in nearby towns showed significant traces of uranium.

According to an environmental study submitted to the NRC, compiled from Kerr-McGee's own records, since the opening of the plant in 1969 'numerous spills and leaks of radioactive materials have contaminated the groundwater and surface water, eventually discharging into the Illinois and Arkansas rivers.' (*The Progressive*, March 1986)

Three months before the accident, the NRC renewed the company's operating licence, which had expired in October 1982. (The company had been operating for almost three years on a legal loophole that allows plants to continue production if they have applied for licence renewal.) As a condition for granting renewal, the NRC stipulated that the company must submit, within six months, detailed reports on the handling of cylinders containing uranium hexafluoride and institute measures that could mitigate the effects of a potential cylinder rupture.

25/26 April 1986

Chernobyl, USSR

NUCLEAR, RADIOACTIVITY, FIRE — code words notifying Moscow emergency centre of fire in Chernobyl Unit 4 reactor.

At 09:00 on Monday 28 April 1986, technicians at the Forsmark reactor, 100 km north of Stockholm, Sweden, recorded abnormally high levels of radiation. Checks on the reactor showed that there were no leaks. But Geiger counter readings from the plant's 600 workers, and soil and vegetation samples from outside the plant, registered levels 14 times higher than normal.

Soon, from other areas in Scandinavia, came similar findings. Something, somewhere, was releasing huge amounts of radioactivity into the atmosphere. The prevailing wind was coming from the south-east — from the USSR. But when the Scandinavians demanded an explanation, they met with evasions from Moscow.

Finally, 12 hours after the Forsmark alert, Moscow TV broadcast a terse statement from the USSR Council of Ministers: 'An accident has taken place at the Chernobyl power station, and one of the reactors was damaged. Measures are being taken to eliminate the consequences of the accident. Those affected by it are being given assistance. A government commission has been set up.'

'Thus began,' commented *Time* magazine (12.5.86), 'by far the gravest crisis in the troubled 32-year history of commercial atomic power.'

The Chernobyl nuclear complex is situated on the Pripyat River, 130 km from the Soviet Union's third largest city, Kiev (population 2.5 million). This part of the Ukraine, known as the 'breadbasket of Russia', is a lightly populated area of farms and small holdings. Until April 1986, the region's largest towns were Pripyat, with a population of 50,000, built to house the complex's construction workers and operators, and Chernobyl, population 12,000.

The Chernobyl complex housed two pairs of RBMK 1000 reactors, generating 1000 MW of electricity each, with a third pair under construction. RBMKs are boiling-water, graphite-moderated high-power reactors. Uranium fuel is cooled by boiling water, and this process produces enough

steam to drive two 500-MW turbines. The fuel rods are
tained within a graphite moderator, which slows c
neutrons and sustains the nuclear chain reaction. Cor
rods inserted into the graphite core manage the reactor's power levels. Although the RBMK design was 25 years old, and Soviet engineers were aware that it had disadvantages — the most serious of which was extreme sensitivity — the Chernobyl Unit 4 reactor had run relatively flawlessly since it was commissioned in 1983.

On the afternoon of 25 April, during a scheduled shut-down of Unit 4, operators began a test on one of the huge 500-MW turbines. Ironically, in view of what was to happen, the technicians were testing safety systems. If the station were to be blacked out, was there enough residual energy in the still-spinning turbine to supply electricity for the crucial 45 seconds it would take for emergency diesel generators to power up?

At 14:00 the reactor power had been reduced to 50 per cent of normal output, and the emergency core cooling system had been disconnected, when the electricity-grid controller in Kiev requested that power continue to be supplied until 23:10. The Chernobyl staff agreed to this but, after 23:10, they did not switch the emergency cooling system back on. This was the first major error in what was to be a chain of serious operating faults: the operators reduced the power output to just one per cent of the normal operating power, far below the level required for the experiment; they withdrew most of the control rods; and they switched off other essential safety systems.

By 01:23 on 26 April, the combination of all these fundamental violations of safety rules had made Unit 4 unstable. Power output was rising. The technicians were rapidly losing control of what was now a runaway reactor. At 01:23:40, part of the reactor suffered a 'prompt neutron power burst'.

Within four seconds, Unit 4 reached 100 times normal power and may have come close to an atomic explosion. Some of the fuel disintegrated and evaporated the cooling water. This caused a steam explosion which blasted apart the 1,000-tonne lid of Unit 4. Flaming reactor debris showered on to adjoining buildings, starting more than 30 fires. More water reacted with the red hot, 1,700-tonne, graphite core to generate hydrogen gas, which detonated in a second explosion, hurling radioactive debris and radio-nuclides a mile into the sky.

By this time, a quarter of the graphite core was alight, sending more radioactivity into the sky. It would take 10 days of truly heroic effort for the burning reactor core to be brought under control. To dampen the fire and contain the radioactivity, helicopters dropped more than 5,000 tonnes of lead, boron and other material on to the exposed core, but up to 26 per cent of the reactor's inventory of radioactivity was lost, nonetheless. So were the lives of 31 firemen and plant workers, most of whom were covered in severe radiation burns.

SARCOPHAGUS

Though the initial disaster had been contained, the problems facing the Soviet authorities were far from over. On 1 May the temperature of the nuclear fuel started to rise again, reaching a maximum of 2,000 °C, due to the heat generated by the radioactive decay of the nuclear fission products remaining in the damaged reactor. To solve this problem, nitrogen was pumped under pressure into the space beneath the reactor vault, and by 6 May the temperature was declining.

With this new emergency contained, work then began on the entombment of the reactor. A group of coal miners were drafted in to build a 'cooling slab' under Unit 4 in order to prevent the hot reactor fuel burning a hole in the reactor base. This alone was a mammoth operation conducted at breakneck speed and under hazardous conditions. By 24 June, four hundred men, working three-hour shifts, had built the 168 m long, reinforced concrete tunnel and installed the monolithic concrete slab.

The plans for the long-term entombment of the plant involved the building of a giant 'sarcophagus' containing more than 7,000 tonnes of steel and 410,000 cu m of concrete. This giant construction was completed in November 1986.

COUNTING THE COST

'The sun was shining brightly and everything was a brilliant white. But in the village, there was no life. Not a single path to be seen across the snow, not a single chimney smoking. That's what Chernobyl means in human terms.' Soviet film-maker Vladimir Gubaryev, after a flight over an evacuated village.

According to the Soviet Minister of Atomic Energy, Nikolai Lukonin, the cost of the Chernobyl disaster to the Soviet

Union alone is 8 billion roubles and still rising. It is, said the article, 'perhaps the most expensive accident on the planet.' Statistics show the extent to which Unit 4 polluted land and threatened lives: 1,000 sq km of land around the Chernobyl reactor was, to some extent, contaminated; 135,000 people and 86,000 head of cattle were evacuated from a 30-km-diameter zone centred on the reactor; one evacuation column — 1,216 large buses and 300 trucks — stretched for 15 km; 60,000 buildings in 500 residential communities required decontamination; in Kiev, 400 wells were bored to provide emergency water supplies as a precautionary measure. Two 6-km emergency water mains were also installed.

According to a Soviet estimate, half of Chernobyl's fallout dropped within 35 km of the reactor complex. The other half fell on more than 20 countries worldwide. The debris and radionuclides thrown into the atmosphere by explosion and fire formed into a cloud which initially passed north-west across Lithuania and Latvia (26-27 April) before becoming stagnant for a time. Then prevailing winds swept some of the cloud across central Scandinavia, continental eastern, western and southern Europe (30 April), and the UK and the Republic of Ireland (2-3 May). Eventually, Chernobyl's distinctive 'calling card' was detected as far afield as the Arabian Peninsula, Siberia and North America.

Differences in topography and weather conditions throughout Europe led to unpredictable contamination levels across the continent. The areas worst affected were those where heavy rain precipitated fallout. In northern Sweden and parts of Finnish Lapland, contamination was so severe — up to 33 times higher than Swedish safety limits — that thousands of reindeer were slaughtered. In southern Germany, where there was particularly heavy rain, levels were several times higher than at the peak of atmospheric nuclear tests in the 1960s.

This wide variation in radioactive fallout throughout the continent was echoed by the variation in advice and safety limits set by European countries. 'Safe' limits on iodine-131 in milk, for example, ranged from 20 Bq/l in the state of Hessen, West Germany, to 2,000 Bq/l in France. The Swiss, monitoring TV stations across Europe, noted that the Germans were being told that milk was safe, but vegetables were not; Italians were advised not to drink milk and to shoot all rabbits; while the French were told almost nothing.

The French government did not even supply a World Health Organisation experts' commission with figures on contamination levels in France. Their first report, dated 6 May 1986, included maps of European countries with contamination levels per region. The French map was empty except for the word 'low'.

The monitoring of the cloud's progress was also inadequate: its arrival over the United Kingdom was signalled, not by government scientists, but by radiologists at London's Charing Cross Hospital, whose monitoring machine registered a rise in radiation levels as the cloud passed overhead.

Nine months after Chernobyl, nuclear experts had still not arrived at a consensus. At a meeting of European parliamentarians in January 1987, the following exchange, reported in *Le Monde* (13.1.87), took place: ' "What distance from the accident did farm animals die?" asked a Scottish MP concerned with lamb. "The Russians have told us nothing on this subject," replied one expert. "And the pollution of the water?" asked an English MP. "Impossible to give a general response," declared another expert; "Certain radioactive elements have been absorbed by the soil, others by streams. Everything depends on the ecosystem and the particular radionuclides." A Luxembourg MP was astonished by the disparities in planned zones of evacuation: "In France it is 10 km, in the United States it is 16 km., in Germany and Switzerland it is 20 km., and at Chernobyl they evacuated to a distance of 30 km." An English MP insisted "But what is the level of becquerels acceptable per kilogram of meat?". "There is no common level in Europe," declared a third expert.'

Agreement between western European countries on a 'safe' limit for radioactive contamination of food and drink was never reached. The EEC issued 'recommended' limits of 500 Bq./kg. 18 months after Chernobyl. This was far lower than many national limits: Albania and France, for instance, set acceptance limits as high as 2,000 Bq./l. for radioactive iodine-131 in dairy milk. If the EEC limit had been adopted at the time of the disaster, a far more damaging picture of European fallout levels would have emerged, and far stricter controls on food and drink would have been necessary throughout the continent.

High-pressure hoses being used to wash down buildings in the town of Pripyat that have been contaminated by radioactive fallout from the accident at Chernobyl. (Credit: Novosti/Science Photo Library)

The true extent to which Chernobyl polluted Europe with windborne fallout is only now becoming apparent. In Autumn 1988, the Munich Environment Institute reported that south German wild mushrooms contained up to 11,400 Bq/kg of radioactivity. The Institute likened eating them to playing 'Russian radiation roulette'. (*Guardian* 29.9.88)

'A study sponsored by the US DOE a year after the Chernobyl disaster and considered, by its authors, to be the most complete and sophisticated at that time concluded that Europe — not the Soviet Union — is likely to be most affected by Unit 4's fallout. (It should be borne in mind that the DOE, operator of the US government's reactors, is a *pro-nuclear* body.) The study's chief author — Marvin Goldman, a professor at the University of California at Davis — and his colleagues collated Soviet data with readings taken outside the USSR, matched them with a computer model, and found that levels of caesium-137, which has a half-life of 30 years, were three times as high as previous studies assumed. Goldman calculated that roughly one million curies of caesium-137 fell within Soviet borders, another one million curies fell on the rest of Europe, and a final one million curies spread throughout the Northern Hemisphere: a total amount comparable to the fallout from all atmospheric weapons tests to date.The total emission of *all* radionuclides from Chernobyl (excluding noble gases) amounted to some 50 million curies.

Goldman maintained (*Science* 8.5.87) that his higher number for caesium-137 increased radiation dose estimates. The death rate from cancer is determined by applying a statistical 'risk factor' to dose estimates. The DOE study used epidemiological data from the US National Institutes of Health to arrive at a risk factor of 2.3 fatal cancers per 10,000 person-rem of radiation, whereas the Soviet Union applied a risk factor of one, a United Nations benchmark set in 1977, to determine post-Chernobyl deaths.

By dividing the new, increased radiation levels by the new, increased risk factor, Goldman arrived at a chilling conclusion: worldwide, deaths from Chernobyl-induced cancers would be around the 40,000 mark. Out of that figure, 'only' 12,000 casualties would be Soviet citizens. More than half the total, some 21,000 people, would be Europeans.

If Goldman's appalling forecast is accurate Europe and the Soviet Union will still have got off lightly. The radioactive cloud missed Kiev and its 2.5 million inhabitants; no rain fell until it reached Sweden, sparing Poland and the Soviet

Union initially; the prevailing winds saved Western Europe from dramatically increased fallout levels; none of the complex's remaining reactors were affected, apart from external contamination; and only a small percentage of the total radioactive inventory was released into the atmosphere.

The US nuclear engineer and author Dr R. E. Webb calculated that, if three quarters of the fission products within Unit 4 had been released, the accident could have been cataclysmic for Europe.

He estimated that the radiation consequences would have been 200-400 times greater. No less than 160,000 sq km of land would have had to be abandoned, and one million sq km would have been lost to agriculture for many decades. 'All things taken together,' he wrote 'the Chernobyl accident could have been much worse.' (1986)

It should be noted that Marvin Goldman's figure represents only the middle of the range of estimates. Other studies, such as that by Dr J. Gofman, using the latest information on mortality amongst atom-bomb survivors, concludes that globally, up to one million people may die from cancers induced by exposure to radiation from Chernobyl.

None of those figues includes consideration of other grave health effects such as non-fatal cancers, mental retardation or genetic abnormalities, which all increase due to the increased exposure of the world population.

THE POST-MORTEM

In August 1986, the IAEA held its first post-mortem on the Chernobyl accident. The Soviet delegation described the steps that led to the disaster:

- The experiment had never been checked with plant safety experts.
- Operators violated no fewer than six safety rules.
- Operators had come to believe so firmly in the safety of the reactor that they 'lost all sense of danger' and became over-confident.
  This led them to make 'deliberate, systematic and numerous violations of procedure'.
- The operators' attitude led them to run the reactor 'as if they were piloting a high-flying plane and opening the doors'.

FALLOUT SANS FRONTIERS

The immediate after-effects of the Chernobyl disaster further undermined Western Europe's nuclear power programmes. In Sweden, anti-nuclear feelings forced a referendum which led to plans to phase-out nuclear power. In Austria, the government announced plans to dismantle the newly completed Zwentendorf reactor, which was not yet in operation. A senior official at the plant, Walter Roznovsky, said, 'It is Chernobyl that has destroyed us. And the future? It is grim, so grim.' (*Guardian* 20.5.86)

In West Germany, the nuclear debate became more heated and gained new momentum. Already fierce, the opposition to the planned reprocessing plant at Wackersdorf and the fast-breeder reactor at Kalkar intensified (the former was finally cancelled in 1989). Ten months after Chernobyl, the Green--Socialist coalition government of Hesse was torn apart by its failure to agree the fate of a nuclear fuel plant at Hanau. (In an irony of the German political system, the conservative, pro-nuclear Christian Democratic Union won instead in the election in April 1988.)

In non-aligned Yugoslavia, there were calls for the scrapping of nuclear power, and on 19 February 1987, the one functioning reactor was closed down indefinitely after it suffered its 36th emergency shutdown in six years.

Radionuclides from reactor accidents respect no national boundaries, a fact that was not lost on Europe's densely-packed populations. After Chernobyl, attention was focused on nuclear plants sited close to national borders — in particular, the four reactors at Cattenom, on France's eastern border, just 18 km from Luxembourg and West Germany.

French national law permits higher levels of routine radioactive discharges than West Germany. However, because the liquid discharges from the Cattenom plant will cross the Franco-German frontier in the Moselle River, Article 37 of the 1957 Euratom Treaty, which governs trans-frontier nuclear contamination, is applicable. In consequence, the regional government of Saarland, which has spent £1 million on a system of radiation monitoring stations, took the French to the European Court. 'If Cattenom were to have an accident on the scale of Chernobyl,' said Saarland Prime Minister Oskar Lafontaine, (*Guardian* 20.5.86), 'the whole of the Saarland, Alsace, Lorraine and parts of the Rheinland would be uninhabitable for ever.' The Cattenom dispute is being repeated on other European borders, with the issue of trans-frontier radioactive pollution being taken up by the EEC's nuclear watchdog body, Euratom.

A month after the special IAEA Post-accident Review meeting in Vienna in August 1986, the IAEA's annual conference hosted the initial ratification of two rapidly-drafted conventions: the Convention on Early Notification of a Nuclear Accident, committing signatory nations to let each other (and the IAEA) know promptly in the event of a nuclear accident — civil or military — and to set out the international implications; and the Convention on Assistance in the Event of a Nuclear Accident. By March 1989, at least 70 nations had signed the conventions.

However, there is no detailed definition in either Convention of what constitutes an accident and it is debatable as to whether accidents of lesser radiological consequence than Chernobyl will be required to be reported.

THE AFTERMATH

17 January 1987: AP report from Moscow that keys needed to bypass safety systems at Chernobyl are now in sealed blocks to prevent human error from causing another accident there, according to Morris Rosen, head of the IAEA's Division of Atomic Safety. (*The Times*, 17.1.87)

14 February 1987: It is reported on this date that Polish bees headed straight back to their hives when they sensed contamination from Chernobyl, and hid for several days after the explosion, according to Henryk Ostach, head of the Polish beekeepers' association. (*The Times*, 14.2.89)

14 March 1987: The Chernobyl accident deposited about one kg of radioactive caesium on Norway, according to the state radiation board — six per cent of the total amount of caesium released by the fire. (*The Times*, 14.3.87)

10 April 1987: AFP report that children in Hamburg's 189 orphanages cannot eat their chocolate Easter eggs this year because they contain too much radioactivity from Chernobyl. The 20,000 eggs were made from milk from the Bavarian region of southern Germany. (*The Times*, 10.4.89)

21 April 1987: Reuter report that five people who suffered acute radiation sickness after Chernobyl have since become parents of healthy children, according to Leonid Ilyin, vice-president of the Soviet Academy of Medical Sciences. (*The Times*, 21.4.87)

25 April 1987: In a report from Kiev, *Times* correspondent Christopher Walker describes how a fleet of 300 blue and

orange water-carriers still regularly hose the streets in order to keep down the contaminated dust from the Chernobyl exclusion zone, 80 miles to the north. When temperatures are below zero, giant vacuum cleaners are used. Kiev is now protected by an extraordinary network of new dams, ramparts and underground walls, built to try to prevent radioactivity seeping into the city's main water supply. Walker notes that the building of these defences has been described by a Western expert as one of the most remarkable civil engineering feats in recent years. All fresh food sold in Kiev's 22 peasant markets must be accompanied by a special certificate showing that it has been checked for radiation. On every main road into the town there are radiation monitoring stations. (*The Times*, 25.4.87)

26 April 1987: First anniversary of Chernobyl disaster. Thousands of demonstrators stage protests in Europe, the Far East and in the Americas. Mexicans condemn the building of the country's first nuclear plants, calling the plants 'little Frankensteins'. (*The Times*, 27.4.87)

27 April 1987: Up to a tenth of the radioactive caesium and other isotopes released from Chernobyl may have fallen on Sweden, according to an analysis of meteorological data published in *Ambio*, the prestigious journal of the Swedish Academy of Sciences (vol xvi, No 1. 1987). (*The Times*, 27.4.87)

5 May 1987: Following allegations that Polish milk powder, exported to Nepal, has been contaminated by radiation from Chernobyl, the 'poisonous' milk is widely boycotted and all further imports are stopped. During the furore it caused, two Nepalese MPs were arrested for criticising government policy and the government itself comes under heavy criticism. (*Independent*, 5.5.87)

30 May 1987: The Moscow weekly, *Nedelya*, reports that Vladimir Shevchenko,the director of a documentary made within days of the accident, *Chernobyl: A Chronicle of Difficult Weeks*, died two months ago, and that two cameramen who worked with him on the film are receiving hospital treatment. (*Independent*, 30.5.87)

23 June 1987: Celestine Bohlen of the *Washington Post* was one of a number of western journalists allowed into the 18-mile 'dead zone' around Chernobyl for the first time since the accident. She reports on the 'red-headed forest' as it has been nicknamed — 70 acres of pine trees, extending north from the damaged reactor, which have now lost all their needles and turned a strange rust colour. Pine trees are

especially sensitive to radiation and can stand no more than a man. (*International Herald Tribune*, 23.6.87)

7-29 July 1987: The trial of six senior officials blamed for the Chernobyl disaster is held in the still contaminated town of Chernobyl, just 11 miles from the damaged reactor. The former director and two of his senior colleagues were sentenced to 10 years hard labour after being found guilty of criminal negligence. Three other former employees of the plant received lesser sentences. (*The Times*, 30.7.87)

25 September 1987: Vladimir Yavorivsky, a Ukranian journalist who was an eyewitness to the immediate aftermath of the accident, reveals, in an article entitled 'The X-ray of truth' in the Soviet weekly, *New Times*, that a local reporter from Pripyat, the town closest to the reactor, had published an advance warning about the possibility of a disaster — which had been ignored; that the disgraced director of the plant had told the Kiev regional civil-defence unit three and a half hours after the accident: 'It is merely a fire on the roof. We will put it out'; and that five days after the disaster, when Kiev was suffering high radioactivity, local party chiefs deliberately ignored the dangers, and allowed the traditional May Day procession, involving thousands of young children, to go ahead. (*The Times*, 25.8.87)

9 October 1987: The European Commission launch an urgent inquiry into the delivery to Ghana of 750 tonnes of EEC milk powder containing apparently dangerous levels of radioactivity from Chernobyl. The incident focuses attention on the continuing uncertainty over what constitutes a safe level of radioactivity in food. (*The Times*, 10.10.87)

14 January 1988: The Politburo reveal that the total cost of the Chernobyl accident was 8 billion roubles (£8 billion), some four times higher than the last estimates made public by the Soviet authorities. This makes it the most expensive accident ever. (*The Times*, 15.1.88)

31 March 1988: *New Scientist* reports that the Soviets solved the problem of what to do with radioactive sheep; they were slaughtered and fed to the mink that are farmed for the fur trade. Radioactive-contaminated milk was made into butter; the process takes long enough for the radioactivity to fall to lower levels before the butter is ready for human consumption. (*New Scientist*, 31.3.88)

24 April 1988: A critical report in *Pravda*, claims that sloppy repairs, drunkenness and nepotism are rampant at the Chernobyl plant, where managers act 'as though there hasn't been an accident'. (*Independent*, 25.4.88)

9 September 1988: A 37 per cent increase in the incidence of leukaemia in children in Scotland aged under four may be linked with the Chernobyl accident, according to research published in the *Lancet*. (*The Times*, 9.9.88)

15 September 1988: The Soviet authorities have decided to turn the 30 km area around the Chernobyl plant into a national park, according to Ilya Likhtaryov, a Soviet scientist from Leningrad University's Institute of Hygiene. (*New Scientist*, 15.9.88)

8 October: *Pravda* reports that official plans have been made to demolish the 800-year-old town of Chernobyl and that 994 people, mostly elderly, have dodged roadblocks and returned to live unofficially in villages surrounding the town. (*International Herald Tribune*, 10.10.88)

3 October 1986

Soviet Submarine Fire North Atlantic

A nuclear-powered Soviet Yankee I class submarine suffered an explosion and fire in one of its missile tubes, killing at least three crewmembers. The submarine, armed with 16 SS-N-6 nuclear missiles (with 1 Mt warheads), was 480 miles east of Bermuda and less than 1,000 miles from the US mainland. The submarine was forced to surface, and called for assistance from a Soviet merchant ship in the area. The Soviet news agency Tass was quick to state: 'After analysing the situation, an expert commission in Moscow reached the conclusion that there is no danger of any unauthorised action of weapons, a nuclear explosion or a radioactive contamination of the environment.' (*International Herald Tribune*, 6.10.86)

The swift official announcement of the accident was seen as a successful move to defuse any tension with the US over the incident in advance of the Reagan-Gorbachev summit at Reykjavik. It also provided the first test of a new international convention, signed 10 days before, which obliged nations to give early warning of a nuclear accident that might threaten a neighbour. The convention, signed by 52 countries, followed the accident at Chernobyl.

The 10,000-tonne, 425-ft submarine continued to make headway under its own power and was then taken in tow by a Soviet freighter. Then, three days after the fire began, liquid fuel in one of the missiles exploded, blasting a hole in an area of the hull behind the conning tower, and the submarine sank in about 18,000 ft of water, some 600 miles north-east of Bermuda. Two hours before the sinking, a US Navy P-3 patrol aircraft had reported that towing had been halted and that the submarine was being evacuated. It is believed that no further crewmembers were injured.

There was much speculation at the time as to how much danger was posed by the nuclear materials in the weapons and reactors on board the sunken submarine. The US Defense Department was quick to allay fears. At a press briefing held shortly after the sinking, a spokesman pointed out that there had been no reported rise in levels of radioactivity from sunken US nuclear submarines, and both Soviet and US spokesmen were quick to state that the submarine's reactors had been shut down before the vessel sank, implying that a shut-down reactor is a safe reactor.

A technical study by Greenpeace put estimated the reactors' radioactive content to be between 10 and 20 million curies of radioactivity, in the form of highly active materials that will remain harmful to life almost indefinitely. This figure represents as much as one third of the radioactivity released into the air by the Chernobyl accident, and 20 times the radioactivity dumped in the Atlantic by the UK, Belgium, the Netherlands and Switzerland between 1971 and 1982.

Unless the submarine is recovered, the radioactive material will eventually be released, and a proportion will be absorbed into the marine food-chains and, through the fisheries, will make its way towards man. The only unknowns are how soon and how much.

10 January 1987

Nuclear Transporter Crash, UK

On this date, an accident involving a 20-ton Mammoth Major warhead transporter provided a rare insight into the transport of nuclear weapons around Britain. The accident occurred eight miles from Salisbury, Wiltshire, on a Saturday afternoon, when a convoy of 10 vehicles was heading for the Royal Navy's top secret arms depot at Dean Hill on the Hampshire-Wiltshire border. At about 15:50 two of the specially-adapted green armoured trucks, fitted with a sliding canopy to facilitate loading by crane, skidded off an icy road. One of them, believed to have been carrying nuclear depth charges, careered down an embankment and overturned in a field. The other came to rest on the edge of the embankment. The crash site and an area one and a half miles around was immediately sealed off by 50 Royal Marines, and police set up road blocks. The following day, the lorry was towed away after being righted by a 50-tonne crane.

Nuclear weapons need to be regularly overhauled to ensure they can be relied on to detonate when used, and their transport to and from the Royal Ordnance factory at Burghfield, near Reading, is organised by the RAF's Special Convoy. The trucks that carry the weapons are escorted by Transit vans or minibuses containing armed guards with military police motorcycle outriders clearing the road ahead. The convoy also contains a safety team of decontamination experts and a fire engine. Bringing up the rear is a mobile command centre and armoury, on hand in case of a terrorist attack.

The main danger posed by the crash was the possible detonation of the conventional explosive in the weapon's trigger device and a subsequent fire. This would have scattered radioactive material over a wide area.

Later reports claimed that the coded radio message — Oldham Two — was transmitted to defence, security and environmental agencies. This is only sent when there is a real danger of a major leak of radioactivity from an explosion or a fire. The message triggers preliminary arrangements for widespread evacuation of the area.

13 September 1987

Goiânia, Brazil

When a private radiotherapy institute in the city of Goiânia (about 180 km south-west of Brasilia) moved to new premises at the end of 1985, they left a caesium-137 radiotherapy unit in their abandoned clinic, without notifying the appropriate licensing authorities. The building was subsequently partly demolished and, two years later, two people entered the premises and tried to dismantle the machine, thinking it had some scrap metal value. Attracted by the shiny stainless steel casing of the radioactive source capsule, they removed it in a wheelbarrow to one of the men's houses nearby. Thus began one of the most serious radiological accidents ever to have occurred.

The capsule contained strongly radioactive caesium chloride salt which is highly soluble and readily dispersible. The day that they removed it, both men began vomiting and one developed diarrhoea, felt dizzy and had a swollen hand. A doctor advised him that he was suffering an allergic reaction caused by eating bad food and he was told to stay at home for a week.

Five days later, on 18 September, his partner began trying to break open the capsule. In the process the caesium salts spilled out. The whole of his house and grounds became so extensively contaminated that it later had to be demolished and the topsoil removed.

The radioactive capsule was then sold to Ivo Ferreira who lived next to the junkyard he managed, and it was transported there in a wheelbarrow by one of his employees. That night, he went into the garage and noticed a blue glow emanating from the source capsule. Thinking how pretty it looked and how the powder might be valuable or have supernatural powers, he took it into his house. Over the next three days he and his wife examined the powder closely and they invited various neighbours, relatives and acquaintances to come and see this curiosity.

A friend of Ivo Ferreira's used a screwdriver to remove fragments of the caesium powder, about the size of rice grains, and gave some to his brother and took the rest home, where he distributed them among his family. Several of them daubed it on their skin.

The wife of Ivo Ferreira was by now vomiting and had diarrhoea. She was also told by a doctor she was suffering from an allergic reaction to something she had eaten and was sent home to rest. She subsequently died. Her mother came over to nurse her for two days and carried a significant amount of contamination home with her.

In the meantime, two of the junkyard employees worked on the scrap to try and extract the lead. They both subsequently died. The lead and remains of the source assembly were sold on to another junkyard. The junkyard man's brother took some fragments of caesium back to his house where they were placed on a table during a meal. His six-year old daughter handled them while eating by hand. She also subsequently died.

By now a significant number of people were physically ill and the junkyard man's wife was convinced the glowing powder was the source of the sickness. Along with one of her husband's employees, she put the remnants of the unit in a bag and travelled by bus to the Vigilancia Sanitaria where they saw one of the doctors, telling him it was 'killing her family'.

That same day the woman and her companion were referred from the local health centre to the Tropical Diseases Hospital where several other people with similar symptoms had been examined. One of the doctors began to suspect that the patients' skin lesions had been caused by radiation.

After consultation with other experts, a local physicist borrowed radiation monitors and quickly discovered that there was evidence of contamination over a wide area of the city. They persuaded incredulous officials at the Secretary of Health for Goiás State that they had an emergency on their hands and a full-scale alert involving the police, fire brigade, ambulance and hospitals was announced. The main areas of contamination were resurveyed, evacuated and cordoned off. Plans were laid to examine contaminated persons in the city's Olympic Stadium; that night 22 people were identified as having been highly exposed.

In all, 249 were found to have been contaminated either internally or externally, of whom four subsequently died. Using a specially equipped helicopter and cars fitted with detectors, the whole city was surveyed for contamination and the seven principal sites were isolated. In total, 85 houses were found to be significantly contaminated and 200 people were evacuated from 41 of them. Seven of the houses had to be demolished and large areas of soil removed. In all, 3,500 cu. m. of low- and medium-level waste — more than 275 lorry loads — was taken to a temporary waste-storage site 20 km from the city. A team of 550 people took about 11 weeks of intensive work to survey and decontaminate the highly contaminated sites and a further three months to deal with the residual contamination in 45 public places and 50 vehicles.

The 1988 IAEA report on the incident notes: 'The accident in Goiânia had a great psychological impact on the Brazilian population owing to its association with the accident at Chernobyl...Many people feared contamination, irradiation and

Gerardo da Silva, one of the people contaminated by the accident at Goiânia, strikes a pose for the photographer. His companion, Edson Faviano, looks on. Shortly after this picture was taken, they both boarded a military aircraft that flew them to Rio de Janeiro for medical treatment. (Credit: Popperfoto)

damage to health; worse still, they feared incurable and fatal diseases...Some of the inhabitants of Goiânia were discriminated against, even by their relatives, and sales of cattle, cereals and other agricultural produce, and of cloth and cotton products — the main economic products of Goiás State — fell by a quarter in the period after the accident.'

The funeral of two of the victims turned into a riot when 500 people threw rocks and tried to block the hearse. Local people believed the irradiated corpses would cause further contamination; the authorities denied this, noting that each corpse was sealed in a 1,400-lb lead-lined coffin. These were encased in cement tombs with seven-in-thick walls.

All things nuclear in Brazil are shrouded in military secrecy and there is an almost total lack of public accountability. The Brazilian Society for the Progress of Science, the country's leading scientific body, concluded in a report issued in November 1987 that there may have been other similar radiation incidents in Brazil that went unnoticed. Top nuclear officials have admitted that at least 50 other pieces of equipment containing radioactive materials have been abandoned across the country; Brazil's Nuclear Energy Commission stopped examining the 236 known to be in use in 1984 because of lack of resources.

16-17 December 1987

Biblis-A Reactor, West Germany

In power plants like Biblis-A, near Frankfurt, the low-pressure-injection system (LPI), which forms part of the emergency core-cooling system, is isolated from the primary cooling circuit of the reactor by a system of valves. During reactor operation it is of vital importance that the LPI-system is reliably isolated. Otherwise it could allow primary coolant to leak out of the containment — a particularly hazardous loss-of-coolant accident, since the leaking coolant would not be available for recirculation. Furthermore, the LPI-system is not designed to withstand high pressures and could be destroyed if exposed to the operating pressure of the reactor. A core-melt accident would be inevitable.

On 16 December 1987, during the start-up of the reactor, the main valve (No. TH22 S006) between the primary circuit and the LPI-system was left open by mistake. Although a red light in the control room indicated that this had happened, the operators assumed that there was something wrong with the signalling light and that the valve was properly closed. The open valve was overlooked for 15 hours, during which time there were two changes of shift in the control room. Only two secondary valves stood between the LPI-system and the high primary pressure; failure of either of these would have led to disaster.

A leak then developed at another, minor valve, and small amounts of hot water entered the cooling water processing plant. This leak was also initially overlooked because the corresponding signalling equipment happened to be out of order. Finally, at 03:03 on 17 December, the increase of temperature in the processing plant was indicated by a signal in the control room. Two hours later — with the reactor already operating at full power — the operators finally realised that valve TH22 S006 was open.

Even in this highly hazardous situation, it took more than two hours for the operators to decide that a plant shutdown was necessary (at 05:18). However, they didn't want to lose production time and about 10 minutes later they changed their minds and made one last attempt to avoid shutting the plant down. The operators' plan was a dangerous gamble: they hoped that by slightly opening one of the secondary valves — the last barrier — they would create a coolant flow in the line between the primary circuit and the LPI-system, which would trigger the closure of valve TH22 S006. In fact it led to the brink of disaster. Valve TH22 S006 did *not* close, and primary coolant started to escape. The secondary valve was shut again after seven seconds and only 150 l of primary coolant were lost. Fortunately the secondary valve did not stick like the first, and a catastrophe was avoided. The operators finally shut the reactor down.

This type of accident was considered practically impossible according to the official safety philosophy. It was not a 'Design Basis Accident'. The plant's owners, RWE, classified the event as a low-priority incident, category 'N' for normal. The authorities of the State of Hesse were notified five days later. The description of the incident they received was inaccurate and no mention was made of the last step, the attempt to shut one valve by opening another.

German officials only learned how serious the event had been following a technical review in March 1988. There was no public disclosure because the West German government rarely discusses safety incidents at individual nuclear plants.

In December 1988 an US trade newsletter, *Nucleonics Week*, triggered off the debate in the West German press. An article quoted unnamed NRC officials as saying that the accident was a 'top-level' event; if it had happened in the USA, an investigator would have been at the scene 'within hours'.

Klaus Toepfer, federal minister for the environment, then confirmed details of the accident and said that 303 incidents affecting safety had occurred at West German nuclear plants in 1987. Of these incidents, 11 were considered worthy of 'rapid attention'. Also in December, the month that it finally became public knowledge, it emerged that the accident was logged with the reporting system for nuclear accidents run by the Nuclear Energy Agency of the Organisation for Economic Co-operation and Development (OECD) in Paris, along with a request that it should remain secret.

Never before has a West German nuclear power plant come so close to a catastrophic accident. This incident has shaken the foundations of German nuclear-safety philosophy, since an accident to which the official German Risk Study had assigned a probability of 1 in 33 million per year had very nearly occurred after less than 100 operating years of German PWRs.

It is less than reassuring that — as the astonished public learned in February 1989 — the computer print-out of the December 1987 incident was thrown away in January 1988. It was therefore impossible to reconstruct the accident sequence and find out why valve TH22 S006 had not been closed by the coolant flow. In February 1989, Biblis-A was permitted to go into operation again, but the plant's troubles were not over. Serious deficiencies in the plant's fire brigade's training and equipment were soon reported, and, on 9 March 1989, another serious incident occurred, putting half of the emergency core cooling system out of action for some time. (After Biblis-A had gone back into operation it transpired that an incident similar to that of December 1987 had occurred before — in 1978.)

1988-89

US Weapons Production Complex

One of the biggest scandals in US nuclear history broke open in the second half of 1988 following two congressional investigations into the US DOE's management of the network of 17 weapons labs and production facilities — located in 12 states and employing 100,000 people — that make up the US weapons production complex.

Controlled by the AEC until 1977, this network of facilities has operated behind a screen of national security since the Manhattan Project began. Little investment has been made in these ageing facilities and, over the last decade, there has been a string of revelations about radioactive pollution, poor management, faulty equipment and bureaucratic cover-ups which has had the cumulative effect of triggering prime-time media interest and broad public concern. The weapons production system is currently in disarray and, for the first time since World War II, the US public is questioning the necessity of maintaining it.

Four of the most critical of these weapons plants, and the ones principally affected by these revelations, are Hanford, Savannah River Plant, Rocky Flats and the Fernald Feed Materials Production Plant.

HANFORD

Some six weeks after the FOIA revelations about radioactive releases from Hanford in the 1940s (see Hanford, p. 81), the accident at Chernobyl focused attention on the last operating reactor at the Hanford site — the N-reactor.

Built in 1963, the N-reactor is the US reactor most similar in design to Chernobyl, as it has a graphite-moderated, water-cooled core and no concrete containment dome. The reactor also has the highest peak power level — 4,000 MW (thermal) — of any reactor in the US and is designed to produce the maximum amount of plutonium per unit of uranium used. It supplied 20 per cent of the nation's plutonium plus electricity for the north-west power utilities.

On 19 May 1986, three nuclear safety experts testified before the House Interior and Insular Affairs subcommittee that a similar accident to Chernobyl was possible; and that, in the event of a reactor emergency, highly radioactive cooling water from a back-up system could be dumped directly into the Columbia River.

In the summer of 1986, the US Energy Secretary, John Herrington, commissioned an independent safety review of the N-reactor by a panel of six nuclear experts, the results of which were made public in December.

Four of the experts recommended remedial safety measures but supported the continued operation of the reactor for another three to five years. Two others, including the nominal chairman, Louis Roddis — a nuclear engineer and former president of Consolidated Edison — recommended the reactor be permanently shut down. Roddis's exact comment was that the DOE should 'shut down the N-reactor unless a positive judgement is made that the requirements for defense material warrant accepting public hazards exceeding those of commercial reactors.' (Schwartz, Greenpeace Chronology, 1988/89)

Throughout the Chernobyl crisis, the DOE had maintained that all their reactors were safe. Yet on receipt of this report, the DOE announced the N-reactor would be shut down on 7 January 1987, for a $50 million safety upgrade programme. In fact $110 million was spent before, on 16 February 1988, the DOE announced that the N-reactor would not be restarted but would be placed in 'cold standby'. The reason given was that plutonium requirements could be met through the Savannah River Plant and by recycling the existing stockpile. Herrington subsequently remarked to a congressional subcommittee: 'We're awash in plutonium. We have more plutonium than we need.' (Schwartz, Greenpeace Chronology, 1988/89)

After irradiation in the N-reactor, uranium was taken in heavily-shielded casks on a rail car to the Plutonium and Uranium Extraction (PUREX) plant on the Hanford site for chemical processing, where it was dissolved in acid to separate out the 'daughter' products of the fission process. On 8 October 1987, the PUREX plant was also closed, following disclosure of a catalogue of employees' safety violations.

SAVANNAH RIVER PLANT

Revelations about a long history of accidents were triggered by an incident at P-reactor, the oldest operating military production reactor in the US and one of the five reactors comprising the vast 192,000 acre Savannah River Plant (SRP), near Aiken in South Carolina. Constructed and managed until April 1989 by Du Pont for the DOE, the reactors produce tritium and plutonium; the site is also used to store nuclear waste.

For three days, operators at the plant had been trying to start a nuclear reaction when, on 10 August, the reactor suddenly produced an unexpected and unexplained power surge. When debriefed by DOE personnel, the operators admitted they had no idea what had caused the mysterious incident; worse, according to one DOE safety investigator's notes, 'THEY DIDN'T CARE!' (*Washington Post* 6.10.88)

When asked by Congress about the incident, the DOE at first denied it, and then made public its own report which revealed a catalogue of 'reactor incidents of greatest significance' dating back 31 years. The DOE report was based on a 19-page memo written in 1985 by G. C. Ridgely, a technical supervisor at Du Pont who had been asked to review the plant's history of operating problems.

According to a report in the *New York Times* (4.10.88): 'The Energy department (DOE) said it was seeking to determine why nothing was done about the memorandum. The department said the failure to disclose the problems illustrated a deeply rooted institutional practice, dating from the days of the Manhattan Project in 1942, which regarded outside disclosure of any incident at a nuclear weapons plant as harmful to national security.'

The accidents documented in the memo, which occurred between 1957 and 1985, were among the most serious ever documented at a US nuclear facility. They include:

1960, January 12: Technicians trying to restart the L-reactor after it had shut down automatically, almost sent it out of control. The power increased 10 times faster than was considered safe.'

1965, January 4: A leak deposited at least 14.5 lb of plutonium sludge on the walls of a concrete air duct inside the plant; the engineers feared that it was very close to going critical, or generating a spontaneous nuclear reaction.

1965, May 10: During full power operations of C-reactor a 'very significant leak' developed that spilled 2,100 gal of cooling water on the reactor vessel floor. As a result, the level of coolant within the reactor fell precipitously and the reactor automatically shut down.

1970, November: Huge amounts of radiation were released in a room adjacent to one of the reactors, where cooling water is filtered. A total of 900 workers spent three months in a highly contaminated environment cleaning up the radiation. This accident was caused by the melting of a vital reactor component known as the source rod. Made of antimony and beryllium, this provides a source of neutrons to start the nuclear chain reaction. The memo reveals that technicians ignored a radiation alarm for two hours.

1970, December: Technicians tried to start a chain reaction inside C-reactor but it automatically shut down because the flow of cooling water was too low and the reactor was overheating. Instead of investigating the problem, the operators tried three more times to start the reactor. The heat generated during the attempts melted the fuel assembly — the first stage of a meltdown of the reactor core.

1982, March: A technician left a water valve open and for 12 hours the undetected flow flooded a large plutonium-processing room to a depth of two feet.

The main DOE report ('Review of the operating experience history through 1987 of Savannah River plant production reactors') also revealed that:

- Chronic equipment failure and poor operating procedures caused the Savannah River reactors to shut down unexpectedly nine to 12 times a year over a 20-year period — a rate twice that of the US civilian nuclear power industry. The highest number of shutdowns was 43 in 1977.
- A former engineer claimed he tried to warn officials in 1982 about leaks from the holding tanks where highly radioactive liquid waste was stored.
- There had been 25 accidents in which workers had been accidentally exposed to radiation. Workers had complained of unsafe practices but were ignored.

The Savannah River site also holds about 70 per cent of the nation's military high-level nuclear waste. The total volume is 34 million gal, of which 90 per cent is a liquid solution, called supernate, containing mainly caesium-137 (160 million curies in total). Ten per cent of the waste is in the form of sludge containing all the plutonium and strontium-90 (790 million curies of beta-gamma radioactivity).

This waste is stored in 51 sub-surface storage tanks; 43 are double-walled tanks of various kinds, eight are single-walled. There have been a number of unintentional releases of high-level waste at these tank farms.

Other liquid wastes were discharged into a series of 68 seepage basins — inlined ponds designed to leak radioactivity into the surrounding earth. As a result the network of shallow aquifers under the vast site is contaminated with radioactive compounds. Deeper aquifers, which may be the source of drinking water for the surrounding area, were found to contain toxic, non-radioactive materials.

Of the five reactors at the plant, two have been permanently shut down; one of these, the C-reactor, was retired in 1985, after cracks were discovered in the reactor vessel. Reactors P, K and L were temporarily shut down on 10 April 1988, after the discovery of cracks in their cooling systems, and have remained shut down since. The earliest they can be repaired and reopened will be 1990.

The importance of the Savannah River reactors is that they were the last dedicated US facilities remaining in operation for the production of plutonium and tritium, a radioactive form of hydrogen that helps produce a more efficient chain reaction in nuclear weapons, thus enabling weapons designers to produce smaller, lighter warheads with the same explosive yield. The problem with tritium, from the weapon makers' point of view, is that it decays at a rate of five and a half per cent annually and at some point — an official secret — decays enough to make the weapon inoperative.

Pentagon officials claim that not having the plutonium- and tritium-producing reactors at Savannah River is tantamount to unilateral nuclear disarmament. Opponents of the weapons programme say that nuclear weapons will still explode without tritium but will just give a smaller blast. Furthermore, they claim that by recycling the gas from other weapons, the US can guarantee its supply for a long period.

If the SRP reactors are restarted, their extreme age makes their continued operating life extremely limited. Accordingly, the DOE is requesting funds to build new production reactors at SRP and INEL in the next decade.

ROCKY FLATS

Problems at the Rocky Flats plant multiplied in 1988. They began on 29 September when a DOE safety inspector and two workers accidentally inhaled radioactive particles after walking into a contaminated area because the warning sign was obscured. Four days after this incident, the same DOE inspector called his headquarters in Washington with a long list of further safety concerns in Building 771. These included: poor radiation monitoring, haphazard placement of warning signs, electrical dangers, poor housekeeping, a backlog of nuclear waste, and complacent attitudes. When further troubles came to light in the next few days, the DOE decided, on 7 October, to shut the building down. (The DOE claimed at the time that this decision was due to the one contamination incident. It was only after an investigation by the GAO that the string of 'very serious' violations that left 'no margin for safety' was revealed.)

Building 771 was reopened on 20 January 1989, following extensive cleaning to remove decades of accumulated plutonium dust. DOE officials were confident that full production in the building could resume by April, but this was by far from being the end of the problems at the plant:

In November 1988, an investigation by the FBI and the DOE's inspector-general revealed that, for 18 years, 30 per cent of the highly skilled engineering and technical staff's time in Building 881 — a top secret establishment where models of bombs and new weapons were made — had been spent instead on making items for private use and sale. These items included medallions, baseball caps, wine presses, clocks and gold- and silver-plated jewellery.

Building 371 was designed in the early 1970s as a safe and efficient successor to Building 771. When it was completed, after eight years construction at a cost of $215 million, its equipment was found to be completely unworkable. After a further seven years of study, the DOE decided to replace all the building's reprocessing equipment; this will take an estimated 10 years and cost about $400 million.

Rocky Flats has also been beset with problems over nuclear-waste disposal. Waste used to be transported to the Idaho National Engineering Lab (INEL) at Idaho Falls, where the intention was to store it temporarily while a permanent underground vault in New Mexico — the Waste Isolation Pilot Project (WIPP) — was being completed. In 1988, because the vault was still not ready, the Governor of Idaho, Cecil D. Andrus, decided that Idaho would no longer play host to Colorado's nuclear waste, and banned its entry into the State. On 22 October 1988, Governor Andrus prevented a railroad car of wastes from Rocky Flats from entering INEL, and it was returned to the plant.

But Colorado doesn't want the waste either. At a public meeting near the Rocky Flats plant on 25 October 1988, Earl Whiteman, the DOE's manager of the plant, startled his audience by informing them that each box car was loaded with 140 55-gal drums of waste and that each drum was permitted by DOE regulations to contain up to 200 g of plutonium. This is twice the amount Rocky Flats says would cause a serious nuclear accident if released into the air.

Under Colorado law, the plant has limited facilites for storing this waste. Colorado Governor Roy Romer vowed to close the plant if its holding capacity of 1,600 cu yd was reached.

On 23 February 1989, Governor Andrus said he had been forced to rescind his ban after the federal government gave him 'no alternative short of closing Rocky Flats and precipitating a national security crisis.' Governor Romer also acquiesced but both governors warned this was only temporary and that the Bush Administration 'must come up with a permanent answer for this waste...or that plant could close.' The new agreement allows Rocky Flats to operate until December 1989, when accumulated wastes will once again be a problem (Schwartz, Greenpeace Chronology).

FERNALD FEED MATERIALS PRODUCTION CENTER

Located near the town of Fernald and 18 miles north-west of Cincinnati, the innocent-sounding Feed Materials Plant actually turns uranium hexafluoride into fuel rods for DOE military reactors and components for nuclear warheads.

Richard Shank, director of Ohio's Environmental Protection Agency, estimates that this plant has, in the last 35 years, released 289,000 lb of uranium oxide dust into the air, deliberately discharged 167,000 lb of wastes into the Great Miami River and stored a further 12.7 million lb of wastes in pits which may be leaking.

Since 1951, when the plant began operating the contractor in charge, National Lead of Ohio, has been dumping liquid radioactive waste into concrete storage tanks on the site. In 1958, National Lead warned the AEC that liquid waste was leaking from cracks in the tanks; their suggested solution was simply to lower the level of liquid below the cracks. The flawed tanks are still in use. (Westinghouse Materials Company took over management operations on 1 January 1986.)

In 1985, some 14,000 residents from the area around the plant filed a class-action lawsuit against National Lead. Aware that radiation exposure is difficult to link conclusively with specific health problems, they are seeking $300 million in damages from lowered property values and the emotional trauma of living near the plant. However, the federal government is immune from such lawsuits and a recent Supreme Court decision ruled that a contractor meeting specifications set by the government is also immune from legal action. So it seems that no one is legally accountable for the plant.

On 6 December 1988, the DOE released a new report which described 155 incidents of contamination — occurring at 16 weapons plants and laboratories — which pose a threat to public health and the environment. In addition to the incidents already mentioned they include the following:

- Los Alamos National Laboratory, New Mexico: Test firings of high explosives have contaminated the test area with uranium.
- INEL, Idhao Falls: The plant, which reprocesses nuclear fuel and operates experimental reactors, has discharged radioactive and toxic waste into unlined waste lagoons. From here it has leaked into the Snake River Aquifer, a giant underground reservoir, the principal source of water for drinking and irrigation in eastern Idaho.
- Other plants discharging toxic chemicals into the environment include: the Lawrence Livermore Laboratory; the Kansas City Plant; the Mound Facility at Amarillo, Texas; the Pinellas Plant at Largo, Florida; the Portsmouth Uranium Enrichment Complex at Piketon, Ohio; the Sandia National Laboratories in Albuquerque, New Mexico, Livermore, California, and north of Las Vegas; the Y-12 Plant at Oak Ridge, Tennessee.

In the four decades since the Manhattan Project, the government has spent about $250 billion (in 1988 dollars) building and equipping the massive nuclear weapons complex. Now it will cost almost that much to clean it up. The DOE estimate the costs to be as follows: $110 billion for long-term clean-ups, the installation of pollution control equipment and the upgrade of waste disposal practices (the estimated cost of cleaning up Hanford alone ranges from $46 billion to as much as $100 billion), and $20 billion to process high-level waste for permanent storage. Plans are to create a new underground repository at Yucca Mountain, Nevada, that will cost $10 billion minimum; $10 billion for the disposal of low-level nuclear waste, including the cost of building the WIPP in New Mexico; $5-10 billion to decommission several hundred redundant facilities.

The GAO believe the costs will be much greater, as does Senator John Glenn, a leading figure at the congressional investigations into this issue, who puts the clean-up 'superfund' at $200 billion.

In addition, the DOE wants to spend a further $50 billion on building new plant in the next decade. This will include building four small reactors at the INEL and one large reactor at Savannah River.

Additional Stories

1980 This year there were 3,804 accidents at the 69 operating US nuclear power plants. Of these, the NRC identified 104 as 'especially significant mishaps'. Two thirds of the plants accounted for all of the accidents, and just seven plants accounted for over half. (**Public Citizen**, 1979-87)

1980 February A two-ton cask, dropped from a crane above the US Rancho Seco reactor because of a frayed cable, missed the reactor vessel by inches. Two weeks later, a seven-ton load was lifted by the same crane, with the same cable; this was in violation of NRC regulations. If it had dropped on the reactor, large amounts of radiation would have escaped, damaging California's most productive agricultural valley. The NRC didn't learn of either incident until 1986 when a journalist raised the question with NRC officials. (**Public Citizen**, 1979-87)

1980 February 26 An electrical short circuit at the Crystal River Unit 3 reactor in Citrus County, Florida, caused the pressure relief valve to open (as happened at Three Mile Island), and resulted in false signals being sent to the integrated control system. The most serious effects of this were a reduction in feedwater flow to the steam generators, the withdrawal of control rods, and an increase in power output. Temperature and pressure in the reactor began to rise, the reactor shut down automatically, the coolant system depressurised, and the high-pressure injection system (HPI) was automatically activated. An alarm indicated that coolant was escaping from the system but, in contrast to events at Three Mile Island, the operators did not turn off the HPI and cooling was maintained. After 20 minutes, power was restored by chance when the foil on a printed-circuit card burnt through.

It took almost seven hours to stabilise the cooling of the reactor, by which time there were 430,000 gal of reactor coolant on the floor of the containment building, giving radiation levels of 50 rem per hour at one time. (**Public Citizen**, 1979-87; 'The safety of nuclear power plants', Uranium Institute, 1988; NRC Annual Report 1980)

1980 August 21 A Soviet Echo I class nuclear-powered attack submarine suffered a serious accident off the coast of Okinawa, Japan. British sailors on the freighter *Gari* observed several bodies on the submarine deck, and crewmembers receiving artificial respiration. It is believed that at least nine men died, probably in a fire in the nuclear propulsion compartment. The Japanese reportedly found evidence of radioactive contamination afterwards. (**Neptune**)

1980 September 15 A B-52 bomber at Grand Forks AFB, North Dakota was on the ground when a fire broke out in one engine; it burnt for three hours, injuring five people. The plane was standing at nuclear deterrent alert status, was full of fuel and had nuclear weapons on board. (**Bradford**; *Washington Star* 17.9.80; *Omaha World-Herald* 17.9.80; *New York Times* 18.9.80; *Air Force Times* 20.10.80)

1980 October 17 Following a series of human errors and equipment failures, personnel at the Indian Point Unit 2 PWR, 40 km north of New York City, discovered that the reactor containment building was flooded by almost 500,000 litres of brackish water from the Hudson River, which had leaked from the plant's cooling systems. The water had filled the containment sump, overflowed through the in-core instrumentation tunnel and filled the reactor cavity, submerging the bottom of the reactor while it was at full operating temperature and pressure.

The accident forced the plant to be shut down until June 1981, costing the utility that owns it $25 million per month to replace the lost power — a cost that was passed on to New York's electricity consumers. The accident occurred on 17 October but the fact that the hot reactor vessel had been sitting in nine ft of river water was not reported to the NRC until a week later — two days after operators had shut the reactor down. The NRC imposed civil penalties for violations associated with the incident. (NRC Power Reactor Events Nov/Dec 1980 vol. 3 No 1; *New Scientist* 4 December 1980)

1980 November At RAF Kinloss in Scotland, a Nimrod aircraft used in a nuclear anti-submarine role crashed with nuclear depth bombs on board. An MOD spokesman claimed that the plane was unarmed. (**Bradford**)

1980 November 19 During a drill at McConnell AFB, Wichita, Kansas, a Titan II ICBM was apparently started on a real launch sequence. One of the launch crew shut the missile down to prevent launch. (**Bradford**)

1980 December The Royal Navy's nuclear-powered attack submarine HMS *Dreadnought* suffered serious machinery damage (reportedly cracks in the secondary cooling system) which required the reactor to be shut down. The submarine was later retired. (**Neptune**)

1980 December 19 The first serious reported accident involving a DOE 'Super Safe Transport' truck occurred on the ice-coated Interstate 25. The 40-ft tractor-trailer truck jack-nifed, but its cargo of plutonium and other radioactive materials survived intact. (*Washington Star* 30.12.80)

1981 This year there were 4,060 mishaps at the 75 operating nuclear power plants in the US, 140 of them identified as especially serious. The NRC ranked 22 plants as 'below average' or 'far below average' in terms of safety. The average power output of all plants was just 59 per cent of the rated capacity. During the year, more than 83,000 nuclear workers received measurable radiation doses. Total exposure was an all-time high of 54,555 person-rem. (**Public Citizen**, 1981)

1981 February 11 The Tennessee Valley Authority's Sequoyah I plant was the first reactor to be licensed by the NRC after the Three Mile Island accident. While the reactor was in a 'cold shutdown condition', the unit operator told an auxiliary operator, who was new to the job, to check that a containment spray valve (in the system for removing heat from the reactor) was closed. The operator made his way to the auxiliary building and, instead of closing the valve, opened it.

As a result, some 40,000 gal of primary cooling water and a further 65,000 gal of water from a storage tank sprayed into the containment building.

Eight workers were doused by this slightly radioactive shower, which burst all the light bulbs and plunged the containment building into darkness. The official report on the incident blamed it on 'lack of adequate oral communication'. (**Public Citizen**, 1981; *New Scientist* 10.12.81; NRC Case Study Report AEOD/C206 October 1982)

1981 February 23 An unarmed US Pershing II missile exploded during a routine training exercise when a US truck on which it was mounted caught fire at Althuette, West Germany. No injuries were reported. (*Washington Post* 25.2.81)

1981 March 8 A radioactive waste storage tank at the Japanese Tsuruga 1 reactor overflowed and leaked into Wakasa Bay.

The plant's operators, JAPCO, kept the incident secret, but it came to light a month later when a team from the Kyoto University School of Agriculture, who had been monitoring contamination in the bay for 10 years, took samples of seaweed near the plant's drain outlet. They found 11 picocuries of cobalt-60 and 40 picocuries of manganese-54, 250 times the level of contamination that JAPCO had claimed was present in the bay.

NATURE V. NUCLEAR

On 25 April 1981, important feedwater cooling systems used to remove heat from the Brunswick 1 and 2 reactors were found to be blocked by oysters, mussels, barnacles and tube worms. On 9 June, at the San Onofre 1 reactor, the flow of cooling water was slowed and a valve malfunctioned because of an infestation of barnacles in the heat exchanger discharge pipe. On 28 August, at the Pilgrim plant, a heat exchanger was rendered inoperable due to the growth of mussels in cooling system pipes. These were just three of a whole string of incidents of a similar kind affecting US nuclear plants.

The tiny Asian Corbicula clam enters the plants as larvae — too small to be caught by a filter screen — in the water supply. Here they thrive, feeding on organisms swept past them in the water flow. It is not an easy problem for utilities to solve. Chlorinating the water might kill the animals but it would also raise the threat of chlorine toxicity to the surrounding area and require a special permit from the EPA.

On 3 September 1984, cannonball and clover jellyfish clogged the mouths of two water intakes at Fort Pierce in Florida. The reactors had to be shut down for 11 days, which cost Florida Power and Light $11 million in bought-in electricity. The jellyfish arrived in a slick two miles wide and seven miles long. They stuck on screens used to filter out flotsam; their weight was such that a six-in-thick steel frame was bent 20 in out of shape. Divers and fishermen working with nets and special trucks fitted with suction devices tried to clear them out, but it was Hurricane Diana that effectively finished the job.

A problem of microbiologically induced corrosion (MIC) of stainless steel piping at the H. B. Robinson Unit 2 reactor forced it to shut down for the whole of 1984 to replace parts of its steam generator and carry out other maintenance work. Six different classifications of micro-organisms, containing over 30 species, have been found to cause MIC. According to an NRC report on the subject, they form a 'synergistic cross-feeding support system' with other bacteria, fungi, and algae to enhance survival under the most adverse conditions. The NRC says: 'They have been known to tolerate a wide range of temperatures (10 to 90 ^{o}C), pH values of 0 to 10.15, oxygen concentrations from zero to almost 100 per cent and extreme hydrostatic pressure.' (**Public Citizen**, 1981; *Newscript* 13.8.80; *International Herald Tribune* 19.8.80; *Economist*, September 1984; NRC Information Notice No. 85-30, 19.4.85; WISE News Communique, 3.3.89)

An inquiry was demanded and the operators finally admitted the scale of an earlier accidental leak in 1975 and disclosed that this latest leakage, the tenth at the plant, had been caused by tons of contaminated water leaking into a drainage system for three hours. 56 workers had been contaminated, as had the sub-contractors involved in the clean-up. Several of these had not worn protective clothing while clearing up the radioactive water with mops and buckets.

JAPCO paid out 1.2 billion yen in compensation to local fishermen and the tourist industry. (J. Takagi, original contribution)

1981 April 9 The USS *George Washington,* the first US sub to carry nuclear ballistic missiles, collided with the 2,350-ton Japanese freighter *Nissho Maru* in the East China Sea, about 110 miles off Sasebo, Japan. The submarine was surfacing when it struck the underside of the freighter, which sank in 15 minutes, killing its captain and first mate. Damage to the submarine was slight. The incident led to severe criticism of the US by the Japanese. Neither the submarine nor a US P-3 Orion aircraft made any attempt to rescue the ship's crew, and it was 24 hours before the US notified Japan of the incident. The US refused to say what the submarine had been doing only 20 miles outside Japanese territorial waters and would neither confirm nor deny that the vessel was carrying nuclear weapons. Several months later, the US government accepted responsibility for the accident. (**Neptune**)

1981 May 26 A Marine Corps EA-6B Prowler aircraft, designed to jam enemy radar by using high-powered transmitters, crashed while attempting to land aboard the nuclear-powered aircraft carrier USS *Nimitz*, 70 miles off Jacksonville, Florida. The pilot reportedly applied power as the plane was landing, veered to one side and crashed into parked aircraft. Ammunition aboard the parked aircraft exploded, starting numerous fires, destroying three F-14 Tomcats and damaging 16 more. 14 people were killed and 48 others were injured. The cost of the damage was put at about $100 million. The incident prompted a debate into whether drug use aboard the *Nimitz* may have been a factor in the crash. (**Neptune**)

1981 July 3 A transformer at the North Anna 2 reactor in the US broke down, spilling oil which then caught fire. Automatic extinguishers were activated but the water mixed with the oil and spread the fire to another room. The fire, which burned for 30 minutes, disabled all the on-site generators which supply electricity in the event of a power cut. Had off-site power been lost, the reactor could not have been shut down safely in an emergency. (**Public Citizen**, 1981)

1981 September According to CIA intelligence reports, a Soviet nuclear-powered submarine in the Baltic suffered a series of explosions, and some crewmembers were sealed into the compartment in which they were on duty. The submarine was towed to Kaliningrad and the men who had been sealed in were taken to hospital in Riga exhibiting signs of terminal radiation sickness. (**Neptune**)

1981 October 27 A Soviet Whiskey-class attack submarine ran aground in Swedish waters, near the Swedish naval base of Karlskrona, 300 miles south of Stockholm. The Swedish government alleged that the vessel was laying mines or carrying out illegal reconnaissance, and stated their belief that the submarine was carrying nuclear weapons.

The Soviet captain claimed that bad weather and a faulty compass had caused him to enter Swedish waters by accident, but the Swedes maintained that good navigation would have been necessary for the submarine to get as far as it had. On 29 October, Swedish warships turned back a Soviet tug, and anti-submarine helicopters chased away an unidentified submarine in Swedish waters.

On 2 November, Swedish tugs refloated the submarine to prevent it being damaged by heavy seas, and three days later the Swedish government announced that the vessel was probably carrying nuclear weapons. The submarine was returned to the Soviets on 6 November, along with a statement from the Swedish government to the effect that the incident would affect Swedish attitudes to any future proposals for a Nordic nuclear-free zone by the USSR. (**Neptune**)

1982 This year there were 4,500 mishaps at the 81 operating US nuclear power plants, 253 being considered particularly significant by the NRC. More than a quarter of the mishaps were due to human error. There were a record 84,322 instances of workers in the nuclear industry receiving measurable amounts of radiation. (**Public Citizen**, 1979-87)

1982 February 1 At the Salem I plant in the US, 23,000 gal of radioactive water from a spent fuel pool spilled into auxiliary buildings, contaminating 16 workers. (**Public Citizen**, 1983)

1982 May 4 The Phénix 250-MW research reactor, at Marcoule in southern France, was shut down following a sodium coolant leak and a small fire. Phénix is the prototype for the controversial Super Phénix reactor. Both reactors can be used to produce plutonium for use in other nuclear reactors or in nuclear weapons. (*The Times* 4.5.82)

1982 May 25 In an incident involving British aircraft in north Germany, a Phantom accidentally fired a Sidewinder missile which locked on to a (nuclear-capable) Jaguar. The pilot ejected and the Jaguar crashed. (**Bradford**)

1982 June 30 At the Soviet-designed Kozloduy Unit 3 VVER reactor in Bulgaria, the level of the cooling water in the pressuriser suddenly dropped, and the pressure in the primary cooling circuit fell sharply. The reactor scrammed, and an alarm showed that water had entered the steam generator. An hour later this was found to be radioactive cooling water, but because of the heat and high radiation levels in the steam generator, it took a further 13 hours to discover that a seal had ruptured.

Just seven months later, on 21 February 1983, in the most serious accident known to have occurred at a VVER-440 reactor, Kozloduy Unit 2 suffered a sequence of events that bore unpleasant similarities to the Three Mile Island accident. The main safety valve controlling pressure in the primary cooling system opened spontaneously, causing the pressure to drop and triggering an automatic shutdown.

The main valve could not be closed manually, and the emergency high-pressure cooling system was activated, injecting a sudden rush of cold water into the hot pressure vessel. The 'thermal shock' that this caused could have ruptured the pressure vessel. Several VVERs are known to have brittle welds because of impurities in the materials used, and the risk is even greater in a reactor whose welds have been bombarded by neutrons over a long period, in this case for more than eight years. The accident was caused by a break in the insulation on a cable supplying power to the control system. (OECD/NEA Incident Reports IAEA-26 and -27)

1982 July 19 A fire started by an electrical fault caused damage estimated at several hundred thousand pounds to the new French-built nuclear power station at Koeberg near Cape Town, South Africa. The fire delayed the start-up of the station by some weeks. No one was hurt in the fire. (*The Times*, 20.7.82)

1982 November 2 At least one civilian was killed and two US soldiers were injured when a 20-ton US Army Pershing I missile transporter crashed into four cars after its brakes failed on a hill outside the village of Walprechtsmeier in West Germany. The missile was not carrying a warhead but the 1,200 villagers were evacuated in case the highly inflammable propellant of the Pershing exploded. (*The Times*, 4.11.82)

1982 November 9 A technician at the US San Onofre plant knocked a power cable out of its socket, causing a drop in water level in both steam generators of the pressurised water reactor. In response, the operators shut down the reactor as a precaution. When the technician plugged the cable back in, too much water flowed to the generators, causing excessive cooling and a drop in pressure. The stresses caused by the sudden drop in temperature could have cracked the reactor vessel, and the lower pressure could have led to steam forming in the vessel, preventing adequate cooling of the core. (**Public Citizen**, 1983)

1982 November 29 While surfacing in the South China Sea, 40 miles east of Subic Bay in the Philippines, the nuclear-powered ballistic missile submarine USS *Thomas Edison* collided with the destroyer USS *Leftwich*. (**Neptune**)

1983 This year, the United States' 84 operating nuclear power plants suffered 5,060 mishaps, 247 of them considered particularly significant. New records were set for the number of workers exposed to radiation (85,646) and for the total exposure (56,507 person-rem). Over one third of all nuclear workers were exposed to 500 millirem, three times the maximum amount allowed for the general public. The NRC imposed a record 49 fines on the nuclear industry. (**Public Citizen**, 1979-87)

1983 February 22 and 25 In one of the most significant US nuclear power plant accidents since Three Mile Island, the Salem I reactor suffered the first complete breakdown of the automatic shutdown system at a US facility. The water level in one of the steam generators fell extremely low and two circuit breakers which should have reacted by triggering an automatic scram and inserting the control rods, thereby shutting the reactor down, failed to respond. (It later turned out that the breakers, which the makers recommend should be oiled every six months, had been lubricated once in seven years and the wrong lubricant had been used. A part of the breaker system was found to have been damaged before it was installed.) Alarms sounded, lights went out briefly in the control room, and the operator decided to shut the reactor down manually.

Operators and supervisors studied the computer printout of what had happened, but there was no indication that the automatic shutdown system had failed, and the plant was started up again three days later.

When the plant reached 12 per cent power output, there was another low-water-level signal from the steam generator. Again the breakers received the signal for an automatic shutdown and again they failed to respond. It was some 30 seconds before an operator realised that the reactor should be shut down manually.

It was later revealed that Salem's equipment lists indicated that the breakers that failed to operate were not considered to be safety related. An NRC statement reads: 'The reactor trip system is fundamental to reactor safety for all nuclear power plant designs. All...accident analyses are predicated on its successful operation.' The NRC later found that at least three other facilities failed to classify the breakers as safety related. Salem 1's owners, Public Service Electric and Gas, were fined $850,000. (NRC Annual Report 1983; NRC Office for Analysis and Evaluation of Operational Data, NUREG 0090 Vol 6, No 1; **Public Citizen**, 1984)

1983 March 10 Before finishing his shift, an engineer at Quad Cities in the US wrote a note for his replacement explaining the procedure for a scheduled shutdown. When the evening shift operator read it, along with a computer printout, he concluded that he was to insert the control rods in an order that was the reverse of the normal procedure. By the time the night shift arrived, the evening shift operators had succeeded in incorrectly installing 33 rods. In order to do this, they had to switch off a safety feature (the rod worth minimiser or RWM) specifically designed to prevent improper insertion.

The night-shift operator switched the RWM back on and it prevented further rods from being inserted, though it did not alert the control room to any errors. (The NRC later concluded that no insertion errors were indicated because there were too many.) The operator discussed the problem with an engineer. They decided that the RWM was faulty, and again bypassed it, succeeding in inserting a further 10 rods in the wrong order and shutting the reactor down.

The NRC expressed deep concern over these events because conditions had been created for a 'rod drop accident' in which a control rod drops from the reactor core with the possibility of causing a localised burst of power which may damage surrounding fuel elements. A fine of $150,000 was imposed against the owning company, Commonwealth Edison. (**Public Citizen**, 1984; NRC Annual Report 1984)

1983 April 15 The auxiliary feedwater system at the US Turkey Point plant Unit 3 reactor was rendered inoperable when a plant worker closed the valves and mislabelled them. This system should supply water to the steam generator if the main system fails, preventing major damage to the reactor core, and NRC guidelines require that these valves be checked every four hours. The auxiliary feedwater system was out of action for five days before the error was noticed. (**Public Citizen**, 1984)

1983 June A Soviet nuclear-powered submarine reportedly sank somewhere east of Petropavlovsk naval base, near the southern tip of the Kamchatka Peninsula, with the loss of most or all of the 90-man crew. The cause of the accident is not known. The vessel was recovered by the Soviet Navy two months later. (**Neptune**)

1983 June 30 At the Argentine Embalse CANDU reactor, a valve failure led to a total loss of feedwater to the steam generator. Operators at the plant spent three hours working in steam and water spray to correct the problem. In the meantime the reactor had to be cooled using the shutdown cooling system. When this was switched on, the shutdown heat exchanger suffered a severe 'water hammer' when it was subjected to the full temperature of the reactor's heat transfer system, causing one pipe to move more than 20 cm. (OECD/NEA Incident Report IAEA-24)

1983 July 1 At the Philippsburg-1 nuclear power plant — a 900-MW boiling water reactor situated between Karlsruhe and Mannheim — leaks in about 20 fuel sub-assemblies raised the concentration of iodine-131 in the coolant water, and subsequently in the off-gas. Filters designed to remove iodine did not perform properly because the iodine reacted with impurities in the coolant water to form compounds that the filters could not retain sufficiently. The emissions of iodine-131 exceeded the permissible limits. (Helmut Hirsch, original contribution)

1983 August 1 An 'unanticipated criticality accident' occurred at the Argentine RA-II Research Reactor in Constituyentes, near Buenos Aires, while a technician was changing the fuel elements. Safety systems automatically stopped the criticality, which was apparently caused by the insertion of two fuel rods without corresponding control rods, but the technician received an estimated dose of 1,400 rad of fast neutrons and 500 rad of gamma radiation. He died two days later. 17 other workers were exposed to radiation. ('The safety of nuclear power plants', Uranium Institute, 1988)

1983 October 1 The French Blayais 1 reactor was running at full power when an operator carried out a routine check on the back-up electrical system. To do this he had to take maximum current from the system, and so he activated the switch-over between the two intermediary coolant systems, stopping system B (the one normally in use) and starting system A. The test went smoothly and he switched back to the normal situation.

However, when the pump in system B started up again, a gasket ruptured and the cooling water began to leak out. The operator paid no attention to the initial alarms and only realised what was happening when the level in the reservoir tank fell to the 'distress' level.

He responded by switching on the A coolant system again. Both systems are connected by valves to a common auxiliary system. Switching on both systems opened all the valves and therefore had the effect of interconnecting the two systems. System A began to empty out through the ruptured gasket in system B. By the time the appropriate valves had been closed, some 12 cu m of water had leaked out, flowing down the electrical conduits, spraying a high voltage electrical installation and almost reaching the back-up safety electrical system.

The accident revealed a fundamental conceptual blunder — it should not have been possible for the two systems to become interconnected. (*Sûreté Nucléaire* September/October 1984)

1984 The NRC revised the Licensee Event Report (LER) system by which mishaps at US nuclear power plants are reported. The new system results in about half as many accidents being reported, but in greater detail. During the year there were 2,435 LERs from the country's 92 nuclear power plants, an average of 29 events per plant. The combined power output of all the nuclear plants was only 57 per cent of their capacity. A record 98,162 workers received measurable doses of radiation, an increase of 13.3 per cent on the previous year, although the number of operating power plants increased by only 9.2 per cent. The NRC imposed 31 fines, totalling $2,054,000. There were 599 scrams, an average of seven per plant. (**Public Citizen**, 1979-87)

1984 January 10 A USAF Alarm Response Team at Warren AFB, near Cheyenne, Wyoming, parked an armoured 'Peacekeeper Vehicle' on top of a Minuteman III silo when a computer malfunction indicated that the missile was about to launch itself. Officials described this as a standard 'precautionary measure'. In the event of a launch, the armoured vehicle would have fallen into the silo, damaging the missile and blocking its path. Details of the incident were not revealed until 1987. The incident was not made public at the time, nor was it reported to SAC or Congress. (*Casper Star-Tribune* 28.10.87; **Bradford**)

1984 March 17 Following repairs at the San Onofre Unit 3 plant, several valves were left in the wrong position, completely disabling the emergency core-cooling system. For 13 days the plant operated at 100 per cent power without the situation being discovered. (**Public Citizen**, 1984-85)

1984 March 21 While the USS *Kitty Hawk* was involved in joint naval exercises with South Korean Forces at night in the Sea of Japan, a Soviet Victor I nuclear-powered submarine that was shadowing the fleet apparently lost track of it. The submarine surfaced to locate the aircraft carrier and collided with it, making a hole below the ship's waterline in a tank full of aircraft fuel. (**Neptune**)

1984 April 14 Each French nuclear reactor has two electrical supply and distribution systems — line A and line B. Each line has a back-up diesel generator and, in some reactors, a turbo generator running off the steam generators. The safety system is thus designed to respond to a total loss of power from the external supply. However, on 14 April 1984, the Bugey 5 reactor suffered a chain of events initiated by a *gradual* loss of power affecting both the *external* and *internal* power supplies. While the reactor was running at almost full power, a rectifier supplying a constant 48 volts to the line A control-command panel failed because of an electronic malfunction. Power to the panel was then automatically supplied by a battery, but as the battery discharged, the voltage steadily fell. The inadequate power supply to the control panel caused the reactor to scram. The panel should then have switched on the auxiliary power supply, but without the necessary voltage, it was unable to do so. For the same reason, the emergency diesel generator for line A did not start up. The only remaining source of electricity was the second diesel generator in line B, and this supplied power until the auxiliary external service was back in action.

In the meantime, a fuse failure in the power supply system had deprived the reactor's protection system of electricity. The core was being cooled by natural circulation in the steam generator circuit while pressure in the primary circuit was being controlled by discharge valves on the pressuriser. Two of these were operated by power from line A and, after opening and closing several times, they ceased to function. The one valve running on line B had to be opened

21 times. Had either of the line A valves stuck open, the consequent breach in the primary circuit could not have been closed and the result would have been an accident comparable to Three Mile Island. (International Reactor Hazard Report. Gruppe Ökologie, Hannover)

1984 April 19 Workers were cleaning instrument guide tubes at the Sequoyah I PWR when a mechanical seal ruptured, spilling at least 16,000 gal of hot radioactive coolant into the containment area. The accident was caused by workers using a tool that they had modified themselves. (**Public Citizen,** 1984-85; 'The safety of nuclear power plants', Uranium Institute, 1988)

1984 May 11 During test operations at the Kalinin Unit 1 VVER reactor in the Soviet Union, incorrect wiring in the control circuits caused a pressuriser relief valve to open accidentally and remain open. The reactor scrammed, the main circulation pumps were turned off and the core was cooled using the emergency core cooling system until the relief valve could be closed. (OECD/NEA Incident Report IAEA-47)

1984 May 17 Fire broke out in the battery well of the nuclear-powered attack submarine USS *Guitarro* 65 miles north-west of San Diego. The fire was still burning when the submarine reached port. (**Neptune**)

1984 July 1 An operator at the French Saint-Laurent-des-Eaux reactor, intending to open valves that connect the primary circuit and the cooling circuit of the B1 reactor, which was in cold shutdown, accidentally sent the command to the B2 reactor, which was still hot. The instruction was to open two valves in series, and the first of these opened. Fortunately, the pressure difference on the two sides of the second valve stopped it from opening and prevented a breach of the primary circuit. A similar accident was to occur at the West German Biblis reactor three years later (see 1987, December 17, p. 298).

On 12 January 1987, the Saint-Laurent A1 reactor had to be shut down when ice in the River Loire blocked the water intake to the reactor cooling system. It took half an hour to restore adequate cooling to the core. The cold weather also caused the A2 reactor to shut down automatically following a drop in the off-site power supply. (Bulletin of Sûreté Nucléaire, July/August 1984; WISE, Paris; La Nouvelle République du Centre-Ouest, 15/17/18/20.1.87)

1984 September In West Germany, a US Pershing II launcher went off the road. The rocket broke and lost fuel and oil. (**Bradford**)

1984 September 19 A Soviet Victor I nuclear-powered attack submarine was damaged in a collision with a Soviet tanker in the Strait of Gibraltar while leaving the Mediterranean. The submarine was reportedly hiding in the 'noise shadow' of the tanker and probably rose suddenly in the thermal gradients where cool and warmer waters mix. The submarine's bow section was ripped open, revealing the sonar and torpedo tube compartments. (**Neptune**)

1984 December 28 An unarmed Soviet Shaddock (SS-N-3) sea-launched cruise missile (SLCM), having been launched in the Barents Sea, strayed into Norwegian airspace and crashed, or was shot down, over Finland. The Soviets reported in February 1985 that they had recovered wreckage of the SLCM in Finland. (**Bradford**; *The Times,* 17.9.86)

1985 There were 2,997 Licensee Event Reports at the 97 operating US nuclear power plants, an average of 33 per plant and an increase of 13 per cent per plant over the previous year. With the total number of scrams at 645 for the year, the average remained at seven per plant. The NRC levied 38 fines, totalling almost $4 million. (**Public Citizen** 1979-87)

1985 March 6 During tests at the 1,300-MW Grohnde PWR near Hanover, all four high-pressure injection pumps in the emergency cooling system were found to be inoperable. One pump was actually filled with gas. In the event of a leakage in the primary cooling circuit, the pumps would not have been able to prevent the core from melting. Federal and state authorities misinformed the public about the incident. They claimed that only one pump was affected and that two operable pumps were sufficient in the event of an accident. The true version was not published until two years later by *Der Spiegel*. (Helmut Hirsch, original contribution)

1985 March The failure of a valve at Hungary's only VVER plant, at Paks, caused the water in the steam generator to fall to its lowest possible level, uncovering the upper tubes. Cold water had to be injected by the emergency feed-water pumps. (OECD/NEA Incident Report IAEA-39)

1985 June 9 While the US Davis-Besse plant on Lake Erie was running at 90 per cent capacity, a spurious electronic signal from the automatic control system caused the main feedwater pump to close. In the fast-moving events that followed, an operator punched the wrong control buttons, shutting off the auxiliary feedwater to the steam generators as well, and as a result the reactor coolant began to overheat. It took half an hour to restore normal cooling, during which time the reactor was temporarily over-cooled and 12 equipment malfunctions occurred.

This sequence of events was similar to the Three Mile Island accident in 1979; a major catastrophe was prevented by an operator switching on a pump normally used for other purposes. The NRC fined the operating company $900,000 for multiple violations associated with the incident. (**Public Citizen** 1984-85; NRC Annual Report 1986; NRC NUREG 0090 vol 8, No 2)

1985 June 10 The Royal Navy's nuclear-powered ballistic missile submarine HMS *Resolution* was en route to test-fire a Polaris missile at the US Navy's Atlantic Test Range when she was struck by the US yacht *Proud Mary*, off the coast of Florida. (**Neptune**; *The Times*, 11.6.85)

1985 June 20 At Helensburgh, Strathclyde in Scotland, two lorries carrying warheads collided with each other. The navy stated that there had been no danger of radioactive release. (**Bradford**)

1985 August 30 At the reprocessing plant in Karlsruhe, West Germany, a barrel, in which 20 kg of inflammable radioactive waste had been mixed with one kg of radioactive sludge, caught fire. Emissions of ruthenium-106 were 10.5 times the daily permissible emission limit. (Helmut Hirsch, original contribution.)

1985 August 31 While the Fukushima I-I reactor, in Japan, was shut down for a routine inspection, fire broke out in the turbine building and destroyed most of the power cables, including those that supply auxiliary equipment in the reactor. Had the plant been operating at the time, the situation could have been extremely dangerous. (J. Takagi, Original contribution.)

1985 October 8 At the Hinkley Point B advanced gas-cooled reactor in the UK, radioactively contaminated water that had leaked from a burst boiler pipe was pumped into tanks, diluted and released into the Bristol Channel. Less than two months later, a one-in hole in the cooling gas circulation system allowed eight tons of contaminated carbon dioxide gas to escape into the containment building and then into the atmosphere over a period of four hours. A further 15 tons of filtered gas were deliberately vented to reduce the pressure. The reactor hall was evacuated and all 500 staff were given potassium iodide pills as a precaution against the effects of radioactive iodine-131. (*Ecologist* vol. 16, No 4/5 1986)

1985 December 8 An unarmed Tomahawk sea-launched cruise missile, launched from a submarine beneath the Gulf of Mexico, veered sharply off course forcing the test flight to be aborted. The missile parachuted to the ground in a populated area near Freeport, north-east of Eglin AFB.

The launch crew had fed information into the computer too quickly, erasing the middle portion of the flight programme. The missile went from the launch part of the programme direct to the landing portion. (*International Herald Tribune* 22.2.86)

1986 There were 2,836 mishaps at US nuclear reactors reported to the NRC, and a total of 678 scrams. The NRC imposed 50 fines for violations of their requirements. The efficiency of US reactors remained the lowest in the world, operating at less than 57 per cent of their capacity.

At the 39 plants built by General Electric, the NRC estimated that in the event of a severe accident there would be a 90 per cent chance of containment vessel failure, which could lead to a serious leak of radiation. (**Public Citizen** 1979-87)

1986 February 11 At the Trawsfynydd reactor, in the Snowdonia National Park in north Wales, 13 tonnes of slightly contaminated carbon dioxide gas were released into the atmosphere when a pressure-relief valve opened accidentally for 14 minutes. The filter, which should have removed solid matter, failed to work, and radioactive particles were released with the gas. (*Ecologist* vol 16, No 4/5 1986)

1986 February 24 According to a Canadian Forces spokesman, an unarmed US cruise missile plunged into the Beaufort Sea moments after it was launched from a B-52 bomber on a test flight over northern Canada. (**Bradford**)

1986 March 13 The nuclear-powered ballistic missile submarine USS *Nathanael Greene* armed with 16 Poseidon missiles and with 168 crew on board, ran aground somewhere in the Irish Sea while on 'hide and seek' manoeuvres. The submarine managed to proceed to the US Navy base at Holy Loch in Scotland under its own power for emergency repairs. It was claimed that the nuclear power plant was not affected. The news of the incident did not come to light until early in May, a few days after reports that another nuclear-powered submarine, the USS *Atlanta*, had run aground on 29 April in the Strait of Gibraltar, also damaging its ballast tanks. (**Neptune**; **Bradford**; *Daily Mail*, 3.5.86; Guardian, 3.5.86; *The Times*, 3.5.86)

1986 March 17 About a tonne of water escaped when one of the boiler tubes of the AGR at Hartlepool, UK, sprang a leak. Some of the water leaked into the compartments housing the electrical motors which power the circulators that drive gas around the reactor to cool it. Six of the 16 motors were put out of action. The reactor was shut down immediately and was out of service for about a month, at a cost to the CEGB of several million pounds. Hartlepool is one of seven AGRs in Britain and, according to geographer Dr Stanley Openshaw, is 'probably the most urban of any nuclear power station in the world and close to the second largest petro-chemical complex in Europe, a uniquely bad site from the point of view of public safety.' (*Guardian,* 26.3.87)

1986 May 4 There were allegations of a cover-up after a small radiation leak from the 300-MW nuclear reactor at Hamm, in North Rhine-Westphalia, West Germany. This experimental graphite-moderated Thorium High-Temperature Reactor (THTR) — the only one of its kind — is fuelled by uranium and thorium pellets embedded in 675,000 graphite spheres the size of tennis balls. Described as an 'inherently safe' design, it cost £1.14 billion to build.

The leak occurred when one spherical fuel element became stuck in the refuelling line of the reactor and attempts to loosen it with a blast of helium coolant resulted in radioactive particles escaping into the atmosphere through defective filters.

Ironically, accurate measurements of the atmospheric contamination were impossible as the area had been contaminated by radiation from the Chernobyl accident a week before. It is possible that this situation was used to 'hide' the seriousness of the leak, news of which only emerged on 30 May, on the same day that the reactor was switched off 'for routine maintenance.' (Helmut Hirsch, original contribution; *Financial Times/The Times/Guardian*, 2.6.86; *Guardian* 3.6.86; *New Scientist* 12.6.86)

1986 May 10 Italy's Caorso nuclear power plant was shut down for several days due to a 'leakage' in a water cooling pipe in a secondary system. There had been 90 mechanical failures at Caorso since 1978. (*International Courier*, 16.5.86)

1986 May 14 At the three-unit Palo Verde nuclear power station in Arizona, power from three of the four transmission lines supplying off-site electricity was lost within minutes. The following day it was discovered that the overhead power cables, which run to the power station from different directions, had been sabotaged 35 miles from the plant. In what must have been a co-ordinated action by several people, straps had been thrown over the cables, earthing them. Local police and the FBI were notified, and a reward of $25,000 was offered for information leading to the conviction of the culprits. (NRC NUREG 0090, vol 9, No 2)

1986 June 23 12 people, including IAEA inspectors, were 'slightly' contaminated by plutonium whilst inspecting a store-room of the Power Reactor and Nuclear Fuel Development Corporation, owned by the Japanese Government, at Tokaimura, north-east of Tokyo. (*The Times* 24.2.86)

1986 July 30 At a NATO rocket site in southern West Germany, the nuclear warhead of a US Pershing 1A missile was knocked off when the driver of a crane accidentally swung its boom in the wrong direction. The warhead fell 10 in on to a work platform without causing serious damage. The missile was in an 'alert' state and capable of being fired within 20 minutes. (*Guardian*/*The Times* 2.8.86)

1986 August 19 Prior to the initial start-up of the French Cattenom reactor, engineers working on a compressed-air valve operating system opened a valve without realising it and left it open. The valve allowed water from the feedwater circuit to run out into the galleries housing the compressed air system, electrical command cables and the electrical supply. Despite the fact that evacuation pumps started up automatically, it took four days before the flooding was noticed. The pumps had been able to remove 1.5 cu m of water per second, but the water, being drawn directly from the Moselle River, was flowing into the galleries at the rate of two cu m per second. EDF calculated the amount of water spilled to be 130,000 tons. Had the reactor been running, a short circuit in the electrical system could have caused a serious accident. (Mycle Schneider, personal communication 22.3.89)

1986 August 29 30,000 gal of unsampled radioactive water leaked into the Irish Sea from the cooling circuits of the Wylfa Magnox nuclear power station on the Anglesey coast in the UK. The CEGB claimed the same amount of water is discharged into the sea after monitoring every two days.

Wylfa, officially opened in 1972, was closed three weeks after it opened when faults were found in its turbine blades. In 1985, hairline cracks were found in the support columns of a crane used to load spent fuel for reprocessing. In June 1986, the station was shut down for 24 hours when a boiler valve malfunctioned. (*Guardian*, 1.9.86)

1986 September A Soviet SS-N-8 submarine-launched ballistic missile, armed with a dummy warhead, was fired from a Delta-2 submarine in the Barents Sea and aimed towards a testing range on the Kamchatka Peninsula in the Soviet Far

East. It misfired and landed in Chinese territory near the Manchurian border, about 180 miles west of the Soviet city of Khabarovsk. The missile was more than 1,500 miles off course. The incident, involving two nuclear powers, could have triggered a grave response. (*New York Times/International Herald Tribune/Guardian/The Times*, 17.9.86)

1986 September Two Soviet-designed nuclear power plants at Loviisa in southern Finland, 30 miles east of Helsinki, were shut down after staff error had led to a leak of 594 cu ft of weak radioactive water. (*The Times*, 8.9.86)

1986 October 31 The nuclear-powered attack submarine USS *Augusta* underwent $2.7 million-worth of repairs at Groton, Connecticut, after colliding either with the sea-bed or with an underwater object in the Atlantic. (**Neptune**)

1986: CHALLENGER AND THE O-RINGS

The news that the destruction of the space shuttle Challenger was caused by a failure in its O-rings was followed later that year by another revelation: that virtually every safety system in every US nuclear power plant uses O-rings to prevent dangerous leaks. Patented in the US in 1939, O-rings look like rubber doughnuts and come in hundreds of sizes. They are engineered seals, similar to washers in taps and gaskets in engines, used to prevent leakage of fluids or pressure in machinery parts. Usually made of flexible synthetic materials, they are also used in dishwashers, waterproof watches, aeroplane hydraulic systems and automobile brake systems.

Daniel Ford, former executive director of the Union of Concerned Scientists (UCS) and Bob Pollard, a former nuclear safety engineer who is now a staff member of the UCS, published their report, 'O-Rings and nuclear plant safety: a technical evaluation', in September 1986. The report revealed that thousands of O-rings are used in US nuclear power plants to provide leak-proof seals in a variety of reactor systems. The authors claim that O-rings are designed to seal tight at temperatures up to 200 °C; but in some accidents, O-rings could be heated to three times that temperature, and a full-scale meltdown could create temperatures more than 10 times higher.

In 1981 the US government found that one common type of O-ring, made of Viton (the material in the Challenger's O-rings), slowly disintegrates when exposed to large amounts of radiation. By 1986 there had been more than 60 reports of O-ring failures in US nuclear plants since 1975. (Daniel Ford and Bob Pollard, 'O-Rings and nuclear plant safety', **Public Citizen**, September 1986)

1986 December The Edwin I. Hatch plant in Georgia suffered the worst high-level-waste storage accident at a US commercial nuclear power plant to date. As a result of mechanical malfunction and human error, 141,000 gal of radioactive water leaked from the plant's spent fuel storage ponds over a period of 11 hours. At least 80,000 gal of the water flowed into the swamp surrounding the plant, within a few hundred yards of the Altamaha River. (**Public Citizen** 1979-87; **Public Citizen**, 'Hatch nuclear waste accident', December 1986)

1986 December 9 At the 14-year-old Surrey 2 nuclear power plant in Virginia, an 18-in steel pipe carrying water and steam at a temperature of 350 °C, exploded. The force of the explosion twisted metal girders, and 30,000 gal of hot water and steam poured into the building, spraying eight workers. Six of the workers were severly burned and four died as a result of the accident. The pipe's walls, originally half an in thick, were paper thin as a result of corrosion, and when a valve accidentally closed, cutting off the steam flow and triggering a reactor shutdown, the pipe burst. Neither the NRC nor the state of Virginia took any action against the utility. (**Public Citizen** 1986; NRC Office of Inspection and Enforcement Information Notice, 18.3.87)

1987 The world's largest fast-breeder reactor, the 1,200-MW Superphénix at Creys-Malville in France, some 40 miles east of Lyon, first came on line in 1987. The reactor soon ran into serious problems. Liquid sodium, used to cool the reactor's fuel rods, was leaking from the fuel transfer and cooling drum at the rate of one and a half gal per hour. Over a period of five weeks, some 20 tonnes of this chemical, which explodes on contact with air or water, seeped out into the secondary cylinder and the leakage finally resulted in a total shutdown.

The fault was eventually discovered but it was estimated that repairs would cost up to £40 million and take between two and four years to complete. These estimates had to be revised in January 1988 when technicians using ultrasound discovered almost 100 fissures in the cooling drum in addition to the main crack. The authorities concluded that the entire drum would have to be replaced at a cost of £100 million. (*New Scientist*, 23.4.87; *The Times*, 8.9.87/16.11.87/2.2.88)

1987 Two Soviet nuclear specialists working in Czechoslovakia admitted that, in the last five years, there had been 356 'deviations from the regulations' at two of the country's oldest nuclear plants — Dukovany and Jaslovske Bohunice. Both stations are powered by Soviet-designed VVER-440 pressurised water reactors. In late May, the Czechoslovak government announced that reactor No. 1 at Bohunice had been closed down after cracks appeared in the steam generating unit. The Czech AEC also revealed that safety inspectors had found 15 defects at Czech nuclear plants in 1986. In three cases they imposed fines for negligence. (*Guardian*, 12.5.87; *New Scientist*, 4.6.87)

1987 By March 1988, 2,810 mishaps in operating US reactors had been reported to the NRC for 1987. About half of these were due to human error. The NRC imposed 46 fines for violations of safety and security requirements. During the year, seven of the 110 nuclear plants licensed to operate at full power produced no electricity at all, and for the ninth consecutive year the average power output from US nuclear plants was less than 60 per cent of capacity.

A congressional report released in December 1987 criticised the NRC's 'cozy' relationship with the nuclear industry, saying that the NRC 'has demonstrated an unhealthy empathy for the needs of the nuclear industry to the detriment of the safety of the American people'. (**Public Citizen** 1979-87)

1987 February 18 In the Irish Sea, off the Isle of Man, a US nuclear-powered submarine dragged the Irish trawler *Summer Morn* backwards by its nets for two and a half hours, until the nets were cut. The trawler's crew hauled in a submarine communication buoy in the torn nets. Nine months later in the same area, the Irish trawler *Angary* was pulled along for a few seconds until a steel chain, tested to 32 tons stress, snapped, and the trawling gear disappeared without trace. The UK Ministry of Defence said that no submarine was operating in the area.

In November 1987, a group called the Celtic League, based on the Isle of Man, suggested that these were not isolated incidents.

Their figures showed that since 1980, 16 trawlers from several countries have vanished without trace in the Irish Sea, with the loss of 32 lives. The League claimed that this was due to the increasing number of US and British submarines operating out of the naval bases at Holy Loch and Faslane on the Scottish coast. The figures for sinkings and loss of life rose in 1989 when the Belgian trawler *Tijl Uilenspiegel* vanished south of the Isle of Man in calm seas with the loss of five men.

The Royal Navy has only accepted responsibility for damage to trawlers and fishing gear in three cases; in some others, it has paid compensation without accepting legal liability. The MOD maintains that no fishermen have lost their lives as the result of collisions with British submarines. (*Independent*, 4.11.88; *The Times*, 6.5.89; **Neptune**)

1987 March The NRC closed the US Peach Bottom plant, 30 miles north of Baltimore, after learning that licensed control-room operators were regularly sleeping while on duty. According to the Baltimore *Sun*, an NRC study published six weeks earlier had identified a fundamental structural flaw in the plant.

The containment structure, which is smaller and closer to the reactor core than that in most US nuclear plants, was found to have a 50 per cent chance of rupturing in the event of a severe accident, thereby allowing a release of radioactivity. The safety of the plant therefore depended heavily upon operators taking prompt action to control any problems in the reactor before a core melt-down occurred. (**Public Citizen** 1979-87; Baltimore *Sun*, 19.4.87)

1987 March 18 A small fire at the Lucas Heights Nuclear Research Laboratory near Sydney, the site of Australia's only nuclear research reactor, slightly contaminated two workers and discharged radioactive gas into the atmosphere. The reactor is used for research and for producing radioisotopes for medical purposes. The only previous accident at the plant since it opened in 1957 was when radioactive effluent drained into a river in 1985 after vandals dodged security patrols and smashed an underground pipe. (UPI report, Greenlink 19.3.87; *The Dominion* 20.3.87)

1987 March 24 At the NUKEM nuclear fuel fabrication plant in Hanau, West Germany, a batch of uranium from the EC Transuranium Institute in Karlsruhe was found to contain plutonium when processed. Workers were contaminated by plutonium, and 75 people had to be monitored. The Federal Ministry for the Environment was not informed for two weeks. (Helmut Hirsch, original contribution)

1987 May 5 A Pershing missile ended up in a ditch at Heilbronn, West Germany, when a US army transporter careered off the road. (Reuter report, *Daily Telegraph*, 6.5.87)

1987 August 16 Two weeks after the event, the Soviet Union effectively admitted at a press conference that an underground nuclear test on the Arctic island of Novaya Zemlya on 2 August could have leaked radiation into the atmosphere.

They claimed that the increase was lower than normal background radiation 'and posed no threat to life'.

In early August, monitoring stations had registered an increase in radiation in the atmosphere over Norway, Sweden and Denmark, Greenland and Alaska, and the US had accused the Soviet Union of violating the Limited Test Ban Treaty. The Soviet Union denied that there had been any fallout.

Five months earlier, in mid-March 1987, radiation monitors in Scandinavia and Western Europe recorded higher than normal levels of iodine-131 and xenon-133 in the atmosphere for several days. Persisting levels of radiation from the Chernobyl accident made it impossible to tell whether caesium was also present, but the composition of the radiation made it unlikely that the source was a nuclear test. Scientists in West Germany and Sweden speculated that there may have been an emission from a Soviet or Eastern European nuclear reactor. (Kessing's Record of World Events, vol. XXXIV, 1988, No. 5.; *Financial Times/Guardian/Washington Post/International Herald Tribune/Wall Street Journal*, 15.4.87; *New Scientist*, 23.4.87; *International Herald Tribune*, 17.8.87; *Independent*, 18.8.87; *The Times*, 18.8.87)

1987 November Two water leakage accidents led to the release of around 10,000 gal of radioactive water from the Rancho Seco nuclear power plant. Some of the contaminated water reached a creek beyond the plant's boundaries. US Federal documents show that radioactive discharges from Rancho Seco between 1980 and 1985 were responsible for what may have been the largest amount of radiation exposure ever to people living near a commercial nuclear plant that was still licensed to operate. (**Public Citizen** 1979-87)

1987 December 22 50 tonnes of heavy water, containing 15 millicuries of radioactivity per l, escaped from the Argentine Atucha I reactor, 800 km east of Embalse. The incident occurred one month after the reactor was restarted following a three-month shutdown for repairs. Less than two weeks earlier, about one tonne of heavy water leaked from the Embalse nuclear power station and flowed into the Rio Tercero Reservoir in Cordoba Province, which provides drinking water for approximately 1.3. million people. This was the third reported accident at the Embalse power station in a year. (WISE 19.2.88)

1988 January 24 A thousand cu m of radioactive carbon dioxide gas leaked from the Dungeness A nuclear power station when an oil seal failed. A year before, four tons of carbon dioxide coolant gas leaked from a valve. Dungeness was the first gas-cooled reactor to be built in Britain. (*Daily Telegraph*, 25.1.88; *The Times*, 25.1.88)

1988 January 26 What the UK MOD subsequently described as a 'minor electrical malfunction' led to the failure of the primary coolant system in the PWR on the nuclear-powered submarine HMS *Resolution* while it was in dock at Faslane in the Firth of Clyde, on Scotland's west coast. Two back-up pumps were not operating and an emergency power supply did not activate. As a result, heat began building up in the reactor core and those on board had between one and a half and seven minutes to bring the situation under control. Disaster was averted by two crewmen racing to start up a diesel generator to power the coolant system. One of the men is believed to have been exposed to radiation. The MOD denied that there had been any danger to the crew or to the general public.

More than 13,000 people live within a five-mile radius of Faslane base. A reactor explosion could spread radioactivity over an area of 2,000 sq miles. (**Neptune**; *Observer*, 14.2.88; *Guardian*, 15.2.88; *Scotsman*, 15.2.88)

1988 February 3 The IAEA in Vienna sent out a test message, describing a fictitious nuclear accident in the USSR, to meteorological stations in Europe via the Global Telecommunication System. In the event of a real accident, the weather monitoring network could be used as an early-warning system. The message was intercepted by the Swedes, interpreted as genuine, and relayed to the world at large. The country's radiation monitoring station went on red alert. The Soviet Ministry of Atomic Power Generation was forced to issue a denial. Share values in the City of London dropped £1.5 billion. (*Daily Mail/The Times*, 4.2.88)

1988 April 27 The French Flamanville No. 2 reactor lost cooling capacity in the spent fuel storage pond when the electrical supply to the pumps was cut off for 24 minutes. (*International Environment Reporter*, 6.8.88)

1988 April 28 As a result of human error, 5,000 curies (0.5 g) of highly dangerous tritium gas escaped from the military nuclear establishment at Bruyères le Chatel, in France. Then, on 15 October, the equivalent of 7,000 curies of tritium was *deliberately* released as part of the EEC fusion programme 'to study the dispersion of tritium and its conversion into tritium water under real conditions,' in the words of EC Commissioner Narjes on 18 February 1988. (*Internation-*

al Environment Reporter, 6.8.88; Mycle Schneider, personal communication, 22.3.89)

1988 June During the first two weeks of June, an anti-submarine warfare helicopter accidentally dropped an unarmed training torpedo on the nuclear-powered attack submarine HMS *Conqueror*, denting its deck plating. (**Neptune**)

1988 July 16 The nuclear-powered attack submarine HMS *Courageous* collided with the yacht *Dalriada* at night in the Irish Sea. The yacht sank, and all four persons aboard were rescued by the frigate HMS *Battleaxe*. (**Neptune**)

1988 September 5 A nuclear plant in Lithuania, which uses reactors similar to those at Chernobyl, caught fire. The blaze started in the cable room of the station's second unit. There were no radiation leaks or injuries but the accident was considered serious, according to *Izvestia*. (Reuter report, *Independent* 6.9.88)

1988 September 18 At the 670-MW West German Stade PWR near Hamburg, an electronic failure caused a valve to close in one of the four main steam lines. Valves in the other lines followed automatically, blocking all four lines. The automatic reactor protection system would have scrammed the reactor and the turbine within seconds, but the operators overruled this by opening the valves, in order to keep the reactor running. However, the valves quickly shut again and the system scrammed. The opening and closing of the valves led to a violent pressure wave in the steam lines, which vibrated with a maximum displacement of almost 20 cm and came close to failure. (The acceptable limit is 10 cm.)

The Stade reactor, unlike more modern power plants, has no precautions against a main steam line break. Certain parts of the lines, outside the containment, cannot be isolated by shutting valves, and there is no second, protective pipe around the steam line — so the operators' action affected a particularly sensitive part of the system. A steam-line break could have led to the failure of a steam generator tube and a release of radioactive steam to the environment. Furthermore, there is the risk of a core meltdown if too much primary coolant is thus lost, or if steam enters the ring space between containment and building walls and disables safety systems there (even without a steam-generator leak).

The event was noticed by the responsible authorities in the State of Lower Saxony at the beginning of November 1988, but was not made public. The Schleswig-Holstein State government voiced suspicions concerning an incident at Stade nuclear power plant at the beginning of December, and the government of Lower Saxony then put out a misleading description of the incident, downplaying its importance and failing to mention the operators' manipulations. The full picture was finally made public by Gruppe Ökologie Hannover on 13 December 1988. (Helmut Hirsch, original contribution)

1988 November A cooling system explosion at Laguna Verde, Mexico's first nuclear power complex, was disguised by the authorities as a 'safety exercise'. When this was disputed in the press, the Federal Electric Commission branded the reporter's action as 'information terrorism', whereupon 15 senior technicians at the plant resigned in protest. Anonymous interviews with plant employees revealed the explosion was caused by poor design: pipes of half the required diameter were used in a closed high-pressure system.

Ever since the plant was first proposed in 1979, it has been plagued with problems and cost overruns. Planned as a prototype, the government was forced to abandon 20 similar plants when Laguna Verde's costs rocketed to £2 billion - seven times the original budget. It consists of two General Electric Mark-II BWRs, and has an inadequate containment system. An internal General Electric report, made public by former GE engineers in 1975, recommended taking the Mark-II reactors off the market. Several utilities have since sued the company, on the grounds that GE knowingly sold them faulty reactors.

More than a decade late, Laguna Verde will only supply a small fraction of the country's energy needs at enormous cost. Situated in a lush cove on Mexico's east coast, it is built on a geological fault eight km from an active volcano and just 30 km from the epicentre of a major earthquake in 1986. Emergency plans in the event of a major accident are seen as a bad joke. The plans apply only to towns and villages within 26 km of the plant and only adults would be evacuated — leaving children to shelter in their schools, which in most cases have unglazed windows and no doors. If the wind blew in the wrong direction, a radioactive cloud would reach Mexico City, 270 km to the east, in three hours. Neither the military nor the medical services are equipped to deal with disaster.

According to a report in *South* magazine (June 1989) there have already been a series of accidents at the plant 'including bent cooling pipes caused by computer error, jammed security doors, overheated electric junction boxes, valve failures, and faults in one of the safety systems designed to prevent a build-up of hydrogen gas.' Mexico has a terrible history of industrial disasters and critics' main concern is that a major accident at Laguna Verde will be caused by operator error or poor maintenance.

The plant has been actively opposed since its inception by a coalition of environmental and church groups. Demonstrations were common until 11 October 1988 when the plant was surrounded by Mexican troops, who have remained in the area since. Gatherings of more than four people are now illegal. (*Independent*, 15.10.88; *Greenpeace USA Magazine*, Jan/Feb 1989 & March/April 1989; *South*, June 1989; *Index on Censorship*, July/August 1989)

1988 November 30 Aboard the nuclear-powered aircraft carrier USS *Nimitz*, in the Arabian Sea, a 20-mm cannon on an A-7 Corsair aircraft accidentally fired during maintenance, killing one person and setting fire to six other aircraft. (**Neptune**)

1988 December 3 Just after 06:00, an explosion at the Burghfield Atomic Weapons Establishment in Berkshire, UK (which assembles and dismantles nuclear warheads) shattered windows a mile from the plant. The blast reportedly occurred during the routine burning of surplus conventional explosives. (*Guardian*, *Independent* 3.12.88)

1988 December 13 During a routine check at the 1,300-MW Brokdorf pressurised-water reactor near Hamburg, during a routine check, it was found (by chance) that four of the eight emergency diesel generators lacked vital components and would have failed if needed. Worse still, the diesel generators concerned had been out of order since the reactor started operation in 1986. (Helmut Hirsch, original contribution)

1988 December 23 Start-up of the Unit 4 reactor at the French Blayais plant was delayed for four days when routine checks revealed that two control rods were jammed. The control rods are designed to slide down into the reactor under their own weight, bringing the nuclear reaction under control in an emergency shutdown. Support plates in the guide tubes had twisted, blocking the passage of the control rods and reducing the effectiveness of a key safety system. (Service Central de Sûreté des Installations Nucléaires report, 31.1.89)

1989 January When a warning siren at Brazil's Angra I reactor was mistakenly triggered during a thunderstorm, mass panic ensued, with cars totally blocking the only highway out of the area. Details of the plant's emergency plans have never been officially published, but a group of visiting scientists discovered the plan allows 15 days to evacuate the local population and lays responsibility for this on the army, civil defence and fire service who, in reality, would be unable to cope. Brazilian houses offer little protection and there are no provisions made in the plan for permanent relocation.

Angra, where a second reactor is already under construction, is built in an area the local Indians call *itaorna*, meaning rotten stone. In March 1985 Angra's radio-ecology laboratory was pushed into the sea by a landslide and there have been several small earthquakes in the region in recent years. (WISE. 10.3.89; *South*, June 1989)

1989 January 6 A violently vibrating recirculation pump at the Fukushima II-3 reactor in Japan was allowed to run for 14 hours, with alarms signalling a fault, before the reactor was shut down. Investigations revealed that a 100-kg bearing ring had broken up, and, by early March, pieces of metal had been found throughout the core, on 122 of the 764 fuel assemblies and at the bottom of the reactor vessel. The largest piece of metal was 10.5 cm long and weighed nine g. The reactor was expected to be out of action for at least nine months. (WISE 14.4.89/28.4.89; *Nuke Info Tokyo*, March/April 1989)

The End of the Nuclear Dream

In these pages, we have followed a few of the many strands that go to make up the story of the Nuclear Age. There can be no doubt that it is a story of truly mythic proportions. But there is a danger that, taken on their own, these tales of disaster and despair may just leave the reader in a state of muted shock; gripped by a paralysis that affects so many of us after 50 years in pursuit of the Nuclear Dream.

Documenting a 'hidden history' such as this may provide illumination, but on its own, it can show no direction. It poses many questions, but points to few answers. It raises deep fears, and, of itself, offers little hope.

The casting of light, the posing of questions and the raising of legitimate concern all have their place. They are the very fabric of open debate. But unless these enquiries are directed by a positive vision, the sobering conclusions to which they lead may simply compound the spirit of numbed acquiescence they are intended to disturb.

This was not the spirit in which this book was conceived. It is not the way that such a book should end. It is not the spirit of Greenpeace.

For when the other strands that make up the story are examined, there are signs that the Nuclear Age may be drawing to a close. The narrow and precipitous nuclear road is not the only way forward. Different paths stretch away to either side. For the past 20 years, Greenpeace and many others have been seeking to point to some of these paths, and to help nudge us slowly in new directions.

In this final chapter, we offer a look through Greenpeace's eyes at the reasons for hope at the end of the Nuclear Age. It is with a glimpse of some of the new roads which are there to be taken that this book should close.

The Nuclear Age began with two promises: a 'Promise of Plenty' and a 'Promise of Security'. After 50 years, our bombs and reactors have failed to deliver either of these. Why have they failed us? What can be put in their place?

THE PROMISE OF NUCLEAR PLENTY

In the aftermath of Hiroshima, the awful power of the atom stood revealed to a shaken world. Many of the scientists who had laboured to bring this Herculean project to fruition were seized with remorse. For many, there was a strong emotional need to atone for their former ill-directed efforts. It was in this charged atmosphere that rash promises of nuclear plenty were first made to a credulous world.

Every aspect of life was to be transformed by the application of this magical new power. Vegetables would be made larger and more appealing. Swift, safe and clean atomic-powered ships and aircraft would revolutionise transport. In the words of one senior proponent, nuclear electricity would become 'too cheap to meter'.[1] The new atomic explosives could be used 'to dig canals, to break open mountain chains, to melt ice barriers and generally to tidy up the awkward parts of the world'.[2]

However, after nearly 50 years, such ideas have been consigned to oblivion. Virtually the only things now transported by nuclear power are nuclear weapons themselves. After Kyshtym, Windscale and Chernobyl, far from reaping the benefit of glorious new varieties of vegetables, we have seen the wholesale contamination of food and the blighting of entire harvests.

As for 'free energy', nuclear power has been described in the financial press as '...the largest managerial disaster in business history, a disaster on a monumental scale...'.[3] It has been the cause of some of the world's most spectacular bankruptcies, and has helped to create some of the world's largest international debts.

Globally, orders for new nuclear power stations now stand at an all-time low. In the USA, for example, every order for a new nuclear power station made since 1974 has subsequently been either cancelled or indefinitely deferred. In the USSR, some 20 nuclear power reactors that were planned or under construction have been cancelled over the past three years.

Austria and the Philippines both decided to scrap their first fully operable nuclear power plants before they even began operation; writing off billions of dollars of investment. Sweden is committed to phasing out all nuclear power completely before the year 2010.

Italy, Spain and Yugoslavia have called a halt to the further expansion of nuclear power. Australia, Austria, Denmark, Greece, Ireland, New Zealand, Portugal, the Philippines and Norway are among those countries which have shelved all plans for future nuclear power stations and have adopted non-nuclear energy strategies.

A handful of governments persist in declaring support for nuclear power. Such has been the scale of their commitment to the nuclear behemoth, that they are finding it very difficult to extricate themselves. But even in France, Japan, the Soviet Union and the United Kingdom, traditional sanctuaries of a beleaguered nuclear industry, the rate of development has slowed down to a snail's pace.

In France, often hailed as the flagship of the world nuclear industry (with nuclear power generating 70 per cent of electricity), the electricity supply industry has built up a massive international debt of some 20 billion pounds and runs at a staggering loss. The government has ordered that the nuclear debt must be drastically reduced. Even here, the end of the Nuclear Dream is approaching.

Nuclear power now delivers some four per cent of the world's energy needs. It has proved to be unacceptably expensive to the majority of countries who initially expressed an interest. After Three Mile Island and Chernobyl, hopes for truly safe nuclear power have been seriously discredited. And the problems posed by the generation of thousands of tonnes of highly radioactive nuclear wastes remain unsolved.

Though the industry is still far from dead, the promise of 'nuclear plenty' lies in tatters. One by one, countries are beginning to look elsewhere to satisfy the need for energy. Despite desperate attempts by this huge and influential industry to bully its way into the twenty-first century, the signs are that, slowly, the world as a whole is beginning to turn to a new and non-nuclear energy future.

But what can take the place of nuclear power? To an older generation which learnt to identify 'progress' with heavy industry, nuclear technology promised a brave new world. The liberation of the power of the atom represented, for a while, the pinnacle of the human scientific endeavour.

With the passing of time, and the realisation that we live in a fragile and finite world, technological progress has come to be associated less and less with the great mechanical leviathans of the past, and increasingly with elegant design, deeper understanding, better communications, 'soft' engineering.

Rather than heralding the dawn of a new era, the Nuclear Age is better seen as the end of an old. The harnessing of nuclear energy represents just one more step in a long historical progression that began with the steam engine. Since the beginning of the industrial revolution some 250 years ago, industrialised societies have obtained much of their energy by boiling water to create steam. Nuclear power stations are no different. They are glorified steam engines. The only change is that, in place of chemical energy, it is nuclear energy that is used to boil water to raise steam. The basic designs of nuclear power stations have not significantly changed since the 1950s. The fundamental engineering principles involved would have been familiar to James Watt.

Now it is the technical insights of the twentieth century which promise a truly new age. Developments in the fields of quantum physics, fluid dynamics, electronics, materials science and information technology herald the energy management and production technologies of the twenty-first century.

Radical improvements in energy efficiency are now made possible by the use of innovative design and modern materials. Automatic control systems allow the same tasks to be undertaken using only a fraction of the energy. The government utility in Sweden, for example, is now developing a strategy by which all nuclear energy can be phased out (presently generating almost half of all electricity) and no further large hydroelectric projects developed, without increasing emissions from fossil fuels; all by increasing efficiency. This is in a country which is already one of the most energy-efficient in the world.[4]

The potential for this new approach is huge. A group of leading international authorities on energy, the 'End Use Global Energy Project', published their findings in 1988. Looking only at existing but under-utilised energy-efficient technologies for transport, industry, commerce and the home, they came to some profound conclusions.

Taking into account an inevitable increase in global population by the year 2020, they found that there is no purely technical reason why poverty cannot be alleviated worldwide, and the quality of life of all rise to the level presently enjoyed only by the few, whilst the total rate of world energy consumption need increase by only some 10 per cent.[5] The alarmist predictions of voracious future energy needs, which have been used to sustain nuclear projects, are simply not tenable.

There is another important benefit from learning to use less to achieve more. For this is the only way that we can begin to escape the Greenhouse Effect. The dumping of huge quantities of carbon dioxide gas in the atmosphere through the burning of fossil fuels and the destruction of the world's forests is contributing to a threat of catastrophic climatic disturbance and global warming. This spectre of environmental destruction on a massive scale has been seized upon by a beleaguered nuclear industry as a reason to switch from fossil fuels to nuclear power.

However, the opposite is the case. Quite apart from the absurdity of proposing to remove one source of environmental damage only to substitute another, spending huge amounts of money on nuclear power could actually make the problem worse. There are a number of reasons for this.

Energy efficiency measures cost at least six times less per unit of energy than the cost of producing the same energy by nuclear power. One pound spent on nuclear power prevents some six times less carbon dioxide being released to the atmosphere from the burning of coal than does the same pound spent on improving energy efficiency. Spending on nuclear power ties up resources that could otherwise be invested in improving energy efficiency, and so compounds the problem of global warming.[6]

Investment in energy efficiency actually saves money, whilst nuclear enterprises lead to enormous debts. Conservation measures may be implemented in a few years, whilst nuclear programmes require decades. The construction of large numbers of nuclear power stations to replace the world's coal stations would cause the crippling debts of developing countries to be multiplied many times by the enormous expenses of nuclear projects. Furthermore, the energy requirement for such a colossal construction programme would itself increase carbon dioxide emissions.

In a hypothetical case where all the world's electricity would be produced from nuclear power, we would face the nightmare prospect of dealing with the pollution and wastes of some 3,000 more nuclear power stations than at present. The risks of future Chernobyls would be multiplied accordingly. Such are the technical obstacles to this scenario that not even the most enthusiastic proponents would advocate so strong a swing to nuclear power. Yet even in this 'best case scenario', the arithmetic of the problem is such that mammoth global nuclear programmes, developed over long periods of time, could reduce the Greenhouse Effect by only some 10 per cent.[7]

Energy efficiency offers far greater success in combating the Greenhouse Effect for a fraction of the economic and environmental cost, and in a fraction of the time. It offers an escape from the industrial treadmill of ever-increasing consumption and environmental destruction. Freed from delusions of a 'progress' demanding constant exponential growth, and liberated from dependence on ever-increasing infusions of energy, we can begin to look to less harsh and disruptive energy sources; the Sun, the Earth, wind and water.

Although no source of power can ever be entirely without impact, these 'renewable' resources offer a more gentle approach to our planet. Modern environmentally-sensitive techniques for harnessing solar and wind power are now at the point where they have become economically competitive as well as relatively environmentally benign. This progress has been achieved without the benefit of the immense subsidies in research and development enjoyed by nuclear power and fossil fuels. It also renders unnecessary the hidden subsidies which go to fund the huge environmental costs of nuclear and fossil power.

'Geothermal energy', from deep within the Earth, and the energy in waves, tides, free-running rivers, and residues from ecologically managed crops are also now beginning to yield benefits through the application of new technologies. Though at an earlier stage, they are gradually being harnessed on a commercial scale. Much work remains to be done, but the possibilities are now in sight.

Detailed studies of the potential for these more benign approaches to renewable energy have been scrutinised by international groups such as the prestigious 'World Commission on Environment and Development'.

They conclude that all the energy needed for the hypothetical world of 'efficient affluence' envisaged by the experts could be obtained solely from renewable sources of supply, carefully selected and managed for minimal impact on the environment.[8]

There are no concrete technical reasons why we should have to choose between such fearful prospects as the Greenhouse Effect or nuclear pollution and proliferation. There is a third way forward.

THE PROMISE OF NUCLEAR SECURITY

While 'atoms for peace' offered a world of affluence, 'atoms for war' promised a future of stability and order to a fractured, chaotic post-war world. Security would be bought at the price of terror. The Quixotic notion of deterrence became elevated to the status of a creed.

After the detonations of nuclear weapons at Hiroshima and Nagasaki, it was declared that the consequences of their future use would be so unthinkable that a state of nuclear armed 'peace' might be maintained into the indefinite future.

Fear of the horrors of nuclear holocaust would stay the hand of any potential adversary.

Innumerable terrible conflicts, including Afghanistan and Vietnam, have helped dispose of this mistaken hope. When small and desperate countries have faced a nuclear armed opponent, such tacit diabolical threats have counted for little. They are no longer credible. The atom bomb has not prevented war.

But have nuclear weapons at least deterred the use of other nuclear weapons? Although somewhat circular, it is this well-worn argument that continues to sustain remaining support for 'nuclear security'. No doubt the same promise might have been made at the introduction of any of the many horrific new weapons devised over the history of war. Though the idea of deterrence is presented as new, the phenomenon itself is as old as weaponry. Once deployed, the crossbow, the cannon, the machine gun and the bomb have all served, to some extent, as a deterrent. Fear of the consequences of their use ensured compliance with the wishes of those who brandished them. However, sooner or later, all such 'deterrents' have eventually been wielded in anger.

Nuclear weapons may be different in scale, but they are no different in kind. They are but the latest products of an ancient military ingenuity. The question now is how long can such a nuclear armed 'peace' persist? Unless the answer is 'forever', hopes of a 'nuclear security' which might have some meaning for our children, begin to look a little shallow.

Yet, once uttered, this perverse logic has been inexorably at work since Hiroshima. For once a particular country sought security in such a paradoxical fashion, others inevitably followed. The USA was soon joined in its vain quest by the Soviet Union, Britain, France and China. Now India, Israel, Pakistan, Argentina, Brazil and South Africa have all acquired an ability to make nuclear weapons under the cover of 'peaceful' nuclear programmes. Iran, Iraq, North Korea and Libya are in pursuit.

To many of these states, the development of nuclear power has been barely more 'peaceful' an enterprise than it was to the nations who pioneered the work after World War II. For as long as the logic of 'nuclear security' is propounded by any one country, others will continue to follow suit. With each new recruit to the nuclear club, the security of all diminishes.

Meanwhile, the arsenals of the established nuclear-armed countries have continued to mushroom in scale and variety.

With each new weapon that is introduced, tension, suspicion and the possibility of accident all increase.

The result of this proliferation of 'nuclear security' is that the world stands in constant danger of annihilation. With more and more weapons in the hands of the superpowers, and with more and more fingers on the button, the world has become less safe. After almost half a century, nuclear weapons have failed to deliver whatever promise of security they may ever have been seen to hold.

It has happened before. The cumbersome, clumsy knights in armour signalled the end of the medieval era. The top-heavy galleon and the ironclad battleship in turn relinquished their place. Each embodied a train of development pursued to the point of absurdity. Nuclear weapons have reached that same critical juncture. Dubbed by one weapons designer as the 'baroque, even rococo'[9] products of our time, nuclear weapons represent the final decadent impulse of a passing age.

Since the Middle Ages, diplomacy and politics have been conducted under the constant threat of explosives. Just as the nuclear reactor represents the apotheosis of the steam engine, so the nuclear weapon embodies the ultimate development of the ancient technology of gunpowder. We have reached the end of the line for this way of thinking.

Nuclear weapons continue the old imperial tradition in which preparation for war takes the place of communication, discussion, and negotiation. In a seemingly limitless world, with industrial civilisation in its infancy, such a petulant approach to politics may have seemed affordable.

Now, as the limits of our natural world become clear, and with a new appreciation of our common destiny on 'Spaceship Earth', it is no longer acceptable that we conduct our affairs under the continual threat of violence. The advent of global communications opens for the first time the possibility of a new age, based upon interdependence and discourse, rather than confrontation and suspicion. Nuclear weapons have no part in such an age.

It is with this realisation that there lie grounds for hope. For the first time, discussion of complete nuclear disarmament has come onto the global political agenda.

A new climate now governs relations between the old nuclear adversaries, and has led to limited but significant new developments, such as the conclusion by Presidents Reagan and Gorbachev of the Treaty on Intermediate Nuclear Forces.

'...the treaty whose text is on this table offers a big chance, at last, to get on the road leading away from catastrophe. It is our duty to take full advantage of that chance and move together toward a nuclear-free world...'[10]

The real importance of such agreements lies not in the numbers of warheads shuffled from one stockpile to another, but in the dawning realisation, even in the citadels of power, that these weapons are working against our mutual security. Just as the nuclear arsenals have been built up, so they can be dismantled. It is now finally becoming recognised that the only rational objective for long term global security is complete nuclear disarmament.

No doubt it was once thought that slavery was so deeply rooted in society that it could never be dislodged; never be 'disinvented'. Now, as a global institution, it has been consigned to history. Likewise, the tide of history is beginning to turn against aspirations to security through nuclear weapons. The battle is far from won, but there are enough encouraging signs to keep hope alive.

The vision of a secure world, without nuclear weapons, is becoming the vision of the twenty-first century. We are beginning to take the first tentative steps towards a new world order. For the first time, the prospect of a global community founded on mutual respect, communication and interdependence is offering itself as a real, indeed the only, sensible alternative to perpetual confrontation.

TOWARDS A NUCLEAR-FREE WORLD

It is here that the visions converge: of a world without nuclear power and a world without nuclear weapons. For 'atoms for peace' and 'atoms for war' have never been separate. One has always brought the other; they are just different sides of the same nuclear coin.

Each face of our Janus-like nuclear creation is irreversibly dependent upon the other. Without the great lure of nuclear weapons, the monstrous expenditure, sacrifice and secrecy of nuclear power would not have been conceivable; at the same time, the nuclear power industry helps to screen the production of nuclear weapons from public scrutiny. After 50 years of the Nuclear Age, their destinies remain entwined.

Now, as we recover from seduction by the easy promise of 'nuclear plenty', we are finding more practical tools to hand. They are all around us, and are here for the taking. As replacements for nuclear power, the options for increased efficiency and renewable energy are vastly more diverse.

But to place faith in technical solutions alone would simply be to repeat the misplaced enthusiasm of the early atomic advocates. Technology by itself cannot solve our problems. Human ingenuity may provide the tools, but it is human imagination which must give us the vision.

With this realisation, the true irony of our present situation becomes clear. For there is serendipity in our current predicament. As we begin to see beyond the glittering chimera of 'nuclear security', we are forced to look to the real world. We find that there are no machines that can take the place of our bombs and missiles. We see our quest for stability and order by such means for what it is: naïve idealism on a breathtaking scale.

It has taken a threat of the proportions of Nuclear Armageddon to make us realise that in a high technology world of five billion people on a small planet, we cannot continue to live by the old rules. We must embrace a new way of thinking. If we are to achieve a global community at peace with itself and its environment, then we must learn the lessons that the Nuclear Age has taught so well. Confrontation and suspicion, greed and conflict lead, ultimately, to disaster.

A nuclear-free world will not be without its problems. Far from it. The twentieth century leaves us with a formidable panoply of threats. Deforestation and desertification, pollution of the seas and disruption of the climate, poverty and injustice all challenge human resourcefulness as never before. The countless billions of pounds squandered annually on nuclear arsenals, and wasted in subsidies to an ailing nuclear industry, could be used instead to help combat these pressing threats.

Such problems will not diminish with the end of the Nuclear Age. It is the way that they are approached and solved, and the resources available to tackle them, that will be different.

The Nuclear Age has taken us a long way down the path of despair. Beginning with false promises and naïve faith, and continuing through disappointment, disillusion, and cynicism, it has left us with a legacy of fear. Fear of the sudden siren blast. Fear of evacuation from our homes at night. Fear of the insidious stealth of cancer and genetic damage. Fear of what a technology out of control can do. Fear of the Holocaust.

Yet it is with these fears that age-old patterns are beginning to crumble. Our loss of innocence has brought new hope and new ideas. The scale of the challenge that now faces us is bringing forth a new vision. Greenpeace, along with many others, is seeking to encourage and nurture that vision.

In the end, it is a question of personal responsibility. We know the problem. We can see the solutions. We have the choice. It remains to be seen whether we will rise to the challenge.

Andy Stirling
Amsterdam, June 1989

Notes

1. Statement by Lewis Strauss, Chair of the US Atomic Energy Commission, at a National Association of Science Writers' Founders Day Dinner in New York (16.9.54).
2. S. Chase, *The Nation,* (22.12.45).
3. J. Cook, 'Nuclear follies', *Forbes Magazine* (11.2.85).
4. T. Johansson, B. Bodlund, R. Williams (eds), *Electricity: Efficient End Use and New Generation Technologies, and Their Planning Implications* (Lund University Press, 1989).
5. J. Goldemberg, T. Johansson, A. Reddy, R. Williams, *Energy for a Sustainable World* (Wiley Eastern Limited, 1988).
Using the most energy-efficient technologies currently available, by 2020 AD a world population of 7 billion could attain average current West European living standards for a global power demand of 11 Terawatts. *The World Bank Development Report 1986* assesses current world demand at 10 Terawatts.
6. W. Keepin and G. Kats, 'Greenhouse warming: comparative analysis of two abatement strategies' (Annual Review of Energy, 1988).
7. The world electricity supply industry is responsible for only about 30 per cent of global fossil fuel consumption. The remainder is largely used in transport which cannot readily be substituted by nuclear power (note 5). Fossil fuel combustion as a whole is responsible for only about two thirds of the nett carbon flow to the atmosphere resulting from human activity. The remainder arises largely through deforestation. (See I. Mintzner, 'A matter of degrees: the potential for controlling the Greenhouse Effect', World Resources Institute, 1988.) Carbon dioxide as a whole is in turn responsible for causing only about one half of the global warming phenomenon (ibid). The end result is that even if all electricity production throughout the world were generated by nuclear power, the Greenhouse Effect would have been reduced by only 30 per cent of two thirds of one half; or 10 per cent.
8. World Commission on Environment and Development, 'Our common future' (1987). The potential for 'environmentally constrained' renewable sources of power, such as small-scale hydroelectric and tidal projects, wind power, solar power, biofuels, geothermal and wave power is assessed in this report to be some 13 Terawatts. This contrasts with the figure for total global power demand of 11 Terawatts arrived at in note 5. The total potential for renewable sources of power, without consideration of environmental impact, is assessed by the International Atomic Energy Agency (not known for their enthusiasm for competing energy technologies) to be around 24 Terawatts. (See 'Nuclear power, the environment and Man', International Atomic Energy Agency, 1982, pg 2.)
9. Herbert York, former Director of US Lawrence Livermore Radiation Laboratories, cited in Mary Kaldor, *The Baroque Arsenal* (Abacus, 1981).
10. President Mikhail Gorbachev at the signing of the INF Treaty, Washington, November 1987.

Stop Press

The following developments came to light in 1989, between the book's completion and its going to press.

British Nuclear Tests

May: A new report by the independent Australian Radiation Laboratory, 'Plutonium contamination in the Maralinga Tjarutja Lands', reveals that fallout from Britain's bomb tests spread far beyond the official test range in a plume that came close to an Aboriginal community at Oak Valley, 75 miles to the north. The report is being withheld until another study is completed in 1990 by British, Australian and US scientists. (*T* 5.7.89)

Windscale, Sellafield, UK

May: Denied legal aid to pursue a claim against BNFL in February, the families of 18 children who live in the vicinity of Sellafield and are suffering from leukaemia are granted legal aid on appeal. They will now be able to go ahead with their claims against Sellafield's owners. The government-appointed Committee On the Medical Aspects of Radiation in the Environment (COMARE) has already concluded that some feature of the plant may be responsible for the high incidence of leukaemia in the area, some 10 times the national average. (COMARE 2nd Report, HMSO 1988; *Obs Mag* 28.5.89.)

Chelyabinsk-40, USSR

June: A brief Tass news agency report officially confirms that there *was* a powerful explosion at the atomic weapons complex in the Ural mountains in 1957. A chemical explosion discharged two million curies of radioactive elements into the atmosphere, creating a plume 65 miles long and five to six miles wide, forcing the evacuation of 10,000 people. Though a £200 million clean-up campaign restored economic activity in more than 80 per cent of the contaminated zone, large areas around the city of Kasli remain contaminated and the water reserves are undrinkable. (*DT* 17.6.89; *Ind* 17.6.89; *NSc* 22.7.89.)

Chernobyl, USSR

May: At an East-West conference on pollution and health in Poznan, Poland, representatives of Green World, described as an 'international group of environmentalists against the Soviet government' allege that: 1. In some of the worst affected areas around Chernobyl, women have been forced to sign declarations that they will not have children; 2. While deaths have officially been put at 18, at the plant alone 50 specialists died; the official diagnosis was given as heart, lung or liver disease - not radiation effects. There are also more than 100 patients in a radiological cleaning centre in Kiev; 3. Hundreds of people from Chernobyl have been relocated all over the Soviet Union; Green World assumes that a third of them have been irradiated; 4. In the worst affected area of Poland, thousands of children are popularly believed to be suffering from leukaemia.

May: According to the *Observer* (21.5.89); a letter in the young communist newspaper *Komsomolskaya Pravda* claims that a radioactive cloud generated by the Chernobyl disaster was deliberately seeded to cause fallout over villages in the Byelorussian and Russian republics in order to save Moscow. In Byelorussia, which lies to the north of Chernobyl and which bore the brunt of the radioactive fallout from the accident, some 18 per cent of the territory, with a population of 500,000, is contaminated, a fifth of it being agricultural land. The republic also provides a new home for the 92,000 evacuees from the exclusion zone around the plant. There is a shortage of medical equipment, personal radiation dosimeters and 'clean' products, and concern for the safety of farm workers in contaminated fields. In the Narodichi district of the Ukraine there are reports of congenital abnormalities among farm animals, rising incidences of sickness among children and an increase in the abortion rate. Doctors have been among the first to leave the contaminated areas.

July: *Izvestia* reports that at 10 of the 44 functioning nuclear power stations in the USSR, there have been 30 stoppages or power reductions in June 1989 alone. Two occurred at the surviving sections of the Chernobyl plant. More than a third of the 'unplanned' incidents were due to personal negligence. (*Obs* 21.5.89; *G* 23.5.89; *NSc* 24.6.89; *DT* 13.7.89; *NSc* 15.7.89.)

Soviet Submarine Fires

April: Fire breaks out aboard an experimental Soviet Mike-class nuclear-powered attack submarine, submerged off the northern coast of Norway, and the vessel is forced to surface. Some of the crew abandon the submarine while others attempt to bring the blaze under control, but the vessel finally sinks, killing 42 of the 69 crewmen.

June: A Soviet Echo-2-class nuclear-powered submarine, armed with cruise missiles and nuclear-tipped torpedos, surfaces about 210 miles north of Norway's Bear Island when the first seal of its reactor loses 'air tightness' and has to be shut down. Smoke is seen billowing from the hull, but the Soviets deny any fire on board. Before the accident, Captain V. Orchinnikov had written in the Soviet publication *Smeena*: 'It will probably surprise you if I say that the nuclear installations in our submarines are operated by people who are not sufficiently trained, and some of them not trained at all. But we still set sail. The operators know and can do only 30 to 50 per cent of what they should know and be able to do.' (Quoted in *Time*, 10.7.89.)

July: Norwegian Defence Command reports that a Soviet nuclear-powered Alfa-class submarine has been sighted 75 miles east of Vardo, on Norway's eastern coast, with smoke pouring from its conning tower. Soviet ships appeared to be assisting the vessel. A Soviet spokesman stated that there had been no accident and that the submarine was playing its part in a military exercise. (*T*, 16.4.89; *Time*, 17.4.89; Greenpeace News Release, 2.5.89; *NYT*, 4.5.89; *Time*, 15.5.89; *T*, 27.6.89; *Ind*, 27.6.89; *Time*, 10.7.89; *IHT*, 17.7.89.)

US Weapons Production Complex

June: The new US Energy Secretary, Admiral James Watkins, gives a press conference at which he outdoes external critics of the DOE by lambasting his own organisation: 'The chickens have finally come home to roost and years of inattention to changing standards and demands regarding the environment, safety and health are vividly exposed to public examination daily. I am certainly not proud or pleased with what I have seen over my first few months in office.' He says that, for decades, the DOE had worked on the philosophy that 'adequate production of defence nuclear materials and a healthy, safe environment were not compatible objectives.' He accused his department of lacking trained personnel, giving one-sided internal briefings and having 'insufficient scientific information.' The latest estimate he gave of the cost of cleaning up the 17 US nuclear arms factories was £11.9 billion over the next four years, a figure which will further load the already substantial budgetary deficit and which is considered optimistic by environmental campaigners.

July: The overseeing committee of the House Energy and Commerce Committee describes the situation in nuclear weapons factories as being a 'crisis of the highest order' and discloses many instances of sloppy operation, arrogant indifference and wilful deception. These include: the use of illegal drugs by workers at Oak Ridge, Lawrence Livermore, Los Alamos and Hanford; managers at the finishing plant at Hanford turning off radiation alarms because they were sometimes set off by high winds; at one Savannah River reactor, a garden hose being the only fire-fighting equipment, in another unit, the sprinkler system being turned off for fear that, if it was activated, the computers and records would get wet. (*Ind* 29.6.89, *Time* 3.7.89)

US/Soviet Accord

June: The US and the Soviet Union sign an accord entitled 'The Prevention of Dangerous Military Activities' in Moscow after a year of secret talks. It aims to defuse local crises and prevent military confrontation arising out of misunderstandings or accidents like the Soviet destruction of the South Korean airliner in 1983. It provides radio frequencies for communication between US and Soviet military commanders in sensitive areas of the world, such as the Gulf, and forbids the use of electronics that could 'cause harm to personnel or damage to the equipment.' (*T* 8.6.89.)

US Nuclear Reactors

June: The residents of Sacramento, California's state capital, vote to close the local Rancho Seco nuclear power station — the first time US voters have scrapped an operating nuclear plant. It will take 20 years to transfer all the radioactive fuel from the 913-MW plant.

Acknowledgements

Such a complex international research project as this would not have been possible without the help of literally hundreds of people and organisations around the world who provided assistance, support and encouragement. In addition to our contributors and consultants we would like to extend special thanks to the following within Greenpeace:

Sebia Hawkins, Damon Moglen, Eric Fersht, Josh Handler, Philip Cade, Gerd Leipold, Julie Miles, Hans Kristensen, Bunny McDiarmid, Rebecca Johnson, John Willis, Antony Froggatt, John Sprange, Michelle Sheather and Rick Lecoyt.

Thanks also to Martin Leeburn at Greenpeace Communications, Steve Sawyer and David McTaggart for making our operation possible.

Our special thanks to our editor Liz Knights, who provided unfailing support and encouragement, to our valiant copy-editor Georgia Garrett and to all at Gollancz who have worked so hard to make this book a success.

We are also pleased to acknowledge the help provided by the following individuals, institutions and groups:

Jackie Walsh (Institute for Policy Studies), Paul Trudel (Canada's Wings Inc.), Alex Gray (Fairplay Information Systems), Henry Hamman (Radio Free Europe), Stewart Ullyott, Roger Meade (Los Alamos Archives), James P. Riccio (Nuclear Information and Resource Service), Kenneth Boley (Public Citizen), Patrick Molloy (Radioactive Waste Campaign), Anne Grinyer and Paul Smoker (Richardson Institute for Peace Studies), Jack Briggs (*Tri-City Herald*), Robert del Tredici, Herman Damveld, Patricia Lawson, Rick Johnson (National Association of Radiation Survivors), Dr Aurora Bilbao, Peter Moulson (University of New Mexico Press), Professor F. M. Szasz, Jean McSorley (Cumbrians Opposed to a Radioactive Environment), Bill Hutchinson, Lanny Sinkin (Christic Institute), Kay Pickering (TMI Alert), Rip Bulkeley, Glenn Alcalay (National Committee for Radiation Victims), Lindy Cater and Jim Thomas (Hanford Education Action League), Jane Balfour Films, Bill Hutchinson (ILER Research Institute), Nigel (Popperfoto), Kerrie Cook (Snake River Alliance), Karen Dorhn Steele (*Spokesman-Review*), Bob Pollard (Union of Concerned Scientists), Joyce Hollyday and Brian Jandon (*Sojourners*).

We wish to thank the *Bulletin of the Atomic Scientists* for kind permission to reproduce Lloyd J. Dumas's quote in our Introduction.

We have already begun accumulating additional material and stories on this vast subject for a proposed second edition of this book. We welcome further information and comments which can be sent to Greenpeace Books, 81c High Street, Lewes, East Sussex, BN7 1XN, UK.

Strenuous efforts have been made by the author and editors to ensure that proper accreditation has been given and that copyright clearance has been obtained. We will happily amend any omissions in future editions.

Principal Sources

The cataloguing of nuclear accidents has a long pedigree, and this book would not have been possible without the work of our distinguished predecessors. The following sources were consulted throughout the production of this book.

Bertell: Rosalie Bertell, *No Immediate Danger* (The Women's Press, 1985).

Bertini: H. W. Bertini and members of the staff of the Nuclear Safety Information Center, report entitled 'Descriptions of selected accidents that have occurred at nuclear reactor facilities', Oak Ridge National Laboratory, Oak Ridge, Tennessee (April 1980). The report was prepared at the request of the President's Commission on the Accident at Three Mile Island, in order to provide the members of the Commission with some insight into the nature and significance of accidents in nuclear facilities.

Bradford: Shaun Gregory and Alistair Edwards 'A handbook of nuclear weapons accidents', Peace Research Report No. 20, Department of Peace Studies, University of Bradford (January 1988). A very substantial list of military accidents involving US, UK and Soviet forces — all of which were believed to have some nuclear connection.

Burleson: Clyde W.Burleson, *The Day the Bomb Fell* (Sphere, 1980); published in the US as *The Day the Bomb Fell on America* (Prentice Hall, 1978).

Caufield: Catherine Caufield, *Multiple Exposures* (Secker and Warburg, 1989).

CDI: Center for Defense Information, Washington, DC. In vol. X, no.5 of their journal *The Defense Monitor*, the CDI issued the DOD list (see **DOD** below) together with substantial comments.

Cook: Judith Cook, *Red Alert* (New English Library, 1986).

Deadly Defense: 'Deadly defense: military radioactive landfills, a citizens' guide', produced by the Radioactive Waste Campaign (New York, 1988), provides further information on the US weapons production complex.

DOD: Department of Defense. In 1977 the US Air Force released summary reports of 26 major US military accidents involving nuclear weapons; an amended list containing details of 32 serious accidents was prepared in 1988.

Hall: Tony Hall, *Nuclear Politics* (Penguin, 1986). The best single volume on the development of nuclear power in the UK.

HSE: Health and Safety Executive. In the UK, licensees of nuclear installations have been required, since 1977, to report certain categories of accident to the HSE, under the Nuclear Installations (Dangerous Occurrences) Regulations. These reports are publicly available from the HSE in London. (Quarterly or yearly lists of accidents are also published by the governments of France, West Germany and the Netherlands.)

We have also consulted a list of serious accidents produced by the Uranium Institute in London, called 'The safety of nuclear power plants: an assessment by an international group of senior nuclear safety experts'(1988). A list of civil and military nuclear accidents was prepared by the late Australian senator, Ruth Coleman. Called 'Nuclear power: accidents and leaks, failures and other incidents, industrial and military', it contains many additional stories. We hope to track down the original sources for these and include them in subsequent editions of this book.

Neptune: William M. Arkin, in collaboration with Joshua M. Handler, 'Naval accidents: 1945-88, Neptune Papers III' Greenpeace/Institute for Policy Studies (June, 1989). William M. Arkin and three co-authors previously published a list of nuclear accidents at sea in the appendix to *The Denuclearisation of the Oceans*, edited by R.B.Byers (Croom Helm, 1986).

Newhouse: John Newhouse, *The Nuclear Age* (Michael Joseph, 1989). This book accompanied an excellent TV series of the same name.

NWDB: Key sources of factual information on the US military were the three volumes of the *Nuclear Weapons*

Data Books by Thomas B. Cochran, William M. Arkin, Milton M. Hoenig and Robert S. Norris (vols. II and III). The volumes are: I — US Nuclear Forces and Capabilities (1984); II — US Nuclear Warhead Production (1987); and III — US Nuclear Warhead Facility Profiles (1987). All are produced under the auspices of the Natural Resources Defense Council and published by Ballinger Publishing Co., Cambridge, Massachusetts. A further six volumes (at least) are being planned, to provide 'a current and accurate encyclopaedia of information about nuclear weapons'.

Patterson: Walt Patterson, *Nuclear Power* (Penguin, 1986). The most readable history of nuclear energy, together with the same author's *The Plutonium Business* (Paladin, 1984) and *Going Critical* (Paladin, 1985).

Pringle and Spigelman: Peter Pringle and James Spigelman, *The Nuclear Barons* (Michael Joseph, 1982). An excellent, general history of the nuclear age.

Public Citizen: The Critical Mass Energy Project of Public Citizen, a 'non-profit research and advocacy organisation founded in 1974 by Ralph Nader to monitor nuclear safety issues and promote safe energy alternatives', has, under the US Freedom of Information Act, systematically examined all available documents concerning accidents at nuclear power plants submitted to the NRC or to Congress. In the US, all nuclear utilities are required by law to submit Licensee Event Reports to the NRC when certain types of accident occur. From these and other sources, the NRC identifies those serious accidents that have posed a risk to public health or safety. These are termed Abnormal Occurrences, and the NRC is required by law to report them to Congress on a quarterly basis. Since 1979, Public Citizen has published full accounts of significant mishaps and Abnormal Occurrences in the form of Nuclear Power Safety Reports, and these have been our primary source of information on US nuclear power plant accidents.

SIPRI: The Stockholm International Peace Research Institute Yearbook, 1977, listed and discussed 32 serious military nuclear accidents between 1945 and 1968.

Wasserman and Solomon: Harvey Wasserman and Norman Solomon, *Killing Our Own* (Delta, 1982).

Main Story Sources

The following abbreviations for newspaper titles have been employed throughout the sources:
DM Daily Mail; *DT* Daily Telegraph; *G* Guardian; *Ind* Independent; *IHT* International Herald Tribune; *JDW* Jane's Defence Weekly; *MG* Manchester Guardian; *MoS* Mail on Sunday; *Nat Geo* National Geographical; *NEI* Nuclear Engineering International; *NSc* New Scientist; *NYT* New York Times; *Obs* Observer; *Obs Mag* Observer Magazine; *Sci Am* Scientific American; *ST* Sunday Times; *STel* Sunday Telegraph; *T* The Times; *WP* Washington Post; *WS* Washington Star; *WT* Washington Times.

Entries in bold refer to Principal Sources (see p. 352).

1945 July 16 Trinity Test, New Mexico

F. M. Szasz, *The Day the Sun Rose Twice* (University of New Mexico Press, 1984).
T.E. Vadney, *The World Since 1945* (Pelican, 1987).
Also: **Caufield**; **Newhouse.**
D. Smollar, 'Socorro unmoved by birth of bomb', *G* (17.7.88).
Our thanks to Professor Szasz for clarifying some vital points.

1946 May 21 Louis Slotin

© Copyright Barbara Moon.
Barbara told us that she had just heard about Slotin when she went to interview Dexter Masterson, then editor of the Consumers Union magazine. Masterson subsequently spent eight years writing a novel based on the Slotin story, *The Accident*, which was recently republished on the first anniversary of Chernobyl (Faber and Faber, 1987). The novel was to have been made into a film star-

ring Humphrey Bogart, but the idea was effectively banned by the government and it was never made. Dexter Masterson died early in 1989, but his work lives on.

1945 August Hiroshima and Nagasaki

Oughterson et al., 'Medical Effects of Atomic Bombs: The Report of the Joint Commission for the Investigation of the Atomic Bomb in Japan', 6 volumes, USAEC, Washington DC (1951)

Note 1: The casualty figures have been rounded up to the nearest thousand. Takeshi Ohkita MD, in his paper 'Acute medical effects at Hiroshima and Nagasaki' (in *Last Aid: The Medical Dimensions of Nuclear War* W.H. Freeman, 1982) comments: 'There is continuing controversy regarding the numbers of persons killed by these bombs. The numbers of civilian casualties estimated by the Joint Commission...do not include military casualties, nor survivors with late effects. No data concerning military casualties, which must have been large at Hiroshima, are available.'

Note 2: The popular notion that the dropping of the atom bombs on Japan was necessary in order to bring an end to the war, and thereby saved thousands of lives, has been shaken by the discovery of a secret intelligence study in the US National Archives. The study concludes that the Soviet Union's decision to invade Manchuria would definitely have forced Japan to surrender, and that no large scale US invasion of Japan would have been necessary. Once the Trinity test had proved successful, the decision was taken to use the bomb and thereby prevent the USSR from being in a position to lay claim to Chinese territory. (Gar Alperovitz, 'Did the US need to drop the bomb on Japan? Evidently not', *IHT* (4.8.89).

1946-48 US Nuclear Tests Part 1

The quoted accounts of Tests Able and Baker come from 'Operation Crossroads, Report to the Honorable Alan Cranston U.S. Senate', GAO (Nov 1985). The detailed analysis of Col. Stafford L. Warren's evidence is contained in Dr A. Makhijani and D. Albright, 'Irradiation of personnel during Operation Crossroads: An evaluation based on official documents', International Radiation Research and Training Institute (IRRTI) (May 1983). Useful background was provided by J. Dibblin, *Day of Two Suns* (Virago, 1988). Full details of the test programme were checked with R. S. Norris et al, 'Known nuclear tests: July 1945 to 31 December 1988', NRDC (Jan 1989). Further details were provided by J. Smith, 'Pressure grows for atomic fall-out inquiry', *ST* (29.5.83); and Associated Press, 'Bikini Atoll radiation was greater than believed', *T* (6.12.87). Glenn Alcalay provided us with expert assistance and comments on this story. Additional reading: D. Bradley, *No Place To Hide* 1946/84, (University Press of New England, 1983).

1949 Hanford, Part 1 The Green Run

J. Thomas, 'A summary of radiation releases from Hanford', Hanford Education Action League (18.5.89). Background information on Hanford came from **NWDB vol. II**.
K. D. Steele, 'Hanford's bitter legacy', *Bulletin of the Atomic Scientists* (Jan/Feb 1988); K. Schneider, 'Seeking victims of radiation near weapon plant', *NYT* (17.10.88); K. D. Steele, '1949 radiation release contaminated vast area', *The Spokesman-Review* (4.5.89); D. Whitney, ''49 iodine releases larger, says DOE', *Tri-City Herald* (4.5.89).

1951-1958 US Nuclear Tests, Part 2

Wasserman and Solomon; **NWDB vol. II**.
G. G. Caldwell M.D. et al, 'Leukemia among participants in military maneuvers at a nuclear bomb test: A preliminary report', *Journal of the American Medical Association*, vol. 244, no. 14 (3.10.80).

'Mortality of nuclear weapons test participants', US National Academy of Sciences (1985).
W. Pincus, 'GIs held manoeuvres in 1957 in swirls of radioactive dust', *IHT* (18.1.78); W. Pincus, '3 scientists see cancer link in US atom-blast exposure', *IHT* (26.1.78); M. Clarke, 'Smoky's fallout', *Newsweek* (6.2.78); M. Cross, 'Most US atom veterans unharmed', *NSc* (13.6.85).
Three books recommended to us by Glenn Alcalay as further reading are: M. Uhl and T. Ensign, *G. I. Guinea Pigs* (Playboy Press, 1980); H. L. Rosenberg, *Atomic Soldiers* (Beacon Press, Boston, 1980); and T. H. Saffer and O. E. Kelly, *Countdown Zero* (G. P. Putnam's Sons, 1982).
DOWNWINDERS
H. Ball, *Justice Downwind* (OUP, 1986).
P. L. Fradkin, *Fallout: An American Nuclear Tragedy* (University of Arizona, 1989).
J. Barnes, 'Victims go into battle over atom town's legacy of death', *ST* (21.11.82); C. Joyce and I. Anderson, 'US found guilty of atom-test deaths', *NSc* (17.5.84); 'Supreme insult for nuclear veterans', *NSc* (21.1.88); Keesing's Record of World Events, vol. XXXIV, no. 7 (1988).
Note: Some 225,000 US military troops participated in both the Nevada and the Pacific tests. According to the Veteran's Administration, as of 23 May 1989, there have been 8,881 compensation claims filed for nuclear-related problems, of these only 182 have been awarded.
Box: THE CONQUEROR
H. Medved, *The Fifty Worst Movies of All Time* (Angus & Robertson, 1978).
K. G. Jackovich and M. Sennet, 'The children of John Wayne, Susan Hayward and Dick Powell fear that fallout killed their parents', *People* (10.11.80); "Death Canyon' row over Reagan's idol', *Obs* (8.3.81).
Note: An interesting footnote to the story is that John Wayne came up with the idea of having the world première of the film in Moscow. However, the Soviet Embassy in Washington screened it first and scuppered the plan. The film was subsequently banned in the USSR. *The Conqueror* became Howard Hughes' favourite film. He was later to spend $12 million buying up every existing print of the film and he refused to allow it to be shown on television. For seventeen years, only Hughes was able to watch it until, in 1974, Paramount secured the rights to re-issue it.

1952 - 1958 British Nuclear Tests

A number of books were published to coincide with the The Report of the Royal Commission into British Nuclear Tests in Australia (Australian Government Publishing Service, Canberra 1985), whose conclusions and recommendations we consulted.
They include: *Fields of Thunder* (Counterpoint, 1985) by TV journalists Denis Blakeway and Sue Lloyd-Roberts; *Clouds of Deceit* (Faber and Faber, 1985) by ex-Sunday Times journalist, Joan Smith; and a powerful book which gives the point of view of the servicemen on Christmas Island — Derek Robinson's, *Just Testing* (Collins Harvill, 1985).
More recently published is Lorna Arnold's, *A Very Special Relationship* (HMSO, 1987), which is currently the definitive official account, based on AWRE, UKAEA, MOD and other official archives.
Christmas Island Cracker, by Air Vice Marshall Wilfred E. Oulton (Thomas Harmsworth Publishing, 1987), is a detailed account of the Christmas Island tests by the Task Force Commander. It contains no mention of the ill-effects of the testing programme.
Darby et al., 'Mortality and cancer incidence in UK participants in UK atmospheric nuclear weapon tests and experimental programmes', NRPB (Jan 1988).
We also cross-checked facts with numerous newspaper accounts and have relied on these to track the recent development of compensation court cases:
'Britain's bomb: the secrets of a brilliant disaster', *ST* (1.12.74); S. Taylor, 'Britain in joint project to monitor contamination', *T* (20.2.81); D. Watts and G. Brock, 'Ill wind at Monte Bello', 'Doctor Marston's disturbing

story', 'A scream like nothing you've ever heard', *T* (18-20 June 1984); S. Connor, 'Echoes of Britain's bomb tests', *NSc* (4.10.84); D. Leigh and P. Lashmar, 'Revealed at last: the deadly secret of Britain's A-bombs', *Obs* (24.3.85); P. Brown, 'The fallout that follows 30 years after', *G* (6.12.85); R. Yallop, 'The cost of mutual trust', *G* (6.12.85); 'The Royal Commission into British nuclear tests in Australia', *T* (6.12.85); B. Palling, 'Britain's million-year nuclear legacy', *G* (13.12.85); R. Milliken, 'Maralinga: Countdown to a new fall-out', *T* (21.8.86); F. Gibb, 'Victim of cancer sues over A-tests', *T* (1.9.86); J. Ford, 'Australia asks Britain to pay for its atomic legacy', *NSc* (2.11.86); H. Mills, 'Ex-soldier wins right to sue over atom test', *Ind* (18.12.86); B. James, 'When the wind blew', *T* (24.2.87); R. Milliken, 'Lingering fallout from British nuclear tests', *Ind* (18.6.87); S. Connor, 'Risk and the radioactive service', *NSc* (4.2.88); W. Bennett, 'Pension victories for A-test victims', *MoS* (17.7.88); 'Britain's bomb-test dust halts Aboriginal homecoming', *T* (6.8.88); S. Taylor, 'Aborigines granted tests compensation', *T* (26.8.88); Reuter, 'Australian wins Maralinga claim', *T* (23.12.88); AFP, '300,000 pounds sterling award in N-test case', *Ind* (23.12.88).
Note: Christmas Island was also used for US nuclear tests as new missile technology became available. Beginning with Operation Dominic in 1962, there were some 40 tests, including one with a Polaris missile fired from a submarine off the coast of California, which exploded a half-megaton bomb over the island. Hayles, Zarsky and Bello, *American Lake*, (Penguin, 1986).

1952 December 12 Chalk River Reactor, Ontario

Bertini; **Patterson**; **Bertell**
Jimmy Carter, later to be US President during the Three Mile Island crisis, was involved in the clean-up at Chalk River and reportedly received three rem in an exposure period of only 90 seconds.

1954-1958 US Nuclear Tests, Part 3

J. Dibblin, *Day of Two Suns* (Virago, 1988).
Additional reading: Ralph E. Lapp, *The Voyage of the Lucky Dragon* (Penguin, 1958).
R. S. Norris, T. B. Cochran and W. M. Arkin, 'Known US nuclear tests, July 1945 to 31 December 1988', Nuclear Weapons Databook Working Paper, NRDC (Jan 1989).
G. Alcalay, 'Cultural impact of the US atomic testing program — Marshall Islands Field Report (4 March - 7 April, 1981)', Rutger's University, in his graduate paper.
The AEC and Brookhaven reports referred to in the text are as follows: R. A. Conard M.D. et al, 'March 1957 medical survey of Rongelap and Utirik people three years after exposure to radioactive fall-out', Brookhaven National Lab., Upton, N.Y. (June 1958); 'Report of the ad hoc committee to evaluate the radiological hazards of resettlement of the Bikini Atoll', AEC (12.8.68).
J. M. Weisgall, 'U.S. bungles, and Bikinians become nuclear nomads', *L.A. Times* (30.3.84); I. Mather, 'Masters of war smash refugees from paradise', *Obs* (19.10.86); M. Weisskopf, 'Islanders left unaware of A-test hazards', *WP* (31.5.88); 'Response from Roger Ray, DOE official in charge of relations with the Marshallese', *WP* (11.6.88); I. Anderson, 'Potassium could cover up Bikini's radioactivity', *NSc* (10.12.88).
KWAJALEIN
Dibblin aside, the principal source on Kwajalein is G. Johnson, *Collision Course at Kwajalein* (Pacific Concerns Resource Center, 1984). Fresh details are provided in D. C. Scott, 'Star wars in the South Pacific', *Christian Science Monitor* (8.6.89).
Note: Kwajalein runs on America's time: it may be a day behind America in reality, being on the other side of the international dateline, but the Pentagon insisted the atoll change in case they fired missiles on the wrong day. (Dibblin, 1988).

1956 July 27 Broken Arrow 1 Lakenheath AFB, UK

DOD/CDI

M. Stuart, 'East Anglia's nuclear escape', *G* (6.11.79); M. Bailey, 'The crash that could have made a desert in England', *Obs* (9.8.81).

1957 May 22 Broken Arrow 2 Kirtland AFB, New Mexico

DOD/CDI

M. White, 'US finally admits its H-bomb error', *G* (29.8.86); Associated Press, 'Whoops! There goes that H-bomb', *IHT* (30-31.8.86); R. Highfield, 'H-bomb disarmer', *DT* (1.9.86).

1957 September Rocky Flats, Colorado

Deadly Defense

C. J. Johnson, 'Cancer in an area contaminated with radionuclides near a nuclear installation', *Ambio* vol. 10, no. 4 (1981).

C. Peterson, 'Rocky Flats: Risks amid a metropolis', *WP* (12.12.88); C. J. Johnson, 'Nuclear safety: The whistle-blower is appalled', *IHT* (20.12.88).

1957 October 11 Windscale, Sellafield, UK, Part 1

Patterson; Bertell; Hall

J. Cutler and R. Edwards, *Britain's Nuclear Nightmare* (Sphere Books Ltd, 1988).

M. Stott and P. Taylor, 'The nuclear controversy', The Town and Country Planning Association in association with the Political Ecology Research Group (1980).

'An assessment of the radiological impact of the Windscale fire, October 1957', NRPB (HMSO, 1982).

'Uranium becomes red hot in atomic pile', *MG* (12.10.57); R. Herbert, 'The day the reactor caught fire', *NSc* (14.10.82); J. Smith, 'Cancer question for the fire investigators at Windscale', *ST* (27.3.83); J. Urquhart, 'Polonium: Windscale's most lethal legacy', *NSc* (31.3.83); P. Brown, 'The lingering curse of Windscale's nuclear fire', *G* 5.10.87); 'Conspiracy of silence over nuclear blaze', *DM* (1.1.88); F. Pearce, 'Penney's Windscale thoughts', *NSc* (7.1.88); 'Robot reveals piles of debris in Windscale reactor', *NSc* (22.10.88); G. Lean, 'Windscale cover-up exposed', *Obs* (1.1.89); 'Radioactive leak from Windscale covered up', *Ind* (2.1.89).

RADIATION DISCHARGES

P. M. E. Sheenan and I. B. Hillary, 'An unusual cluster of Down's Syndrome, born to past students at an Irish boarding school' *British Medical Journal* vol. 287 (12.11.83).

N. Timmins, 'Leukaemia cases linked to deliberate rise in Sellafield radiation outflow', *T* (25.8.84); A. Spackman and D. Connett, 'New Sellafield scandal', *ST* (16.2 86).

Further details and comments provided by Jean McSorley of CORE (Cumbrians Opposed to a Radioactive Environment).

1957/1958 Chelyabinsk-40, USSR

Pringle and Spigelman

The two official reports cited in the text are: D. M. Soran, D. B. Stillman, 'An analysis of the alleged Kyshtym disaster', Los Alamos National Laboratory (Jan 1982); and J. R. Trabalka, L. D. Eyman and S. I. Auerbach, 'Analysis of the 1957-8 Soviet nuclear accident', Oak Ridge National Laboratory (Dec 1979).

Dr Z. Medvedev, 'Two decades of dissidence', *NSc* (4.11.76);

Dr Z. Medvedev, 'Facts behind the Soviet nuclear disaster', *NSc* (30.6.77); Dr Z. Medvedev, 'Winged messengers of disaster', *NSc* (10.11.77); Dr Z. Medvedev, 'Nuclear disaster in the Urals', *NSc* (11.10.77); 'The wasteful truth about the Soviet nuclear disaster', *NSc* (10.1.80); J. R. Trabalka et al, 'Analysis of the 1957-58 Soviet nuclear accident', *Science*, vol. 209 (18.7.80); S. White and C. Joyce, 'Urals disaster: Explosion or just pollution?', *NSc* (22.4.82); J. Wilhelm, 'Wasteland in the Urals', *Discover* (June 1982); J. Wilhelm, Soviets may be making nuclear fuel in contaminated area', Associated Press

(1.12.88); 'The Nuclear Accident in Russia' *Sixty Minutes* broadcast (9.11.80); N. Usui, 'Velikhov terms public opinion major issue for Soviet Union', *Nucleonics Week* (8.12.88)
Note: The ill-fated final flight of a U-2 spy plane over Soviet territory in May 1960 undoubtedly had, as one of its many targets, the Kyshtym complex. Pilot Francis Gary Powers had just passed over the area when his plane was shot down.

1958 March 11 Broken Arrow 3 Florence, South Carolina
DOD/CDI; Burleson

1960 February 13 French Nuclear Tests, Part 1, Algeria and Moruroa
B. and M-T. Danielsson, *Poisoned Reign*, (Penguin, 1986).
'French Polynesia: The nuclear tests (A chronology 1767 - 1985)', Compiled by E. Shaw and R. Wilson, Greenpeace Report (1981).
A. S. Burrows et al, 'French nuclear testing, 1960-88' (NWD 89-1), Natural Resources Defense Council (24.2.89).
D. Marsh, 'France tests its atomic might', *NSc* (14.2.85).

1960 June 7/8 BOMARC Missile Fire New Jersey
DOD/CDI
'N. J. Governor looks into 1960 missile accident', *Air Force Times* (22.7.85); 'USAF to release data on 1960 plutonium leak', *JDW* (10.8.85).

1961 January 3 SL-1 Reactor, Idaho Falls
Bertini
NEI, March 1961, vol. 6, no. 58
J. R. Horan and W. P. Gammill, 'The health physics of the SL-1 accident', in *Health Physics* (Pergamon Press 1963).
We went to great effort to track down the makers of *SL-1: A New Way To Die*, a documentary film on the accident, screened by the BBC in May 1986, but with no success. Anna Gyorgy in *No Nukes* mentions a three-part film, *The SL-1 Accident*, available free of charge from DOE's film library. Special thanks to Kerry Cooke, of the Snake River Alliance, for her help.

1961 January 29 Broken Arrow 4, Goldsboro, Carolina
DOD/CDI
G. Hauer, 'The story behind the Pentagon's broken arrows', *Mother Jones* (April 1981).

1962, October 7 Nukey Poo Reactor, Antarctica
Bertini
Barney Brewster, *Antarctica: Wilderness at Risk*, in conjunction with Friends of the Earth (A.H. and A.W. Reed Ltd, 1982).
For photos and further details, see *The Greenpeace Book of Antarctica* (Dorling Kindersley, 1989).

1963 April 10 USS Thresher, North Atlantic
Bradford; Pringle & Spigelman; Neptune
J. Bentley, *The Thresher Disaster: The Most Tragic Dive in Submarine History* (Doubleday, 1975).
D. Miller and J. Jordan, *Modern Submarine Warfare* (Salamander, 1987).
J. H. Wakelin Jr., 'Thresher: lesson and challenge'; Lt. Comdr. D. L. Keach USN, 'Down to Thresher by Bathyscaphe'; E. A. Link, 'Tomorrow on the deep frontier' — all in *Nat Geo* (June 1964).
D. E. Kaplan, 'A Chernobyl at sea haunts countries with nuclear subs', *Toronto Globe and Mail* (8.6.87).

1964 Onwards Chinese Nuclear Tests
Newhouse
'Chinese fallout brings back memories to the US', *NSc* (14.10.76);
'Chinese bomb affects Tokai reactor', *NEI* (Nov 1976); 'Nuclear tests, cancer link concerns China', *IHT* (23.8.81);
'Nuclear protests by students create dilemma for leadership', *T* (31.12.85);

J. Gittings, 'China calls halt to surface explosions', *G* (22.3.86); 'China admits nuclear deaths', *T* (12.5.86).

1965-67 Operation Hat, Himalayas

H. Kohn, 'The Nanda Devi caper', *Outside* (May 1978); R. Wigg, 'Delhi knew of nuclear device spying on China', *T* (18.4.78); 'Spy satellite on a mountaintop', *NSc* (20.4.78).

1965 December 5 USS Ticonderoga, North Pacific

Neptune
D. E. Sanger, 'US-Japan ties worsen on news that warhead was lost 24 years ago in Pacific', *NYT* (9.5.89); P. Pringle, '48 nuclear warheads have been lost at sea', *Ind* (10.5.89); Reuter, 'Japanese fury at US nuclear accident', *Ind* (11.5.89); The one-megaton revelation', *Time* (22.5.89).

1966 January 17 Broken Arrow 5 Palomares, Spain

Palomares Summary Report, Defense Nuclear Agency, Field Command, Technical and Analysis Directorate (17.1.75) (Obtained through the FOIA by the NWDB Project, NRDC, Washington DC.)
Two book-length accounts of the Palomares accident are: F. Lewis, *One of Our H-bombs is Missing* (McGraw-Hill); T. Szulc, *The Bombs of Palomares* (Gollancz, 1967).
P. M. Pinilla et al., 'Epidemiological study of mortality in Palomares', a paper presented to a conference organised by the British Nuclear Energy Society in London, 11-14.5.87.
E. Schumacher, 'Where H-bombs fell, Spaniards still worry', *NYT* (28.12.85); 'Two decades of plutonium poison at Palomares', *G* (18.12.86).
A 72-minute documentary film — *Broken Arrow 29* — about the Palomares incident was produced by The International Broadcasting Trust for Channel Four Television in the UK and released in 1986. Produced, written and directed by Dina Hecht, the film focuses on continuing efforts by the local people to gain compensation for the effects of the accident.

1966 October 5 Fermi Reactor, Detroit

Curtis, Hogan and Horowitz, *Nuclear Lessons* (Turnstone Press, 1980).
J. G. Fuller, *We Almost Lost Detroit* (Berkeley Books, 1984).
Bertini; **Patterson**
Our thanks to Bill Hutchinson for his help on this story.

Nuclear Powered Ships

Pringle and Spigelman; **Patterson**
Curtis, Hogan and Horowitz, *Nuclear Lessons* (Turnstone Press, 1980).
A. Villiers and J. E. Fletcher, 'Savannah — World's first nuclear merchant ship', *Nat Geo* (August 1962).
R. F. Pocock, 'Marine nuclear propulsion', *Marine Engineering Review* (March 1981).
L. W. Brigham, 'Arctic icebreakers: US, Canadian, and Soviet', *Oceanus* vol. 29, no 1. (Spring 1986).
E. Gray, 'How safe are the nukes?' *Sea Classic Int.* (Autumn 1986).
Assoc. Press, 'Nuclear meltdown averted on Soviet icebreaker in port', *DT* (6.3.89); Reuter, 'Meltdown crisis on Soviet ship', *G* (6.3.89); 'Nuclear ship protest', *T* (8.3.89); M. Dobbs, 'Amid the Glasnost a greening emerges', *IHT* (9.3.89); WISE Communique, (24.3.89).
J. Takagi, original contribution, 1988.
Fairplay Information Systems (FISYS), London, personal communication (22.6.89).

1968 January 21 Broken Arrow 6, Thule, Greenland

The main source for this story is the official report contained in USAF Nuclear Safety, vol. 65 (Part 2) Special Edition Jan/Feb/Mar 1970.
Other sources are:
DOD/CDI
D. Ford, *The Button* (Unwin, 1986); P. Feldman, 'Cancer strikes 18 years after nuclear clear-up', *G* (11.12.86); J. Isherwood, 'Greenland tells US to produce atom bomb film', *DT* (17.12.87); 'Greenland demands film

of A-bomb it says lies off coast', *WT* (18.12.87); 'Workers childless after cleaning up H-bomb accident', *DT* (19.12.87); P. B. de Selding, 'A Broken Arrow's dark legacy', *The Nation* (25.6.88).

1968 May 21 USS Scorpion, North Atlantic

DOD/CDI; **Neptune**; **NWDB vol. I**
R. Varner and W. Collier, *A Matter of Risk* (Hodder and Stoughton, 1978).
H. S. Bradsher, 'Scorpion, sub that sank in '68, was carrying nuclear weapons', *WS* (28.5.81); H. Lucas, 'Lost US submarine caused by torpedo mishap', *JDW* (5.1.85).

1969 November Operation Holystone

Neptune
D. Ball, 'Nuclear war at sea', *International Security*, vol. 10, no 3. (Winter 1985/86).

1970-79 French Nuclear Tests, Part 2

B. and M. Danielsson, *Poisoned Reign* (Penguin, 1986).
A. S. Burrows et al, 'French nuclear testing 1960-88, (NWD 89-1)', Natural Resources Defense Council.
Further details about David McTaggart's voyages to Moruroa are provided in our book *The Greenpeace Story* (Dorling Kindersley, 1989). The fullest account is in McTaggart's own book, co-written with Robert Hunter, entitled *Greenpeace III: Journey Into the Bomb* (Collins, 1978).

1970 December 18 US Nuclear Tests, Part 4

R. S. Norris, T. B. Cochran and W. M. Arkin, 'Known US nuclear tests, July 1945 to 31 December 1988' (Natural Resources Defense Council, 1989).
ERDA News, (Energy Research & Development Administration), (18.6.76); 'Nuclear device falls, 11 hurt', *WS* (30.10.75); UPI, 'Radioactive gas leaks into air after Nevada nuclear test', *WS* (27.9.80); R. E. Brim and P. Condon, 'Another A-bomb cover-up', *Washington Monthly* (7.1.81); UPI, 'Two workmen tested for nuke exposure', *Las Vegas* (12.9.81); 'How UK stages desert A-tests at $10 million a bang', *Obs* (24.4.83); Reuter, '12 hurt in US nuclear test', *T* (17.2.84); 'Day the nuclear roof caved in', *NSc* (23.2.84); 'A nuclear unthreat', *The Economist* (29.3.86); I. Mather, K. Hindley, 'Stacking the odds against a test ban', *T* (20.8.86); Keesing's Record of World Events vol. XXXIV, 1988 no. 5.
PREVIOUS VENTS
J. Kay, 'Radiation on the wing', *Arizona Daily Star* (17.11.81).
Box: MIGHTY OAK
W. Pincus, 'Nevada tunnel too 'hot' to enter after A-test', *WP* (14.5.86).
Notes. On 23 October 1975, for a test codenamed 'Peninsula', a nuclear canister containing a device with a yield of less than 20 kt accidentally fell the last 40 ft to the bottom of a 650 ft deep shaft at the Nevada Test Site. Eleven workers were injured, some by electrical cables. Two required hospitalisation; one broke his leg. The canister lay there for four years while another shaft was drilled close to it. It was then blown up by another nuclear explosion.

1973-79 Windscale, Sellafield, UK, Part 2

J. Cutler and R. Edwards, *Britain's Nuclear Nightmare* (Sphere Books, 1988).
Also: **Hall**; **HSE**; **Patterson**
M. Stott and P Taylor, *The Nuclear Controversy* (The Town and Country Planning Association in association with The Political Ecology Research Group, 1980).
A. Dunn, 'Windscale radioactivity leak "presents no immediate danger" ', *G* (29.2.80); M. Briscoe, 'A-plant safety scandal', *DM* (1.8.80); 'Big radioactive leak reported in Britain', *IHT* (2.8.80); N. Hawkes, 'Nuclear leak escaped survey', *Obs* (3.8.80). Our thanks to Jean McSorley of CORE.

1974 November 13 Karen Silkwood

R. Rashke, *The Killing of Karen Silkwood* (Houghton Mifflin, 1981).

Additional reading: H. Kohn, *Who Killed Karen Silkwood?* (NEL, 1983).
'Silkwood case ends in victory', Convergence/The Christic Institute (Spring 1987).
C. Ware, 'The Silkwood coalition', *New West* (18.6.79); H. L. Rosenberg, 'Crusader', *US* (28.4.81); P. Tibenham, 'The Karen Silkwood affair', *The Listener* (19.4.84); M. Hosenball, 'Silkwood family awarded $1.38m', *ST* (24.8.86).
Our grateful thanks to Lanny Sinkin and all at the Christic Institute.

1974-75 Project Jennifer

R. Varner and W. Collier, *A Matter of Risk* (Hodder & Stoughton, 1979).
J. T. Richelson and D. Ball, *The Ties That Bind* (Allen and Unwin, 1985).
Additional reading: Clyde W. Burleson, *The Jennifer Project* (Sphere); W. Shawcross, 'Did Operation Jennifer really go wrong?', *ST* (23.3.75); 'The great submarine snatch', *Time* (31.3.75).
Note 1: The person who had possession of the stolen Hughes documents, which included 3,000 pages of Hughes' handwritten memoranda, was eventually tracked down by Michael Drosnin, who published a detailed account of their contents in 1985 in *Citizen Hughes* (Hutchinson, 1985).
Note 2. The *Glomar Explorer* was subsequently used by US and Dutch firms to search for minerals on the seabed, but this project was abandoned in 1980. *Newscript* (10.3.80).

1975 March 22 Browns Ferry Fire

Power Reactor Events, NRC, vol. 2. no. 5 (Sept 1980).
R.L. Scott, 'Browns Ferry nuclear power plant fire', *Nuclear Safety* vol. 17, no. 5 (Sept/Oct 1976).
Also: **Public Citizen**
WISE Communique (28.5.81).

1975 November 21 USS Belknap

Neptune
M. Stevens, L. Jenkins and J. B. Copeland, 'Collision course', *Newsweek* (8.12.75).

1976 Hanford, Part 2, Atomic Man

P. Loeb, *Nuclear Culture* (New Society Pub, Philadelphia, 1986).
T. Crowell 'McCluskey's recovery amazes doctors', *Tri-City Herald* (27.4.78); L. Williams, 'Victim of 1976 Z plant accident at Hanford dies', *City Herald* (8.8.87).

1978 January 24 Cosmos 954

Principal reports on the use of nuclear power in space are: B. M. Jasani, 'A note on ocean surveillance from space', *Ocean Yearbook 5*, Eds. E. Mann Borgese and N. Ginsburg, (University of Chicago Press, 1985); S. Aftergood, 'Background on space nuclear power', Committee to Bridge the Gap (May 1988); R. Bulkeley, 'Nuclear power in space: A technology beyond control?', Paper included in: *Military Technology, Armaments Dynamics and Disarmament*, Ed. H. G. Brauch (Macmillan, 1989).
Also: **Burleson**
Recent developments and the story of Cosmos 1900 are in:
'Orbiting debris threatens space missions', European Space Agency, (ESA), Features, no.8 (3.8.88); M. Fagan, 'Satellite may rain 50kg of uranium on Earth', *Ind* (10.8.88); S. Soldatenkova, 'Alarm over Cosmos-1900', *Moscow News* (4.9.88); P. Wright, 'How Britain can cope with a burnt-out satellite', *T* (15.9.88); M. Fagan, 'Satellite likely to hit Earth next month', *Ind* (15.9.88); H. Gavaghan, 'Soviet satellite nears Earth with unspent nuclear fuel', *NSc* (29.9.88); 'Earth is safe from Cosmos', *ST* (2.10.88); W. Broad, 'Moscow's new orbiting nuclear reactor may go on sale to the West', *T* (16.1.89); 'Nuclear power space ban sought', *T* (19.1.89); 'Reactors in space could drop in on Earth', *NSc* (28.1.89); 'Stable orbit', Science and the Citzen Section, *Sci Am* (Feb 1989); G. E. Brown Jr, 'Courting disaster in orbit', *Bulletin of the Atomic Scientists* (April 1989).

1978 December 30-31 Beloyarsk, USSR

Reuter, 'Nuclear Fire Revealed' *ST* (23.10.88); Andrew Wilson, 'Nuclear Inferno' *Obs* (8.1.89).

1979 March 28 Three Mile Island

M. Stephens, *Three Mile Island* (Junction Books, 1980).
R. Del Tredici, *The People of Three Mile Island* (Sierra Club Books, 1980).
J. Doroshow, 'A decade of delay, deceit and danger', TMI Alert (1989).
W. Scobie, 'Hollywood cashes in on Harrisburg fallout', *Obs* (15.4.79); L. Torrey, 'The week they almost lost Pennsylvania', *NSc* (19.4.79); M. Gray, 'What really happened at Three Mile Island', *Rolling Stone* (17.5.79); H. Wasserman, 'Three Mile Island did it', *Harrowsmith* (May/June 87); W. Booth, 'Postmortem on Three Mile Island', *Science* vol. 238 (4.12.87); J. Hollyday, 'In the valley of the shadow', *Sojourners* (Mar 89); C. Flavin, 'Ten years of fallout', *Worldwatch*, Worldwatch Institute (Mar/Apr 89); N. Hawkes, 'Still active after all these years', *Obs Mag* (9.4.89).
Additional material from an original interview with Kay Pickering of TMI Alert.

1980 onward French Nuclear Tests, Part 3

J. Dibblin, *Day of Two Suns* (Virago, 1988).
Andrew S. Burrows et al., 'French Nuclear Testing 1960-88', NWD 89-1 Natural Resources Defense Council.
'French Polynesia: the nuclear tests (A chronology 1767-1985)' compiled by E. Shaw and R. Wilson (Greenpeace Report, 1987).
Jim Borg, 'Cost of French Nuclear Testing', *JDW* (25.3.89); 'Telling Omissions', *Greenpeace Quarterly* (Winter 1989).
Box: CIGUATERA
Henry Gee, 'Nuclear testing blamed for poisoning islander's food', *T* (2.2.89); 'Nuclear tests trigger food poisonings on Pacific islands', *NSc* (11.3.89).

Note: An up-to-date view of the various nationalist struggles in the South Pacific is provided by David Robie in his article 'Rising Storm in the Pacific' (*Greenpeace US magazine*, Jan/Feb 1989) and in his new book *Blood on Their Banner*.

1980 onward The Indian Nuclear Programme

Ed. Dhirendra Sharma, *The Indian Atom: Power and Proliferation* (Philosophy and Social Action, New Delhi, 1986).
The State of India's Environment (1984-85): The Second Citizen's Report (Centre for Science and Environment, 1986).
World Nuclear Industry Handbook 1989.
Also: **Pringle and Spigelman**.
N. Hedge, 'Nuclear Energy in South Asia: At What Cost?', *Utusan Konsumer* (March 1987); 'Indian Plant Blaze' *G* (1.5.86); M. Hamlyn and P. Wright, 'Workers Flee Heavy Water Plant', *T* (1.5.86); A. Tucker, 'N-Plant Could Close', *G* (10.9.86); T. Singh, 'Indians Claim New Plant is a "Chernobyl" ', *ST* (16.10.88); 'Too Hot To Handle', *South* (June 1989).

1980, June 3 NORAD Computer Glitches

P. Pringle and W. Arkin, *S!OP, The Secret US Plan for Nuclear War* (W.W. Norton & Co., 1983).
D. Ford, *The Button* (Unwin, 1985).
A. Grinyer & P. Smoker, *It Couldn't Happen Here* (Richardson Insitute, June 1987).
'Recent False Alerts From The Nation's Missile Attack Warning System', Report of Senator G. Hart & Senator B. Goldwater to the Committee on Armed Services, US Senate (9.10.80).
R. Thaxton, 'Nuclear False Alarm Gives A Grim Warning' *Obs* (2.3.80); L. Torrey, 'The Computer That Keeps Crying Wolf', *NSc* (26.6.80).
Note: M. Berger in 'Space sentries' (*Omni*, December 1986) provides an interesting glimpse inside Cheyenne

Mountain. Inside the catacomb are 15 three-storey buildings; each resembles a tall, windowless mobile home, constructed of reinforced steel and mounted on enormous springs. There is a workforce of 1,000 daytime employees and 300 night-shift workers. They dine in a cafeteria named Granite Inn. There is a barbershop, a small general store, a gym and a basketball court. Totally self-sufficient, the mountain has its own generators, air-filtration systems and four water reservoirs. NORAD is also home to Spadoc, the US military's Space Command Defense Operations Center. The mountain complex can be 'buttoned up' by closing two enormous 25-ton doors, operated by a hydraulic system that can close them in just 45 seconds. There are enough supplies for the employees to survive for 30 days.
Box: WIMEX
Babst et al., 'Accidental Nuclear War: The Growing Peril' (Canadian Peace Research Reviews, 1985).
F. Grove/Knight News Service, 'Pentagon Calls Super-Computer a "Disaster" ', *San Francisco Chronicle* (4.11.79); 'Hush-Up on Accidental War', *NSc* (5.6.86); M. Thompson, 'GAO Blasts Defense Attack-Warning System', *Daily Press* (16.12.88).

1980 September 18/19 Titan II Missile Fire, Arkansas

DOD/CDI

M. Leapman, 'Radiation leak averted in Titan missile explosion', *T* (20.9.80); 'Inspection of warhead to take place in Texas', *Muskegon Chronicle* (23.9.80); L. Torrey, 'The night they nearly lost Damascus', *NSc* (9.10.80); W. Pincus, 'Air Force missile blast survivors wonder if buddy died in vain', *WP* (22.10.80); E.Ulsamer, 'The US Senate honors heroic Titan crews', *Air Force* (December 1980).
According to Andrew Cockburn in his book *The Threat* (Hutchinson, 1983), the same year the Titan at McConnell AFB exploded, SAC showed key US technicians a satellite photo of a Soviet silo where exactly the same kind of accident had occurred in 1977.
1965 ACCIDENT:
'Toll of a Titan', *Time* (20.8.65); 'Nuclear-Age Tomb', *Newsweek* (23.8.65).
1978 ACCIDENT:
Airman (May 1979); *Air Force Times* (12.5.80).

1981 January 6 Cap la Hague, France

Pringle and Spigelman

J. Leclerc, *The Nuclear Age: The World of Nuclear Power Plants* (Hachette, 1986).
Surété Nucléaire, no.13, (jan-fev 1980); A. Tucker, 'French nuclear plant halted by fire', *G* (18.4.80); P. Webster, 'Nuclear plant inquest', *G* (19.4.80); Shoja Etemad, 'The luck which saved Cap la Hague', *G* (19.4.80); 'Test on water after leak at A-plant', *T* (4.10.80); I. Murray, 'Nuclear waste catches fire at atom plant', *T* (8.1.81); 'Nuclear fire explained', *NSc* (2.4.81); *Greenpeace France Newsletter* no.6 (printemps 1981); M. Bennitt, 'Five injured in French atom leak', *DM* (22.5.86); 'France discovers the nuclear scare', *NSc* (29.5.86).
Special thanks to Mycle Schneider of WISE.

1981 November 2 LX-09 Explosive

A. Wilson, *The Disarmer's Handbook* (Penguin, 1985).
D. Campbell and N. Solomon, 'Accidents will happen', *New Statesman* (27.1.81); N. Solomon, 'Pentagon Denies LX-09 Safety Hazard', *Pacific News Service* (8.10.81); B. Wilson, 'LX-09, the hidden threat inside Poseidon missiles', *Glasgow Herald* (20.11.81).

1982 January 25 Ginna Reactor, New York

Public Citizen

C. Joyce, 'Gremlins in nuclear plant may strike again', *NSc* (4.2.82); I. Davis, 'US Nuclear Stations Exposed', *T* (1.4.82).
Box: TUBE LEAKS
R. Udell, 'Tube Leaks', (Public Citizen, 1982).
C. Joyce, 'Gremlins in nuclear plant may strike again' *NSc* (4.2.82); I.

Davis, 'US nuclear stations exposed', *T* (1.4.82).
Note: The French state electricity utility EDF has recently decided to change the steam generators on 25 900-MW reactors over the next few years. The costs are estimated to be around FF350 million per reactor (each of which has three steam generators). France has no previous experience in the replacement of these radioactive giants. (Mycle Schneider, personal communication May 1989.)

1983 Radioactive Scrap, Mexico

R. Curtis and E. Hogan, *Nuclear Lessons* (Turnstone Press, 1980).
'Radioactive scrap ends up in table legs', *NSc* (23.2.84); 'Mexican children play on radioactive pick-up truck', *NSc* (1.3.84); E. Marshall-Juarez, 'An unprecedented radiation accident', *Science*, vol. 223, (16.3.84); W. Scobie, 'How thieves caused worst American radiation spill', *Obs* (23.3.84); Hidden peril in a load of scrap', *MoS* (6.5.84); P. Chapman, 'Hot rods still at large in Mexico', *NSc* (13.12.84).

1983, November 14 Sellafield, UK Part 3

K. A. Gourlay, *Poisoners of the Seas* (Zed Books, 1988).
M. Brown and J. May, *The Greenpeace Story* (Dorling Kindersley, 1989).
The official reports referred to in the text are:
'Report on Sellafield Beach Contamination Incident', UK HSE (14.2.84); 'An Incident Leading to the Contamination of the Beaches near to the BNFL Windscale and Calder Works, Sellafield, November 1983', DoE (HMSO, January 1984).
1981 ACCIDENTS:
Patterson
J. Cutler and R. Edwards, *Britain's Nuclear Nightmare* (Sphere, 1988).
N. Timmins, 'Milk alert after Windscale leak', *T* (9.10.81); N. Timmins, 'Windscale leak affects two of 12 farms in the area', *T* (10.10.81.); G. Lean, 'Edge of darkness', *Obs* (23.2.86).
1985 ACCIDENTS:
J. Cutler and R. Edwards, *Britain's Nuclear Nightmare* (Sphere, 1988).
G. Lean, 'Edge of darkness', *Obs* (23.2.86); M. Morris, 'Nuclear mist sounds alert at Sellafield', *G* (6.2.86); 'Nuclear Sights', *ST* (1.2.88); T. Douglas, 'Selling Sellafield as 'open and honest', *Ind* (24.5.89); A. Bell, 'Edge of darkness', *Time Out* (24/31.5.89).

1984, August 25 Mont Louis Sinking, North Sea

K. A. Gourlay, *The Poisoners of the Seas* (Zed Books, 1988).
M. Brown and J. May, *The Greenpeace Story* (Dorling Kindersley, 1989).
M. Bond, 'The transport of nuclear materials to the USSR' (Greenpeace, May 1987).
'Nuclear alert in Channel', *DM* (27.8.84); R. Morris, 'Sunk Channel ship carried hazardous nuclear cargo', *T* (27.8.84); R. Morris, 'Sunken nuclear cargo "safe" ', *T* (28.8.84); 'French attacked for atom cargo silence', *T* (29.8.84); T. Samstag, 'But who carries the nuclear can?', *T* (13.9.84); P. Wilkinson (Greenpeace UK Campaign Director), 'Mont Louis Dangers', Letter to *The Times* (14.9.84).

1985 January 11 HERO Accident, West Germany

A. Grinyer and P. Smoker, *It Couldn't Happen Here* (Richardson Institute, June 1987).
P. Axelrod, D. Babst and R. Aldridge, *A Shocking Event: A Nuclear Catastrophe Waiting to Happen* (Tampere Peace Research Institute, 1988).
Note: According to *Time* (27.2.89), a giant $90-million high-powered early-warning radar owned by the USAF Space Command is only one and a half miles from the approach end of the runway at Robins AFB in Georgia. To guard against the radar triggering electromagnetic explosive devices on military aircraft, it is manually shut down for up to 90 seconds when a

plane comes in to land. The USAF has now compiled a secret list of 300 powerful radio transmitters in the US that their pilots must avoid by a certain distance.

1986 January 5 Gore, Oklahoma

T. Fishlock, 'Radioactive leak kills man in US', *T* (6.1.86); T. Fishlock, 'Scales fault led to death at American atom plant', *T* (7.1.86); 'Training Failures Cited in US Nuclear Plant Leak', *T* (11.1.86); M. White, 'Bhopal Echoes in Nuclear Accident', *G* (11.1.86); D. Bernstein and C. Blitt, 'Lethal Dose', *The Progressive* (March 1986).

1986 April 25/26 Chernobyl, USSR

Dr R. F. Mould, *Chernobyl, The Real Story* (Pergamon Press, 1988).
'Summary Report on the Post-Accident Review Meeting on the Chernobyl Accident', IAEA (1986).
'The Radiological Impact of the Chernobyl Accident in OECD Countries', NEA/OECD (1987).
Dr R. E. Webb, 'Chernobyl: What could have happened', *The Ecologist* (1986); J. Greenwald et al., 'Meltdown', *Time* (12.5.86); L. Martz et al., 'The Chernobyl meltdown', *Newsweek* (12.5.86); J. Steele, 'A reactor on every frontier', *G* (20.5.86); G. Lean, 'The monster in our midst', *Obs* (31.8.86); B. Ramberg, 'Learning from Chernobyl', *Foreign Affairs* (Winter 1986/70); W. C. Patterson, 'Nuclear watchdog finds its role', *NSc* (23.4.87); 'Chernobyl is Ukranian for "Wormwood" ', *NSc* (9.4.87); R. Milne, 'Lessons for the Soviets', *NSc* (23.4.87); J. Simmonds, 'Europe calculates the health risk', *NSc* (23.4.87); P. Quinn-Judge, 'Nuclear power is full steam ahead', *Christian Science Monitor* (27.4.87); E. Marshall, 'Recalculating the cost of Chernobyl', *Science*, vol. 236 (8.5.87); 'Helicopter map radiation', *NSc* (27.8.87).
ADDITIONAL READING:
There is a vast and growing literature on Chernobyl. Useful additions to the literature include the following: H. Hamman and S. Parrott, *Mayday at Chernobyl* (Hodder and Stoughton, 1987); Vladimir Gubaryev, *Sarcophagus* (Penguin, 1987); D. R. Marples, *The Social Impact of the Chernobyl Disaster* (Macmillan, 1988); I. Shchevbak, *Chernobyl: A Documentary Story* (Macmillan, 1989). There are also two novels: F. Pohl, *Chernobyl* (Bantam, 1987); and J. Voznesenskaya, *The Star Chernobyl* (Methuen, 1988).
Two Soviet documentaries, which may or may not be shown in the West, are *Threshold* (Director Rolan Sergienko) and *Microphone* (Director George Shklyarevsky). Further details may be found in *Index on Censorship* vol. 18, nos. 6 and 7 (July/August 1989).

1986 October 3 Soviet Submarine Fire, Atlantic Ocean

Neptune

'Three die in Soviet missile submarine blaze off US', *STel* (5.10.86); C. Walker, 'Quick admission will help summit', *T* (6.10.86); C. Bohlen, 'Soviet nuclear submarine burns off Bermuda', *IHT* (6.10.86); 'Defense Department Briefing on the Soviet Submarine Mishap', *Federal News Service* (6.10.86); G. C. Wilson, 'Submarine sinks as Russians try to tow it to port', *IHT* (7.10.86); A. Brummer, 'US rules out danger as Soviet sub sinks', *G* (7.10.86); R. Highfield, 'Scientists undecided about A-sub hazards', *DT* (7.10.86); A. Tucker, 'The new enemy below', *G* (7.10.86);
Note: Soviet Yankee-class submarines are of a similar layout and vintage to Britain's Polaris submarines. Nineteen of the Yankee-I vessels were built at the Severodvinsk 402 and Komosomolsk-na-Amur yards between 1967 and 1971.

1987 January 10 Nuclear Transporter Crash, UK

M. Urban, 'The lethal routine that went wrong', *Ind* (12.1.87); P. Davenport, 'Nuclear weapons transporter skids into field', *T* (12.1.87); P. Brown, 'Stricken nuclear convoy rescued', *G* (12.1.87); R. Kay, 'Operation nuke

crash', *DM* (12.1.87); M. Urban, 'MPs call for inquiry on nuclear convoy', *Ind* (12.1.87); E. Plaice, 'Britain's 41 minutes on nuclear brink', *Today* (19.1.87).
Note: On 17 September 1988, a 23-year-old medical student died when his MG Midget sports car rebounded off a coach into the path of a Mammoth Major transporter that was leading a convoy of four missile carriers. The transporter, believed to have been carrying four nuclear depth charges, was slightly damaged. J. Craig, 'Inquiry call over nuclear lorry crash', *ST* (18.9.88).

1987 September 13 Goiânia, Brazil

'The Radiological Accident in Goiânia', International Atomic Energy Agency (1988).
M. Margolis, 'Anger as radiation victims are buried', *T* (27.10.87); R. House, 'Brazil attempts to bury its nuclear catastrophe', *Ind* (28.10.87); G. Lean and L. Byrne, 'Brazil N-chiefs face disaster charges', *Obs* (8.11.87).

1987 December 16-17 Biblis-A Reactor West Germany

Original contribution by Helmut Hirsch.

1988/89 US Weapons Production Complex

S. Schwartz, 'The DOE Nuclear Weapons Production Complex: A Selected Chronology', *Greenpeace USA* (1988/89).
'Nuclear materials production', Federation of American Scientists Public Interest Report, vol. 41. no. 10 (December 1988).
'Making Warheads: nine articles on different aspects of US nuclear weapons production', Various authors, *Bulletin of the Atomic Scientists*, vol. 44. no. 1. (Jan/Feb 1988); K. Schneider and M. R. Gordon, 'Reactor shutdown may harm nuclear readiness, US says', *IHT* (10.10.88); E. Magnuson, 'They Lied to Us', cover story, *Time* (31.10.88); C. Peterson, 'Weapons plants' costly cleanup challenges Bush', *WP* (26.11.88); K. Schneider, 'Wide threat seen in contamination at nuclear units', *NYT* (7.12.88); P. Pringle, 'Savouring the 'sweetness' of the tritium factor', *Ind* (29.12.88); 'The Nuclear Mess', *The Economist* (14.1.89).
HANFORD:
L. B. Stammer, 'Critical safety issues overlooked at nuclear plant, 3 professors charge', *LA Times* (20.5.86); M. L. Wald, 'Safety lapses paralyze nuclear bomb complex', *NYT* (23.10.86); E. Marshall, 'End game for the N reactor?', *Science*, vol. 235 (2.1.87); R. L. Stanfield, 'Security leaks', *National Journal* (21.2.87).
SAVANNAH RIVER:
Deadly Defense.
E. Marshall, 'The buried cost of the Savannah River plant', *Science*, vol. 233 (8.8.86).
K. Schneider, 'Accidents at US nuclear plant were kept secret up to 31 years', *NYT* (1.10.88); 'Du Pont rejects contentions it hid reactor problems', *NYT* (3.10.88); 'Chronic failures at nuclear plant are disclosed by the Energy Department', *NYT* (6.10.88); C. Peterson, 'Report cites hazard in arms reactor mechanism, operator attitude', *WP* (6.10.88); 'Enquiry ordered at nuclear arms site', *NYT* (7.10.88).
ROCKY FLATS:
F. Butterfield, 'Report finds perils at atom plant greater than Energy Department said', *NYT* (27 October 1988); S. Davis, 'Top secret — the US nuclear gift factory', *ST* (20.11.88); C. Peterson, 'Rocky Flats: risks amid a metropolis' / 'Design flaws leave new building inoperative', *WP* (12.12.88).
FERNALD:
K. B. Noble, 'Amalgam of agony and anger downwind from uranium site', *NYT* (19.10.88); K. B. Noble, 'US concedes risks to health at atomic plant', *NYT* (29.10.88); E. Magnuson, 'A US nuclear scandal', cover story, *Time* (31.10.88).

Further Reading

An immense amount of further technical detail on the subject, together with a valuable collection of diagrams and pictures, is provided in Chuck Hansen's *US Nuclear Weapons: The Secret History* (Aerofax Inc., Arlington, Texas, 1988), distributed by Orion Books, a division of Crown Publishers, New York, and in Europe by Midland Counties Publications, Leicester, UK. See also, Richard Rhodes, *The Making of the Atomic Bomb* (Simon and Schuster, 1986). Another valuable work is Kosta Tsipis' *Understanding Weapons in a Nuclear Age* (Simon and Schuster).
'Accidental nuclear war', edited by Paul Smoker and Morris Bradley is an excellent edition of the quarterly journal *Current Research on Peace and Violence* (Tampere Peace Research Institute, Finland, vol. IX, no. 1-2, 1988).
Readers seeking a fresh insight into this vast and complex subject are recommended to read Brian Easlea's thought-provoking *Fathering The Unthinkable: Masculinity, Scientists and Nuclear Arms Race* (Pluto Press, 1983); *Nuclear War Atlas* by William Binge (Basil Blackwell, 1988); *By The Bomb's Early Light* by Paul Boyer (Pantheon, 1985), an examination of American thought and culture at the dawn of the atomic age; *Fictions of Nuclear Disaster* by David Dowling (Macmillan, 1987); *Einstein's Monsters* by Martin Amis (Penguin, 1988); and *The Nuclear Age* by Tim O'Brien, a novel that ranks as one of the best on this disturbing subject. Most recent, accessible and telling of all is *Barefoot Gen*, a cartoon story of Hiroshima by Keiji Nakazawa (Penguin, 1989).
Virtually the only photographic work on the subject is the excellent *At Work In The Fields Of The Bomb* by Robert del Tredici (Harrap/Harper and Row, 1987).
Those readers wishing to stay up to date on developments should subscribe to the *Bulletin of the Atomic Scientists* (6042, S. Kimbark, Chicago, Illinois 60637, USA) and *The Ecologist* (Worthyvale Manor, Camelford, Cornwall PL32 9TT).

A variety of valuable publications on nuclear energy are available from Public Citizen's Critical Mass Energy Project (215 Pennsylvania Ave. SE, Washington DC 20003), from the Nuclear Information and Resource Service (1424 16th Street NW, Suite 601, Washington DC 20036), and from the World Information Service on Energy (WISE) (PO Box 5627, 1007 AP, Amsterdam, The Netherlands).

Greenpeace Publications

Greenpeace has published a wide range of reports on both civil and military nuclear subjects, a selection of which appears below.

Nuclear Energy

'Combating the greenhouse effect: no role for nuclear power', Leggett and Kelly (1989).
'The failure of British nuclear emergency plans to meet accepted international standards', proof of evidence of P. J. Sands in support of Greenpeace at Hinkley Point Public Inquiry (April 1989).
'The swelling tide of world opinion against nuclear power', proof of evidence of C. Flavin in support of Greenpeace at Hinkley Point Public Inquiry (March 1989).
'Opinion of law on nuclear liability and insurance', proof of evidence of P. J. Sands in support of Greenpeace at Hinkley Point Public Inquiry (December 1988).
'Statement of case by Greenpeace UK to Hinkley Point Public Inquiry' (September 1988).
'Survey of radiation monitoring organisations in Western Europe', Greenpeace International (1988).
'Shut Them Down: a four-year timetable for closing all Britain's nuclear reactors', Greenpeace UK (September 1986).

Nuclear Safety

'The activities of the International Atomic Energy Agency in the field of nuclear safety', Dr Helmut Hirsch for Greenpeace International (December 1987).

'Too Hot to Handle: an interim report on the under-insurance of British nuclear reactors', Greenpeace UK (1986, republished 1987).
'VVER-440: The main power reactor in eastern europe', Greenpeace Germany (April 1987).
'International nuclear reactor hazard study', Greenpeace International (September 1986).
'Chernobyl UK: how an accident on the scale of Chernobyl could occur in a British gas cooled nuclear reactor', P. Cade and P. Bunyard (1986).

Radioactive Waste

'Exposing the faults: the geological case against the plans by UK NIREX to dispose of radioactive waste', P. J. Richardson (1989). 'The controversy over ocean dumping of radioactive wastes', Greenpeace International (October 1988).
'International liability for damage resulting form the dumping of radioactive wastes at sea', Greenpeace International (October 1988).
'Seabed emplacement of radioactive wastes', submission to London Dumping Convention (LDC) by Greenpeace International (October 1988).
'Radioactive waste management: the environmental approach', Friends of the Earth (1987).
W. Jackson Davies PhD, 'Transportation of radioactive spent nuclear fuel: quantitative site specific accident analysis for the port of Anzio, Italy', Greenpeace International (September 1986).
'An investigation into the hazards associated with the maritime transport of spent nuclear reactor fuel to the British Isles', Political Ecology Research Group for Greenpeace UK (November 1985).

Spent Fuel Reprocessing

'Investigation of allegations of incidents and malpractice at British Nuclear Fuels, Sellafield', Large and Associates, report for Greenpeace UK (May 1988).
'Precautionary approach versus assimilative capacity: the lessons to be learned from the British practice of discharging radioactivity into the marine environment', Submission by Greenpeace to the North Sea Ministers' Conference (November 1987).
'Radioactive discharges into European coastal waters', submission by Greenpeace International to Paris Commission, Amsterdam (March 1986).

Naval Nuclear Technology

'The nuclear arms race at sea', Neptune Papers I, Greenpeace International and the Institute of Policy Studies (October 1987).
'Nuclear warships and naval nuclear weapons: a complete inventory', Neptune Papers II, Greenpeace International and the Institute of Policy Studies (May 1988).
'Naval accidents 1945-1988', Neptune Papers III, Greenpeace International and the Institute of Policy Studies (June 1989).
'Survey of radioactivity in sediments in the vicinity of naval establishments in the UK', Greenpeace UK (1988).
'Nuclear accidents on military vessels in Australian ports: site specific analyses for Sydney and Fremantle/Perth', Dr W. Jackson Davies for Greenpeace International (December 1986).

Index

Notes: **Bold** page numbers refer to illustrations and maps; *key terms* on nuclear matters are listed on page 50; main *abbreviations and acronyms* used in the index are listed on page 51.